Dietary Supplement Good Manufacturing Practices

Preparing for Compliance

Dietary Supplement Good Manufacturing Practices
Preparing for Compliance

William J. Mead

CRC Press
Taylor & Francis Group
Boca Raton London New York

CRC Press is an imprint of the
Taylor & Francis Group, an **informa** business

First published in paperback 2024

First published in 2012 by Informa Healthcare

Published 2024 by CRC Press
2385 NW Executive Center Drive, Suite 320, Boca Raton FL 33431

and by CRC Press
4 Park Square, Milton Park, Abingdon, Oxon, OX14 4RN

CRC Press is an imprint of Taylor & Francis Group, LLC

© 2012, 2024 Taylor & Francis Group, LLC

Reasonable efforts have been made to publish reliable data and information, but the author and publisher cannot assume responsibility for the validity of all materials or the consequences of their use. The authors and publishers have attempted to trace the copyright holders of all material reproduced in this publication and apologize to copyright holders if permission to publish in this form has not been obtained. If any copyright material has not been acknowledged please write and let us know so we may rectify in any future reprint.

Except as permitted under U.S. Copyright Law, no part of this book may be reprinted, reproduced, transmitted, or utilized in any form by any electronic, mechanical, or other means, now known or hereafter invented, including photocopying, microfilming, and recording, or in any information storage or retrieval system, without written permission from the publishers.

For permission to photocopy or use material electronically from this work, access www.copyright.com or contact the Copyright Clearance Center, Inc. (CCC), 222 Rosewood Drive, Danvers, MA 01923, 978-750-8400. For works that are not available on CCC please contact mpkbookspermissions@tandf.co.uk

Trademark notice: Product or corporate names may be trademarks or registered trademarks and are used only for identification and explanation without intent to infringe.

Publisher's Note
The publisher has gone to great lengths to ensure the quality of this reprint but points out that some imperfections in the original copies may be apparent.

Library of Congress Cataloging-in-Publication Data

Mead, William J.
   Dietary supplement good manufacturing practices : preparing for compliance / William J. Mead.
      p. cm.
   Includes bibliographical references and index.
   ISBN: 978-1-4200-7740-7 (hardback)

   1. Dietary supplements--Standards. 2. Dietary supplements industry--Standards. 3. Manufacturing processes--Standards. I. Title.
   RM258.5.M43 2011
   615.1--dc23

                                          2011029665

A CIP record for this book is available from the British Library.

ISBN: 978-1-4200-7740-7 (hbk)
ISBN: 978-1-03-292252-2 (pbk)
ISBN: 978-0-429-14442-4 (ebk)

DOI: 10.3109/9781420077414

Visit the Taylor & Francis Web site at
http://www.taylorandfrancis.com

and the CRC Press Web site at
http://www.crcpress.com

Typeset by MPS Limited, a Macmillan Company

Foreword

Having been involved in the development of Good Manufacturing Practices (GMPs) for dietary supplements since the passage of the Dietary Supplement Health and Education Act of 1994, I am keenly aware of the complexity of the issues involved. I worked with committees of industry experts to draft model GMPs that were submitted to the Food and Drug Administration (FDA) in 1995 and published by the agency as an Advance Notice of Proposed Rulemaking in 1997. Formal regulations establishing dietary supplement GMPs were not published in proposed form until 2003 and not finalized until 2007, capping a herculean effort on the part of FDA, the affected industry, and other stakeholders. The rules became effective for large companies in 2008, for smaller companies in 2009, and for the smallest companies in 2010.

So why publish a guide to GMPs now, in 2012? The answer is simple: because FDA inspections reveal that many companies still lack a full understanding of the requirements and are not in full compliance. Thus, there is a continuing and urgent need for additional education and guidance. Bill Mead is an expert in quality issues, with a lifetime of solid experience upon which to base a highly professional, exceedingly thorough, but amazingly readable book to support and expand any company's comprehension and application of the GMP regulations. The references at the end of each chapter point the way to even more depth of knowledge. This is a book that every dietary supplement company should have, not only as a reference, but as an integral component of its GMP training program.

Annette Dickinson, Ph.D.
September 2011

Preface

One of the fundamental facts of business in any industry is that consistent excellence in quality is essential both to attract and to maintain customers. In the dietary supplement industry, the great majority of firms do indeed provide high-quality goods. Unfortunately, a few rogue firms have taken shortcuts for reasons of cost cutting, and their quality has at times been less than ideal. These few shortcomings have had considerable media coverage, resulting in some lack of confidence in dietary supplement products on the part of many consumers. This has hurt the image of the entire dietary supplement industry to some extent.

Since dietary supplements are legally classified as foods, all such products have been required to be manufactured under the *food* Good Manufacturing Practice (GMP) regulations found in Part 110 of Title 21 of the Code of Federal Regulations, albeit these clearly do not adequately cover the types of products being marketed as dietary supplements.

To correct this situation and to level the playing field, the leading trade associations representing the majority of firms making dietary supplements jointly formed (under the leadership of the Council for Responsible Nutrition) what was called the Industry Standards Working Group, and in 1995 this group drafted their concept of what dietary supplement GMPs should cover. They submitted their document to the Food and Drug Administration (FDA) in November of the same year.

On February 6, 1997, the FDA published the industry's draft in the *Federal Register* as an Advance Notice of Proposed Rulemaking (ANPR) as the first step in eventually promulgating GMP regulations covering manufacturing, packing, or holding dietary supplement products. Interested parties were invited to comment on this proposal. The objectives were to sooner-or-later publish final regulations, taking into consideration the comments from industry trade associations, consumers and consumer advocacy groups, companies, academia, health care professionals, and others, to achieve workable ground rules to prevent product adulteration and misbranding, to ensure that the products actually contain the amount of the ingredients claimed on the labels, and to ensure that all applicable quality specifications are reasonably met.

More than 100 letters of comment on the ANPR were received. In 1998, the FDA formed a GMP working group under the Food Advisory Committee to study these comments, and also made several plant visits to better understand the issues; in addition various stakeholder meetings were held to get still additional input. Obviously, the FDA has its own expertise and background in that GMPs have long been used for foods, drugs, devices, and biologics.

The GMP working group presented a draft of their report in mid-1999 at a public meeting and invited further comments, then in the *Federal Register* of May 13, 2003, the FDA published a revision of the ANPR, this time as a Proposed Regulation, and again public comments were solicited, and carefully evaluated. Following this, the Final Rule appeared in the *Federal Register* of June 25, 2007 (see Appendix).

In short, the FDA concluded that dietary supplement GMPs are indeed needed, and would be of particular importance to small and medium-sized firms that in many instances might not be fully aware of the steps necessary to consistently assure product quality. Moreover, the FDA concluded that having such regulations would be an important step in helping prevent irresponsible firms from making and selling substandard products. Having GMPs in place would help ensure that consumers get dietary supplement products having the expected strength and purity purported in the products' labeling. The regulations also require manufacturers to maintain documented evidence that their methods and processes are properly under control on a consistent basis.

It is important to note that the FDA does have the legal authority to inspect factories, warehouses, and other establishments in which dietary supplement products are manufactured, processed, packed, or held. Similarly, the FDA does have the legal authority to issue GMP regulations for dietary supplements, and the courts have held that such regulations are not merely suggestions as to what should be done, but instead they have the full force and effect of law. In other words, firms involved in making dietary supplements are *required* to conform to the GMPs, or face the possibility of serious legal consequences.

There are many definitions of the term *quality*, but in effect it means that the products made must regularly and consistently meet both the consumers' needs and expectations and, in this instance, also the regulatory requirements of the FDA since its mission is to protect the health of the American public. To continuously achieve quality requires the firm to have in place and faithfully follow a comprehensive *quality system*. There are many forms of such systems, including Total Quality Management, ISO-9000, Six Sigma, Zero Defects, Hazard Analysis Critical Control Point system, and others, all of which have merit and are in no way mutually exclusive.

However, in 1962, in response to an emergency triggered by an attempt to gain approval to market a certain drug in the United States that was found to have caused the birth of many deformed babies, Congress passed what was known as the Kefauver–Harris Amendments to the U.S. Food, Drug, and Cosmetic Act. These amendments had far-reaching consequences by expanding the regulatory authority of the FDA. One of the items added to the act at that time stated that a drug or device shall be deemed adulterated (and therefore subject to regulatory action) if the methods used in, or the facilities or controls used in, its manufacture, processing, or holding do not conform to or are not operated or administered in conformity with *current good manufacturing practice*. It is from this wording that the term "GMP" arose. However, the act did not further define GMPs, leaving this to the FDA to publish implementing regulations, thus providing FDA the authority to prevent the manufacture and distribution of faulty drugs. The first drug GMP regulations were issued in 1963, and have since been revised several times, aiming at greater specificity. With time, the FDA expanded the GMP concept to foods and medical devices, albeit the regulations were tailored to each type of product. Not long thereafter, most other major countries saw the merits of GMPs, and followed suit with their own versions of them. Even the World Health Organization has issued GMPs. The fact is that GMPs are excellent quality systems, and are widely used as such on a global basis. Therefore, with many years of good experience with GMPs, it is not at all surprising that the FDA elected to follow this same basic form of quality system for dietary supplement products.

The regulations outline broad goals of *what* must be accomplished, deliberately and wisely avoiding specifics as to precisely how these goals should be met. In other words, the "how to" points are left largely up to the discretion of each firm. This gives the industry considerable latitude and flexibility, and also provides for future changes as technology advances without necessitating rewriting the regulations frequently to keep them up-to-date.

It is the purpose and intent of this book not only to elucidate the actual final dietary supplement GMP regulations but also to augment them by explaining FDA's intents, and also to supply workable and practical suggestions as to ways a firm can best meet the goals, based on experience with GMP compliance techniques worked out over the years in the food, drug, and medical device industries. Obviously, many firms are already quite knowledgeable about GMP compliance methods, but not all companies have this background. Also, as new people come into the dietary supplement manufacturing industry, they should find it helpful to have a source of information on the significance and meaning of the GMPs, in addition to practical suggestions as to how to comply with FDA's requirements.

ACKNOWLEDGMENTS

I want to thank my family, and in particular my wife Linda, for putting up with the innumerable hours I spent on our computer as well as for the inevitable unruly stacks of papers and documents always present in our home office space during the writing of this book. It is indeed a blessing to have a tolerant and understanding mate.

I also want to extend sincere thanks to Annette Dickinson, Ph.D., whom I have yet to meet face-to-face but who read the manuscripts of each chapter as I produced them and sent them to her via e-mail. She made helpful comments and constructive criticism, which I greatly appreciate. Annette was President of the Council for Responsible Nutrition (CRN), one of the leading dietary supplement trade associations, during the years when FDA was wrestling with writing the DS GMPs, giving her unique insight into the nuances of the development of the regulations. She is a recognized expert in the regulatory, technical, health benefits, and marketing aspects of the dietary supplement industry.

William J. Mead

Contents

Foreword vii
Preface ix

1. The basics of good manufacturing practice *1*
2. Regulatory overview *8*
3. Personnel matters *17*
4. Physical plant and grounds *25*
5. Equipment and utensils *40*
6. Cleaning and sanitation *52*
7. Maintenance and GMP *63*
8. Calibration *68*
9. Production and process controls *73*
10. Specifications *78*
11. Sampling *92*
12. Deviations and corrective actions *103*
13. Incoming components, packaging materials, and labels *107*
14. Master manufacturing record *112*
15. Batch production record *117*
16. Manufacturing operations *121*
17. Packaging and labeling operations *137*
18. Quality control responsibilities *145*
19. Laboratory operations *153*
20. Returned goods *168*

21. Product complaint handling *170*

22. Holding and distributing *174*

23. Handling recalls *179*

24. Top management responsibility *182*

25. Record keeping, documentation, and SOPs *186*

26. Change control *194*

27. Adverse event reporting and record keeping *197*

28. Adulteration and contamination *200*

29. Supply chain integrity *211*

30. Audits *216*

31. Outsourcing *223*

32. The Food and Drug Administration *230*

33. FDA inspections *240*

Appendix FDA dietary supplement GMP regulations 249
Index 289

1 The basics of good manufacturing practice

THE PRIME IMPORTANCE OF QUALITY
The Dietary Supplement Good Manufacturing Practice (DS GMP) regulations are essentially about a means of achieving consistent quality in products made. There are many definitions of quality, which is a somewhat subjective term, but all address meeting consumers' expectations. This means that the product as purchased will, in fact, be what is claimed on the label, will be fit for use, and will not be contaminated with anything that may be harmful. The Food and Drug Administration (FDA) defines dietary supplement quality, stating that it means that the product meets its established specifications for identity, purity, strength, and composition, and has been manufactured, labeled, and held under conditions to prevent adulteration.

From a strategic point of view for product manufacturers, excellence in quality is essential for the continuation and growth of the firm in that consumers will avoid purchasing products that they perceive may be of inferior quality. Thus, quality is an important feature in competitiveness. However, achievement of consistent quality is not easy to measure in that it is based on long-run consumer satisfaction.

To achieve the goals of quality, it is necessary to set and meet meaningful specifications and to have control over all facets of the manufacturing process. For dietary supplement products, this is no longer optional but is instead required both by the desire of consumers and by enforceable FDA regulations.

Quality is a learned cultural system and philosophy that must be indoctrinated into all departments and all employees, not just into those directly involved with manufacturing and quality control. It must become a way of life. This is important in all industries in this competitive global era, but is particularly significant in those firms regulated by the FDA. Achieving consistent quality involves thoroughly understanding what is required. It also depends on shared values and establishing trust within the company.

QUALITY MANAGEMENT SYSTEMS
To consistently meet the quality requirements, it is essential to establish and effectively use a quality management system (QMS) that consists not only of specifications for each step in the manufacturing process but also in having a formalized set of policies, methods, written procedures, documentation, training clear lines of organizational responsibilities, and top-management support. There are many approaches to structuring a QMS, but since the early 1960s, health care products manufacturing (both in the United States and globally) has fared well with the GMP form of QMS, now widely used in the food, drug, cosmetic, medical device, and biological products industries. On the basis of this favorable experience, the FDA has now promulgated mandatory GMP regulations for the dietary supplement industry. The basic concepts of GMP are uniform throughout all FDA-regulated manufacturing, including (but not limited to) dietary supplements. Therefore, much can be learned from the

accumulated experience with the GMPs that have long existed for drugs, medical devices, and various categories of food products.

Since there is great diversity in the types of dietary supplement products, the scale of manufacturing, the equipment and steps involved, and the controls required to produce items with consist excellence in quality, there is no "one size fits all" for compliance. Therefore, careful design of the details of the firm's QMS is required. However, FDA understands and appreciates such diversity and agrees that compliance techniques need to be tailored to the existing circumstances.

Making internal decisions about how best to comply with the GMPs should be based in part on a thorough understanding of the meanings stated in the regulations and what is known about FDA's interpretation of given sections, based in part on good science and in part on common sense. But, in deciding about actions, the interpretation of which could be controversial, it is prudent to document in writing how and why the decisions were reached. That at least helps explain the rationale that was used in reaching such decisions in the event of future disagreements with the FDA, which in turn can help defuse conflicts.

HAZARD ANALYSIS AND CRITICAL CONTROL POINTS

In developing the dietary supplement GMPs, the FDA considered various QMS approaches, one of which was Hazard Analysis and Critical Control Points, the acronym for which is HACCP (pronounced "hassip"). This is a science-based quality tool that originated in the 1960s in connection with the development of foods and beverages for astronauts in the then new space exploration program to help minimize the probability of food-borne illness. It proved to be so successful that it was soon endorsed by the National Academy of Sciences, Codex Alimentarius, and the National Advisory Committee on Microbiological Criteria for Foods. It has since been adopted as a regulatory requirement for identifying and preventing hazards from contamination in various categories of foods. HACCP is favored by both the FDA and the U.S. Department of Agriculture in food production, not only because it works so well but also because it helps simplify government oversight since the required record-keeping facilitates inspectors' ability to see how well a firm is complying with food safety laws, plus it places the responsibility for ensuring food safety squarely on the manufacturer and/or distributor. Applying these techniques helps improve the efficiency and effectiveness of inspections.

HACCP involves seven principles: hazard analysis, identification of *critical control points* (CCPs), establishment of preventive measures with critical limits for each control point, establishment of procedures to monitor the CCPs, providing for corrective actions to be taken when monitoring indicates that a critical limit has not been met, establishment of ways to verify that the system is working as intended, and effective record-keeping. The hazard analysis portion identifies potential hazards and measures to control them. The hazards include unwanted chemicals like pesticides or heavy metals, microbiological contamination, toxins, and physical contamination such as glass or metal fragments. The CCPs are points in the production process, warehousing and transportation, and distribution at which the potential hazard can be controlled or eliminated.

Some of the comments FDA received in developing the DS GMPs favored the HACCP approach, but FDA concluded that while clearly useful in preventing certain specific types of problems, it fails to ensure that a product necessarily conforms to established specifications, and therefore is not in this instance a suitable substitute for the GMP approach. The FDA has said that dietary supplement manufacturers may *voluntarily* apply HACCP as *part* of their quality system, but the rest of the GMP requirements must still be met. However, certain ingredients (e.g., those derived from fish and fishery products such as fish oils and fish cartilage, and those derived from bovine materials) used in some DS products may be subject to compliance with other FDA regulations that do *require* HACCP.

Since the GMPs represent the *minimum* requirements, it is clear that firms are at liberty to do *more* than what is mandated, and therefore incorporating HACCP as a *part* of the overall QMS may be prudent in some situations. HACCP is used as a part of the QMS by some firms in the pharmaceutical and medical device industries. The World Health Organization has also provided guidance on the use of HACCP to help ensure the quality of pharmaceuticals, and this could obviously prove helpful in the manufacture of dietary supplement products.

QUALITY MANUALS
Although not required by the GMPs, some companies use a quality manual summarizing their QMS. This is a concept borrowed from ISO-9000, and can indeed be useful. Such manuals typically include a table of contents, a statement of the company's policy of full and complete adherence with the dictates of the GMP regulations, lists the firm's quality procedures, describes the organizational structure and responsibilities of individuals and departments, clarifies patterns of communications, includes definitions, and may reference specific standard operating procedures.

There is usually also a revision history of the manual, and a page of approval signatures. Since such manuals are optional and not mandated, there is flexibility in the structure and content, but it is a useful way of summarizing the firm's commitment to and methods of compliance with the GMPs. Such manuals can also be useful in training.

The inclusion of a glossary of definitions helps reduce the possibility of confusion of terms. For example, it can be prudent to specify what the company means when referring to a "carton" or a "case," and other such terms, including the names of ingredients, that otherwise might not be uniformly used throughout the company. This helps reduce misunderstandings.

THE GMP PRINCIPLE
When a consumer buys a package of a dietary supplement, he or she cannot easily determine the quality of the product that is contained within the package. The consumer must trust that the firm named on the package label has taken the necessary steps to assure that the product is, in fact, as represented and is neither contaminated nor adulterated in a way that might be harmful. Thus, the company's reputation plays a vital role in the purchasing decisions made by consumers.

In the past, there has been considerable variation from company to company as to the standards used in manufacturing and quality control. While the great majority of firms have deliberately and carefully followed good QMS, a few unscrupulous manufacturers have taken shortcuts to save on costs. The relatively small number of resulting inferior products led to rather sensational negative publicity in the media, causing some damage to the image of the entire dietary supplement industry. However, now that legally enforceable GMPs exist for the entire industry, the competitive playing field has been leveled since all manufacturers are required to meet specified minimum standards. The previously rogue firms will either need to change their ways or go out of business. In general, the dietary supplement industry favors having the regulations in place. Therefore, the GMPs are looked upon as being good for both the dietary supplement industry and the consuming public.

The basic premise of the GMPs is that quality must be *built into* the product, and that simply testing each batch is inadequate to achieve the desired results. The GMPs require achieving control over each step of the manufacturing process, and being sure that the possibility of contamination is minimized in part by using raw materials of high quality, in part by being sure that the buildings, grounds, and equipment are properly designed, constructed, maintained, cleaned, and sanitized, and in part by requiring employees working directly with the products to follow good personal hygienic practices. Moreover, employees and their supervisors must clearly understand what is expected of them and how their duties are to be performed. Every step in the manufacturing process must be well designed and implemented, and properly documented. Master manufacturing and batch production records must be established and used. **Quality** control steps must be followed, and both incoming supplies and finished products must be properly sampled, examined, and stored. Consumer complaints must be followed up and corrective action taken when needed. Controls at all steps of the manufacturing process are needed to ensure quality is consistently achieved.

Still one more concept that is basic to GMPs, but is not specifically spelled out in the regulations, is that of *continuous improvement*. This implies being constantly seeking and alert for finding ways to make the manufacturing processes and quality control methods better. This is usually accomplished in small incremental steps over time. It typically involves reducing variation (for which statistical process control methods can be useful), teamwork, employee involvement, and benchmarking to learn from the experience of others.

As the name implies, GMPs are specifically related to *manufacturing and quality control*, and are not directly involved with the inherent safety or efficacy of the ingredients and the formula, and are not directed toward claims or label copy, all of which are covered by other appropriate regulations.

THE TERM *CURRENT* GMP

The regulations refer to *current* good manufacturing practice (not practices), which is sometimes abbreviated by GMP, cGMP, or CGMP. These are all synonymous.

The word "current" implies that the regulations are constantly evolving as technology and equipment change over time. But, since changing the actual regulations is a tedious and time-consuming process, the wording of the regulations does not necessarily always encompass what is considered to be both "current" and "good." This poses something of a dilemma. It is the FDA and the courts, not the industry, that determine what is deemed to be both current and good, based on what is considered to be both feasible and valuable. Obviously, the FDA has extensive experience in GMP, in part from their many inspections of manufacturing facilities, in part from other consumer health-care industries they regulate, and in part from the outcome of regulatory actions taken and from court rulings in litigation. Current and good therefore are not necessarily synonymous with "best" or "state of the art."

The GMPs deliberately contain some rather vague terms, such as *suitable, adequate, appropriate,* and *necessary*. This yields flexibility for industry and also permits changes in interpretation of the written regulations over time. As stated in reply to Comment 60 in the preamble to the DS GMPs, the FDA declined to define such terms, saying in effect that this depends on individual circumstances based on particular conditions and that therefore each firm must evaluate and properly apply these terms to each particular operation.

Obviously, this can result in some disagreements between the FDA and the regulated firms. There are mechanisms to resolve such conflicts when they do occur. However, it is prudent for the regulated companies to try to stay abreast of FDA's point of view as to what practices are indeed considered current and good. There are many ways of doing this, including information from industry associations and trade publications, monitoring other firms' Establishment Inspection Reports and Warning Letters, keeping up-to-date on appropriate FDA guidance documents, and attending seminars on GMP compliance, or through consultants or law firms. There are also internal FDA documents such as the *Investigations Operations Manual* (IOM) and compliance policy guides that can be obtained, some of which are available online and others through Freedom of Information. Clearly, the FDA greatly prefers voluntary compliance to regulatory actions, so representatives from the FDA do frequently participate in seminars and other activities to help keep the industry informed.

Over the years of having GMPs in other FDA-regulated industries, the question has often been raised as to whether or not the regulations should be made much more detailed and explicit as to just *how* they must be applied. The FDA has wisely adopted the concept of having GMPs define in general *what* must be accomplished as opposed to precisely *how* this must be done, leaving it largely up to each individual firm to decide the "how to" issues.

Since dietary supplements are legally viewed as a subset of foods, the drug GMPs do not apply *per se*. However, the food GMPs do not adequately cover all of the unique manufacturing processes involved with tablets, capsules, powders, liquid extracts, and certain other dosage forms commonly used for dietary supplement products. While basically following the food GMP approaches rather than the drug GMPs, the dietary supplement GMPs do necessarily somewhat overlap with certain aspects of the drug GMPs. Congress specifically mandated FDA to model the DS GMPs after the existing *food* GMPs, not the *drug* GMPs. Since the final DS GMPs have appeared, some in the industry have questioned whether or not the FDA properly followed this requirement. However, in Comment 18 in the preamble to the DS GMPs, the FDA provided a lengthy and careful explanation of how they attempted to accomplish this goal. Short of a challenge in the courts, this is now a settled matter.

All quality systems, including GMPs, are constantly evolving. For example, the drug GMPs today are in reality quite different than they were a few years ago, with current emphasis on various aspects of quality risk management. We can expect similar changes over time with the DS GMPs.

GMPS ARE ALSO GOOD BUSINESS PRACTICE

Apart from the regulatory aspects of GMPs, they do represent one facet of ways of properly conducting business. Quality is indeed important in attracting repeat sales, which is obviously a basic business consideration, and GMP compliance is an effective and time-tested way of achieving quality. Compliance with the full GMP concept and philosophy helps bring manufacturing operations into a state of control, reduces variation, minimizes errors and mix-ups, helps prevent the need for rework, reduces scrap, and helps avoid costly recalls.

GMPs lead to a better understanding of processes and methods, which in turn leads to the development of improved manufacturing techniques yielding better productivity and reduced downtime. All of these factors improve overall company performance. Thus, it can truly be said that GMPs make good business sense in that better products and better customer service help improve a firm's competitive situation by increasing consumer trust in specific brands.

Another consideration is the possible reduction in product liability exposure. In the event of a mishap that might lead to litigation, since the plaintiff's attorneys would have access to FDA records of inspections and/or enforcement actions through Freedom of Information, any negative information about GMP failures could have a significant negative impact on the outcome of the case.

One of the goals of applying GMPs is to assure doing things correctly the *first time*, thus avoiding the necessity for rework and reprocessing. This requires careful attention to details. Whenever a firm attempts to "cut corners" to reduce costs, the risk of failure increases dramatically. The GMPs have been carefully crafted to help avoid the probability of things going wrong during manufacturing. They provide practical ways of assuring excellence of quality, if they are, in fact, meticulously followed. This requires fostering a corporate culture, a philosophy, and a way of life, for all involved, based on honesty and integrity coupled with a sincere interest in achieving consistent quality. It also requires a positive willingness to follow "rules," and an eagerness to meet customer expectations. It is more than just meeting FDA's mandatory requirements, but is instead about conducting business in the best feasible manner.

This is, in part, why many firms already were in full compliance with the DS GMPs, having realized the importance and advantages long before the final rule was published. It took FDA many years to finalize these regulations, but most members of the industry have long realized the importance of excellence in quality and marketing safe healthful products, have not resisted the necessary steps, often with the help of trade associations and other third parties, and have taken the required actions to be in compliance long before the issuance of the final regulations. But, it is also important to keep in mind that Congress recognized the need to establish industry standards and therefore authorized the FDA to issue and enforce the DS GMPs, thereby creating a level playing field for all firms in the industry and reducing the number of consumer complaints and illnesses and the number of recalls. To repeat, better process understanding clearly leads to better decision-making on the part of manufacturers, improved quality and less product variability, greater manufacturing efficiency, lower risk, improved manufacturing performance, better protection of the public health, and enhanced consumer image of DS products in general.

THE COST OF QUALITY

There is no doubt that achieving excellence in quality does involve additional costs. The FDA recognized this and discussed it at length in the preamble to the final regulations. In fact, some small, marginal firms may be unable to afford to take the required steps and may, therefore, be forced out of business. Such costs may include upgrading the building, grounds and equipment, establishing and maintaining a laboratory (or contracting for outside laboratory services), establishing and maintaining a complete QMS, sampling, training, calibration, etc.

However, this is a two-sided coin. Once the investments have been made to bring the firm into compliance, operations tend to become more efficient, and thereby the cost of goods produced should actually change in a favorable way. Therefore, one way of looking at the costs of GMP compliance is that they are, in fact, an investment that will pay off in the long run since the long-term benefits of successfully implementing a robust GMP-based QMS will outweigh the costs.

SUGGESTED READINGS

Bass IS. Dietary supplement GMPs: the conundrum that is DSHEA. FDLI Update, September–October 2007:30–32.
Crowley R, FitzGerald LH. The impact of cGMP compliance on consumer confidence in dietary supplement products. Toxicology 2006; 221(3):9–16.
Dickinson A. Good manufacturing practices for dietary supplements: a necessary step toward assuring product quality. FDLI Update, June 2000:12–15.
Gryzlak BM, Wallace RB, Zimmerman MB, et al. National surveillance of herbal dietary supplement exposures: the poison control center experience. Pharmacoepidemiol Drug Saf 2007; 16(9):947–957.
Messplay GC, Heisy C. FDA finalizes dietary supplement cGMPs. Contract Pharma, October 2007:18–20.
Tannenbaum E, Stock C, DiFelice A. A healthy step: applying GMPs to dietary supplements. Nutritional Outlook, January/February 2008:39–49.
Wood DC. The Executive Guide to Understanding and Implementing Quality Cost Programs: Reduce Operating Expenses and Increase Revenue. Milwaukee, WI: American Society for Quality, 2007.

2. Regulatory overview

THE ORIGIN OF FEDERAL STATUTES

When the need for a new law is perceived, any member of either house of Congress may introduce a proposal, which at that stage is called a "bill." Such bills are assigned to the appropriate committee in that house, where they are discussed and debated, often involving public hearings. The bill may undergo significant changes in this process, but will eventually either be rejected or voted on in a positive way in the committee, and then sent to the entire house for consideration, further debate, and often amendments. In the end, that house will vote on the bill as modified in the process, and if approved, the bill is sent to the other house of Congress where it goes through a similar process. Obviously, politics and the power of lobbyists play roles as do pressures from the constituents of the various members of Congress. But, if the second house approves the bill, often with differences from the version submitted by the first house, a "conference committee" made up of the members of both houses tries to reconcile the differences, after which final votes are taken on the revised bill. If passed, it goes to the President who will either sign it or veto it. If signed, or if the veto is overridden by Congress, it becomes a Public Law cited by the letters PL with an identifying number. This is also known as an Act, and becomes part of what is called the United States Code (U.S.C.), which is where all federal statutes are compiled after they have been finalized. The U.S.C. is divided into 50 volumes, called titles, each of which contains related laws. All of the federal laws pertaining to the FDA are in Title 21 of the U.S.C., which is divided into chapters and these in turn are divided into sections.

The basic law related to the FDA and its activities is the Federal Food, Drug, and Cosmetic Act (FD&C Act) of 1938 as amended (and it has been amended many times), which is found in Chapter 9 starting at Section 301 of Title 21 of the U.S.C. The section numbers as used in the FD&C Act differ from those in the U.S.C., but a cross-reference system exists to correlate the section numbers.

The *original* Food and Drugs Act was signed into law by President Theodore Roosevelt in 1906, in part as a result of a novel that had been published detailing the filthy conditions existing in certain meat-packing facilities.

Then, in 1937, a tragic error occurred when a drug used to treat streptococcal infections, Sulfanilamide, was produced in liquid form using diethylene glycol as the solvent. The resulting elixir was poisonous, killing more than 100 people. The end result was a rewriting of the 1906 law into what became the 1938 FD&C Act, which among other things strengthened the safety requirements for drug products. Then, in 1961, a drug in use in Europe to treat sleep disorders and morning sickness in pregnant women, called Thalidomide, was under consideration for FDA approval for its sale in the United States. However, it was discovered by an employee of the FDA that this product caused serious birth defects, so the application was disapproved. This aroused concern in Congress that the FD&C Act was deficient regarding safety requirements for new drugs, which resulted in the Kefauver–Harris Amendments of 1962. These amendments, among other provisions, for the first time created the requirement that drug products be made in accordance with "current good manufacturing practice." This was the beginning of the GMP concept, which was soon adopted also for other classes of products under FDA control, in addition to many other countries.

PROHIBITED ACTS AND PENALTIES

Section 331 of the FD&C Act provides severe penalties FDA may impose if products are found to be *adulterated* or *misbranded*.

Adulteration of dietary supplements is explained in Section 402 of the FD&C Act, and there are two different meanings of the term. The first of these is if the product presents an unreasonable risk of illness or injury, or both. This may be because of the formula of the product *per se*, or because of contamination from accidental inclusion of an unintended substance, deliberate addition of undeclared substances, the presence of such substances as filth, pesticides, metal or glass fragments, naturally occurring toxicants, heavy metals, or undesirable microorganisms. Thus, the product must be manufactured and handled from the receipt of ingredients and packaging components through all the steps of processing, packaging, labeling, storage, and distribution in a manner that prevents adulteration. Adulteration can be either intentional or unintentional. Intentional adulteration usually is the result of knowingly adding an ingredient (often a drug) to intensify a pharmacologic effect. Most instances of adulteration are unintentional, however, resulting from inadvertent contamination either of the raw materials or during the manufacturing process. Some instances of legally defined adulteration result from the use of an unapproved new dietary ingredient. But, in any event, pursuant to the Dietary Supplement Health and Education Act of 1994 (DSHEA), the FDA bears the burden of proof to establish that a DS product is indeed legally adulterated as defined in the statute.

The second meaning of adulteration is if the product has been "prepared, packed or held under conditions that do *not* meet current good practice regulations." This is stated in Section 402(g)(1) of the FD&C Act. This means that a product need not actually be *imperfect* in any way to be considered to be legally adulterated and therefore subject to regulatory action by the FDA. In other words, the failure to follow current good manufacturing practice can cause the product to be considered to be adulterated and therefore subject to regulatory actions by the FDA, such as costly recalls, seizure, injunctions and negative publicity for the firm involved, and the possibility of fines or even jail sentences for the firm and/or responsible individuals. Obviously, this is a very strong reason why it is extremely urgent to be well aware of, and fully in compliance with, the GMPs.

For further discussion, see also chapter 28, "Adulteration and Contamination."

The other prohibited act is *misbranding*, described in Section 403 of the FD&C Act. Products are misbranded if the label or labeling is false or misleading in any particular. Also, the label must conform to the applicable laws and regulations dealing with claims and format, provide full ingredient labeling, and contain other required statements and directions for use. Specific labeling requirements have been established by the Fair Packaging and Labeling Act, the Nutrition Labeling and Education Act of 1990, specific provisions contained in DSHEA, in applicable FDA regulations. The format and content of dietary supplement labeling and claims is outside the scope of this book except that GMPs require the master manufacturing record to specify the required label, and require companies to examine incoming labels to assure they conform to specifications, and that the correct label is properly applied to the product.

FOODS FOR SPECIAL DIETARY USE

At the time the FD&C Act of 1938 was enacted, dietary supplements were already well established in the marketplace, and Congress carefully considered whether such products should be part of the food category or part of the drug category. The decision was to consider dietary supplements as a subcategory of foods, and a separate section of the Act, 403(j), was created to deal with foods for special dietary uses—a category, which included dietary supplements. The FDA finalized a definition of special dietary uses and regulations regarding the labeling of such products in 1941.

STANDARDS OF IDENTITY

In the late 1960s and early 1970s, the FDA attempted to establish Standards of Identity for dietary supplements of vitamins and minerals to "rationalize" the formulations available in the marketplace. The agency sought to accomplish this in the same way it had established a standard formula for enriched bread, by using its authority in Section 401 of the FD&C Act to establish a Standard of Identity for a particular class of foods. Despite vigorous opposition from much of the industry and the public, the FDA finalized a rule in August 1973 that would have severely limited the quantities and combinations of vitamins and minerals that could be sold as dietary supplements. The rule was subsequently overturned by the courts as being too restrictive, and legislation was passed in 1976 prohibiting the FDA from limiting quantities of vitamins and minerals for reasons other than safety. It had been the agency's intent to establish a narrow standard for vitamin and mineral supplements sold as foods for special dietary use, but to permit higher level and additional combinations to be sold as over-the-counter (OTC) drugs, provided such formulations were endorsed by an expert panel as part of the massive OTC drug review. An expert panel did, in fact, prepare such a monograph, which was published in 1979, but it was subsequently withdrawn by the agency.

THE DIETARY SUPPLEMENT HEALTH AND EDUCATION ACT OF 1994

In the early 1990s, FDA commissioner David Kessler convened a special internal task force to provide advice on how dietary supplements should be regulated. The task force took a restrictive view of the category, and in its report was the basis for an Advance Notice of Proposed Rulemaking, published by the FDA in June 1993, suggesting that the quantity of vitamins and minerals in dietary supplements should be limited, that amino acids were unapproved food additives, and that herbal products were perhaps inherently therapeutic and possibly unsafe. This aroused considerable displeasure among the consuming public and also within the industry. As a result, Congress passed the 1994

DSHEA, which again confirms that dietary supplements are a subclass of foods and are *not* drugs. Among other provisions, DSHEA authorized the FDA, if they should decide to do so, to publish and enforce GMP regulations for dietary supplement products "modeled after" the food GMPs.

DSHEA defines a dietary supplement to be an ingested product (as distinct from a product to be topically applied, for example) intended to supplement the diet consisting of one or more of the following "dietary" ingredients:

- A vitamin
- A mineral
- An herb or other botanical
- An amino acid
- Another dietary substance for human use to supplement the diet by increasing total dietary intake
- A concentrate, metabolite, constituent, extract, or combination of these ingredients

This definition is now in Section 201(ff) of the FD&C Act.
Dietary supplements may not contain dietary ingredients that present a significant or unreasonable risk of illness or injury, under Section 402(f) of the FD&C Act, or that may render the supplement injurious to health under Section 402(a)(1).

DSHEA also establishes some general labeling requirements for dietary supplement products, defines "grandfathered" dietary ingredients and "new dietary ingredients," and permits a class of label claims called statements of nutritional support (more commonly known as "structure/function" claims) under certain conditions.

DSHEA allows the FDA to take action to remove dietary supplements from the market if they pose a significant or unreasonable risk, but the burden of proof is on the FDA to establish that the products are indeed unsafe. The FDA used this provision as the basis for banning ephedra-based products marketed for weight loss and as sports nutrition supplements, which had been the subject of numerous reports of adverse events.

THE ORIGIN OF REGULATIONS

The statutes written by Congress express in a general way what must be done. They tend to be somewhat vague, and usually do not go into great detail, but instead leave this up to the agency responsible for enforcement of the specific laws. This requires interpretation of the meanings and intent of Congress, on the part of the agency, through the publication of implementing regulations. Section 701(a) of the FD&C Act, in particular, authorizes the FDA to publish such regulations, but this must be done in accordance with the Administrative Procedure Act (APA), signed into law in 1946, which is in sections 511–599 of Chapter 5 of Title 5 of the U.S.C. This statute specifies in detail the process agencies must follow in writing regulations. The establishment of regulations by this procedure is called *rulemaking*, and a "regulation" is technically a "rule."

There are two categories of rulemaking, *interpretive* and *substantive*. Interpretive rules are essentially statements of the judgment of the agency regarding how they believe Congress intended them to establish means of carrying out the dictates of the Act. However, substantive regulations have the full force of law.

Many years ago the courts settled the question as to the category into which GMP regulations fall, and it is now well understood that the GMPs are *substantive*, meaning that they are mandatory and that the FDA has specific power to strongly enforce these regulations.

The APA requires the agency to first publish in the *Federal Register* a notice of a proposal to create a regulation. This is called an Advance Notice of Proposed Rulemaking (ANPR). The purpose of this is to solicit written comments from the public prior to drafting the actual proposed rule. Such comments typically come from companies, trade associations, individuals, academia, etc., and are extremely helpful to the agency in formulating the regulation. A time limit is established for such comments, and the mechanism for commenting is explained in the ANPR. Over the years, this process has proved to be both fair and reasonable, although time-consuming, which is why it took the FDA more than a decade to publish the final GMPs for dietary supplement products.

The FDA is required to consider all the comments received and to summarize both the comments and the agency's decisions triggered by them. This information is then included in the preamble to the next step, publication of the Proposed Rule (PR). Again, comments are solicited in the same manner. This is referred to as a "notice-and-comment" procedure. The next step, if indeed there is one, is the publication of the Final Rule (FR). This may come soon after the comment period expires, or years later, or never. But, again, the responses to the comments received must appear in the preamble, which yields valuable information on the FDA's thinking on the matter. Since the preamble appears only in the *Federal Register* printing of the final regulation [not in the version that ends up in the *Code of Federal Regulations* (CFR)], it is prudent to retain and study the preamble.

When a final rule is published, the effective date and the compliance date of it are spelled out in the *Federal Register* notice of the final rule. In the instance of the DS GMPs, the effective date was August 24, 2007, and the compliance date is June 25, 2008, for firms with 500 or more full-time employees; June 25, 2009, for firms having more than 20 but fewer than 500 full-time employees; and June 25, 2010, for firms having fewer than 20 employees.

Regulations are legally binding, both on the industry and on the FDA. Regulations are enforced in the courts as legal requirements. Disputes may arise regarding regulations and FDA's enforcement of them, however, which may end up in court. In many such instances, judges tend to be reluctant to substitute their judgment for that of the FDA, since the agency is generally considered to have the greatest knowledge and experience regarding the issues under their purview.

The comments the FDA receives during the notice-and-comment process can and often do influence the agency's decisions and actions. The FDA allows ample time for such comments, and does, in fact, carefully consider the comments when it draws up a final rule. Therefore, it is a good idea to submit comments, when appropriate, during the development of regulations. Moreover, the FDA is required by two other laws (the Regulatory Flexibility Act passed in 1990 and the Small Business Regulatory Fairness and Flexibility Act passed in 1996) to actively seek out input from small businesses on new regulations, to ensure that this point of view is considered along with the views of the "giant" firms and the trade associations.

THE FEDERAL REGISTER

The *Federal Register* is an official publication of federal rules, proposed rules, notices from federal agencies, and executive orders. It is published daily (except weekends and federal holidays) by the Office of the Federal Register of the National Archives and Records Administration. It was created by the Federal Register Act of 1934, and has been described as the legal newspaper of the executive branch of the federal government. Following what is published in the *Federal Register* is a way of keeping up-to-date with the regulatory scene. In fact, it is one of the most important sources of information on what the FDA is doing. It is available in hard copy by subscription, and on the Internet, both daily and with available archives of past issues.

THE CODE OF FEDERAL REGULATIONS

The CFR is the codification of final rules that have been published in the *Federal Register*. It consists of 50 titles, each covering a specific area of federal regulations, with Title 21 being devoted to the rules administered by the FDA.

The CFR is updated once a year, on April 1 for Title 21. Specific titles, any or all, can be purchased in hard copy, but are also available on the Internet as the Electronic Code of Federal Regulations (e-CFR), with a convenient search feature. There is also a list of CFR Sections Affected (LSA), which lists proposed, new, and amended federal regulations that have been published in the *Federal Register* since the most recent update of a CFR title.

Each CFR title is divided into chapters, which are further subdivided into parts, sections, paragraphs, and subparagraphs. Citation of specific portions of regulations are typically in the form of, for example, 21 CFR 111.15(b)(1), where "21 CFR" means Title 21 of the Code of Federal Regulations, "111.15" means Part 111, Section 15, and "(b)(1)" means paragraph b, subparagraph 1.

In this book, it is assumed that *all* citations are from 21 CFR, and these will be cited through the use of the symbol "§," for example, §111.15(b)(1). The dietary supplement GMPs are found in Part 111, while the food GMPs are in Part 110 and the drug GMPs are in Part 211 of 21 CFR. These can all be accessed from FDA's Web site.

GUIDANCE DOCUMENTS

Since changing the GMPs takes a very long time, the FDA typically makes available informal guidance documents to help explain the agency's current thinking on specific topics. These are helpful both to the industry and to FDA's personnel. However, these guidance documents may be useful, and they are not regulations and do not have the force of law. They are not legally binding on either the agency or the industry, and it is not mandatory that they be followed. They are merely FDA's way of sharing information with industry as to acceptable ways of complying with specific portions of the regulations. If a firm has a procedure or method it elects to follow instead of what is stated in a guideline, it is at liberty to do so, although it should be prepared to defend that decision.

The FDA welcomes suggestions on new or revised guidances, and they solicit comments on drafts that are published. They encourage industry's involvement in their development of guidance documents.

The FDA has an internal set of policies and procedures for developing, issuing, and using guidance documents, which are called Good Guidance

Practices, as explained in §10.115. It also maintains a list of available guidance documents on the Internet.

DIRECT FINAL RULES AND INTERIM FINAL RULES

A direct final rule is a shortcut for noncontroversial rules. This route is used when the agency believes that the rule is unlikely to generate adverse comments or suggestions that it be withdrawn or changed. When published, a direct final rule has a preamble explaining the rule, and also a deadline for submitting any comments. If any adverse comments are received by the end of the comment period, the direct final rule must be withdrawn and if the agency decides to proceed, it must go through the usual notice-and-comment procedure. If no adverse comments are received by the end of the comment period, the direct final rule becomes effective on the specified date. This is a way of expediting rulemaking for routine and noncontroversial matters.

An interim final rule (IFR) can be used when the agency believes there is good cause to make the rule immediately effective even while there may still be some issues under consideration.

The APA provides for what is called a "good cause exception" to the usual notice-and-comment process if the agency finds that such requirements are impractical, unnecessary, or contrary to the public interest. But, based on comments received, the agency must eventually decide to either issue a *revised* final rule or confirm the IFR as final.

CITIZEN PETITIONS

Another route to regulations is what is called a "Citizen Petition," as provided for in Section 701(e)(1) of the FD&C Act and described in detail in 21 CFR 10.30. This allows any interested person, company, trade association, consumer groups, etc. (or a law firm, if it is desired to petition anonymously) to submit a request to the FDA to issue, amend, or revoke a regulation. In other words, one can petition the FDA to either do or not do something, and the process is relatively easy to follow, although the submitter should be careful in preparing the submission. However, the important concept is that this allows anyone the opportunity to make specific requests of the agency. The FDA then considers the petition, and must respond within 180 days by either approving or denying it (in whole or in part), or providing a tentative response indicating why the FDA has been unable to reach a decision.

Petitions to the FDA are public, and anyone can submit their comments to the FDA either supporting the petition or not. Citizen Petitions have been submitted to the FDA requesting certain changes in the DS GMPs.

Both the notice-and-comment process required in formulating regulations and the Citizen Petition process assure that anyone can be involved and participate in rule-making. The system is indeed open and fair.

AVAILABILITY OF ACTS AND REGULATIONS

The entire text of the FD&C Act and the several other Acts enforced by the FDA, plus the entire 21 CFR as well as FDA's guidance documents and many other pertinent references are available on FDA's Web site at http://www.fda.gov.

WHO IS SUBJECT TO THE DS GMPS

Any firm that manufactures, packages, labels, or holds dietary supplements must comply with the GMPs, as specified in 21 CFR 111.1. This includes both domestic firms and those located in other countries that export DS products to the United States. This includes firms that make DS products but then sends them to another firm for packaging and/or labeling. It also includes contract manufactures involved in the production of DS products.

The FDA considered requiring the firms that make the raw materials (components) used in manufacturing DS products to be subject to the DS GMPs. However, many suppliers of such ingredients also provide these same items to other industries too, not to just the dietary supplement manufacturers, so ultimately the FDA declined to make the suppliers subject to the DS GMPs. Thus, manufacturers of *dietary ingredients* and excipients are not necessarily subject to the GMPs. It is up to the manufacturer of the finished DS product to establish appropriate specifications for all components and then take steps to be certain that incoming materials meet these specifications. This can be accomplished through a certificate of analysis from a certified vendor, coupled with positive identification of each ingredient, as further discussed in more detail elsewhere in this book.

A firm that does only *part* of the manufacturing operations of DS, such as just the final packaging and/or labeling, is subject to only those portions of the regulations that apply to those specific operations. However, that firm would also need to comply with the sections related to personnel, the physical plant, quality control, and other requirements that apply to that firm's operations, but *not* with other requirements with which that firm is not involved (e.g., that firm would not be expected to do testing of the finished product since it did not make the finished product).

In general, the firm named on the label must be assured that any firms doing part of the manufacturing are, in fact, in compliance. The firm that releases the product for distribution does need to be certain that all steps did, in fact, comply.

Similarly, a firm that purchases a packaged and labeled DS product and then warehouses that product for eventual distribution to other firms is subject to the GMP requirements related to its specific operations.

Retail stores that merely stock and sell finished DS products are not subject to the GMPs.

In short, the GMPs apply to firms based on the functions and operations that company is performing related to the manufacture, packaging, labeling, and warehousing of DS products. If there are any questions, or if there is uncertainty as to whether all or parts of the DS GMPs are applicable in any given situation, it is advisable to contact the FDA's District Office for clarification. Moreover, in borderline situations where it may not be clear as to what (if any) sections of the GMPs are applicable, the FDA can exercise its enforcement discretion in arriving at a decision, and therefore consultation with the FDA is prudent.

SUGGESTED READINGS

Bass IS, Young AL. The Dietary Supplement Health and Education Act: A Legislative History and Analysis. Washington, DC: Food and Drug Law Institute, 1996.

Cross-reference of Section Numbers: FD&C Act and United States Code. Available at: http://www.fda.gov/RegulatoryInformation/Legislation/FederalFoodDrugandCosmeticActFDCAct/ucm086299.htm

Piña KR, Pines WL, eds. A Practical Guide to Food and Drug Law and Regulations. 2nd ed. Washington, DC: Food and Drug Law Institute, 2002.

Schnoll L. The Regulatory Almanac—A Guide to Good Manufacturing, Clinical, and Laboratory Practices. Chico, CA: Paton Press, 2000.

Sharp J. Good Manufacturing Practice: Philosophy and Applications. Buffalo Grove, IL: Interpharm Press, 1991.

3 Personnel matters

IMPORTANCE OF HUMAN RELATIONS MANAGEMENT
People are arguably *the* most important factor in any company, large or small. The selection, recruitment, training, development, motivation, and supervision of personnel at all levels are complex but extremely significant topics. These relate to organizational effectiveness, productivity, customer satisfaction, and quality. Unlike machinery and equipment, each individual is unique. People not only differ from one another but also vary in performance from day to day or even hour to hour. Personal issues and concerns can and do impact the actions of people, and these need to be given due consideration. Human capital plays a vital role in the long-term competitive situation of the firm. One important facet of this has to do with the impact of personnel issues on GMP compliance.

PERSONNEL QUALIFICATION REQUIREMENTS
The GMPs, in §111.12(a), state that it is necessary to have *qualified* employees who manufacture, package, label, or hold dietary supplements. This means that the individuals who actually perform any or all of these functions must be suitably qualified. This does not speak about the *number* of people required to do the work, but obviously implies the need to have enough employees to ensure compliance with the regulations. Clearly, the number needed will vary depending on the tasks that need to be done, leaving this up to the judgment of the management of the firm. Similarly, the definition of "qualified" is not explicitly stated, except that in §111.12(c) the regulations require that each person engaging in these activities must have the *education*, *training*, or *experience* to perform his or her assigned functions. Clearly, some positions require an appropriate educational background as well as specific on-the-job training (and "training" is indeed a form of "education"). "Experience" as used here implies the knowledge gained over time as an individual becomes increasingly familiar with a set of tasks and/or the use of specific equipment or procedures. Again, then, what is required for given employees to be qualified is subjective and up to the judgment of management.

SUPERVISOR QUALIFICATION REQUIREMENTS
Similarly, §111.13 requires that each of the individuals who *supervise* the manufacturing, packaging, labeling, or holding of dietary supplements must also be qualified to do so. As in the instance of other personnel, it is up to management to determine the appropriate amount and balance of education, training, and/or experience. This would also apply even to executives of the firm, if they supervise employees involved with manufacturing, packaging, labeling, or holding dietary supplements. The term "supervise" is a bit ambiguous, but should be construed in the usual commonsense meaning. The role of supervisors is critical in assigning the right employees to specific tasks, making sure that those people know how to properly perform their assigned jobs and do

them well, and ensuring that GMP compliance matters are properly handled. Moreover, the leadership and motivational qualities of supervisory personnel are also of great importance in accomplishing these goals.

Although this is not specified in the GMPs, it is a good business practice to have a policy indicating who is qualified and authorized to temporarily fill-in to cover for key supervisors during periods when they are absent.

TRAINING

One facet of "qualified personnel" of course involves training. The DS GMPs do not specify precisely *what* training is required for each individual, because of the great diversity in operations from company to company, leaving this to the judgment of each firm's management. However, based on long experience with other FDA-regulated industries using GMP-based quality systems, it is clear that the training must cover both GMPs *per se* and the details of the performance of each person's specific tasks. In other words, each employee needs to understand what GMPs are and why compliance with them is urgent, and must also understand how his or her specific job must be performed. Whenever a person is assigned a new set of tasks, these too must be thoroughly understood.

Moreover, the training must be repeated at appropriate time intervals to assure that the workers still remember the concepts and details.

While not specifically stated in the regulations, the training requirement applies to temporary employees as well as to permanent personnel. If all or some of the employees do not understand English well, the training should be done in the language they do understand. Training should be given to new employees before they actually start performing their tasks.

Despite the fact that the required content of the training is up to the firm's management, §111.14(b)(2) mandates *documentation* of training, including the dates, the type of training conducted, and the names of the person or persons trained. It is advisable to include the name of the person conducting the training, the time spent, and the topics covered. This documentation is critical since successful quality management systems are inextricably tied to appropriate training programs, and documentation is needed to help assure that each individual has, in fact, received the required training. The documentation also makes it possible to track which employees have been trained in which operations.

In addition to documenting training, it is usual and prudent (but not specifically mandated) to keep a file on each employee, including information obtained during the hiring process as well as performance reports and other data. Such files should include information on the person's educational and experience background at the time of employment, in addition to the details of training received during his or her time with the firm. In addition, some individuals may also acquire skills and/or education outside of their employment, so it is a good idea to encourage employees to periodically update such information in their files. This could include courses taken, seminars attended, and other such activities that would expand the record of their education, training, and experience.

There are many possible approaches to providing training. Obviously, classroom lectures are one form. These can be conducted by in-house personnel, or by outside consultants or firms specializing in such training. Or, employees can be sent to outside courses. Training aids, appropriate audio-visual media, computer- and Internet-based training, are also available. Handouts, posters, articles in the firm's newsletter if there is one, or GMP reminders passed out with paychecks can be used to reinforce more formal methods of training. But training needs to be an ongoing dynamic process.

It is important to recognize that adults learn in different ways than children do, so training needs to be tailored accordingly. The "memorize and recite" technique usually is not very effective with adults where interactive discussion tends to work well. But there is likely to be a variety of learning types within any group of employees, so that the training approaches used often require adaptation. The objective is to teach the basic GMP concepts and the reasons for them, coupled with practical ways of accomplishing specific tasks in full compliance with both the letter and the spirit of the regulations.

For specific tasks, on-the-job training, with instruction given by a supervisor or an experienced fellow worker (the "buddy" system), is a useful approach, and for this reference to and use of the appropriate Standard Operating Procedures (SOPs) can be a helpful tool.

Whatever combination of training techniques is used, it is important to check for effectiveness. This can be accomplished by the use of written tests, verbal questioning, observation of performance, and other methods. Reminders and observations on the part of supervisors can be extremely helpful. Positive reinforcement by acknowledging that tasks are indeed being done correctly and in accordance with GMPs shows caring and also helps assure continual improvement. Evaluation of training through observation and coaching helps people respond in a positive way.

While not specifically stated in the regulations, it is a good idea to give *all* company employees training in the basic concepts of, and reasons for, the GMPs, and not limiting training to those directly involved in manufacturing operations.

Furthermore, §111.8 requires having written procedures (SOPs) covering all the facets of personnel matters covered in Subpart B of the DS GMPs, which of course includes but is not limited to training, while §111.14 specifies that the written procedures and the documentation must be kept.

PERSONAL HYGIENE

Humans carry countless microorganisms on their skin, in their hair, in their saliva, and in their respiratory tracts. These organisms are normal inhabitants of human bodies and can easily be transferred, directly or indirectly, to products at various stages, thereby possibly causing contamination that could be ruled adulteration.

Most of the organisms resident in and on people are essentially harmless or even helpful, but they do not belong in dietary supplement products. And, of course some such organisms potentially could be dangerous. Even individuals with scrupulous personal habits of hygiene can inadvertently cause contamination. Therefore, the GMPs require specific measures to minimize the potential for contamination.

In §111.10(a), it is stated that measures must be taken to exclude people from working at certain specific tasks if they have health conditions that could lead to product contamination. This would include individuals who by medical examination, the person's own acknowledgement, or by supervisory observation, are either shown to have or appear to have, an illness, infection, an open sore or wound, or any other such abnormal possible source of microbial contamination. This does not necessarily mean that people who are sick or have an open sore must not work at all while they have such conditions, but it does mean that they must not carry out tasks where they could expose raw materials, packaging components, in-process work, products, or contact surfaces to contamination. They could, for example, do office work or certain warehouse duties, as long as they do not infect others who in turn might expose materials or contact surfaces that could result in product contamination. Management has the flexibility of deciding whether it would be safe and appropriate for a person with a health issue to be allowed to work in an alternate area where product contamination could not occur. Employees must know to inform their supervisor if they have (or if there is a reasonable possibility that they may have) a health condition that could result in product contamination, according to §111.10(a)(2), and the supervisor must then act responsibly. It is important that all employees understand and follow this rule.

HYGIENIC PRACTICES

The GMPs, in §111.10(b), spell out nine specific practices that must be followed to the extent necessary to protect against contamination by anyone working in an operation where adulteration of a component, product, or contact surface could occur. The phrase "to the extent necessary" does give management some flexibility in this regard.

In the GMPs, a "contact surface" is defined as being any surface that contacts a component or dietary supplement product, that is, a surface where contamination of the product could occur.

The first of these hygienic practices is the wearing of outer garments in a manner that protects against contamination of components, dietary supplements or any contact surface. This does not necessarily require workers in the plant to wear uniforms or smocks, although that is one good solution in part because it helps reinforce in the workers' minds the importance of keeping clean. Cleanliness of dress also tends to encourage other personal habits of cleanliness. Moreover, putting on clean outer garments after employees arrive at work does help minimize the potential for contamination. Importantly, however, *any* clothing worn by employees in processing and production areas *must* be kept clean. Another factor worth consideration is the fact that humans constantly shed skin cells at a surprisingly rapid rate, and these cells often carry microorganisms on them which can contaminate, while clean outer garments help reduce this potential.

The second hygienic practice cited in the GMPs is maintaining adequate *personal* cleanliness. This tends to be a cultural factor, and comes from what has been taught at home and school. It is not a subject that is easily broached, and changing adults' points of view and personal habits is not easy, but it is a risk area that does warrant attention. It is surprising how many people are essentially ignorant about the basic principles of personal hygiene. In addition to bodily cleanliness and good grooming, it includes risk behaviors such as not covering one's nose when sneezing, scratching one's head near product contact surfaces, and other obvious unsanitary actions. Effective supervision can be a useful tool in helping teach habits of cleanliness, in addition to company rules and regulations and reminder publications and posters.

The third hygienic practice addressed is washing hands thoroughly (and sanitizing, if necessary, to protect against contamination with microorganisms) in an adequate hand-washing facility, both before starting work and at any time the hands may have become soiled or contaminated. The urgency of this cannot be overemphasized. Hands may appear to be clean when, in fact, they are really not. Hands easily become soiled in using the bathroom, sneezing, from dirt and body oils or perspiration, greeting others by shaking hands, and touching unsanitary items. The contamination on unclean hands is easily transferred to product contact surfaces of packaging machinery or other equipment, to products and containers, and in countless other ways. Many people do not truly understand *how* hands should be washed, so instruction on this is worthwhile. The recommended procedure is to use very warm running potable water to moisten the hands, then apply enough soap to get a good lather, rub the hands together for at least 20 seconds and also wash the wrists, then rinse thoroughly in running water. Then dry the hands. Another step that can be useful is to also use a fingernail brush since microorganisms often exist on and under the nails. There is an ongoing debate as to whether plain soap or antibacterial soap is preferable. When it is inconvenient to use soap and water, alcohol-based hand sanitizers are useful, although it is usually difficult to remove visible soil by using such products. A number of commercial types of hand sanitizer products exist, which the regulations say should be used "if necessary." But, while the regulations stress the importance of clean hands, they are silent on the details of just how this must be accomplished.

It has been shown that the method of drying hands after washing is nearly as important as is the washing step, with disposable paper towels performing significantly better than warm air dryers. Keeping fingernails clean and well trimmed is also an important part of good hygienic practice, albeit this is not specifically addressed in the GMPs.

The fourth topic on personal hygiene in §111.10 is removing all unsecured jewelry and other objects that might fall into components, dietary supplements, equipment, or packaging, and removing hand jewelry such as rings that cannot be adequately sanitized during periods in which components are manipulated by hand. If hand jewelry cannot be removed, it must be covered by tape or other material that is maintained in an intact, clean, and sanitary condition and effectively protects against contamination of components, dietary supplements, or contact surfaces. This is both to prevent possible contamination since rings, bracelets, necklaces, earrings, etc., can harbor microorganisms, and as a safety measure in that jewelry and other such items can get caught in moving machinery, and thus endanger the employee. It is advisable for employees not to carry pens, pencils, glasses, etc., in shirt pockets where they could fall out and cause problems.

The fifth item on personal hygiene has to do with gloves used in handling components or dietary supplements. If used, these must be made of an impermeable material, and must be in an intact, clean, and sanitary condition. The regulations do not require the use of gloves in any specific operations, although they are required *when necessary* to prevent contamination. In general, gloves are needed for anyone who must directly touch the product or contact surfaces. Disposable gloves are frequently used in such instances, but if reusable gloves are used, they must be well cleaned and sanitized.

The sixth required hygienic practice requires, *where appropriate*, wearing in an effective manner hair nets, caps, beard covers, or other effective hair restraints.

The concept here is to prevent hairs from getting into product. Moreover, hair generally carries a heavy load of bacteria and other microorganisms, and therefore hair should be kept clean, although this is not specified in the regulations. Since caps are allowable as head coverings, when they are used as the hair restraint, while not mentioned in the regulations, it is advisable that the hair be neatly trimmed at the sides and back, so that it is at least mostly covered. Many firms permit sideburns, as long as they too are neatly trimmed. Again, the statement "where appropriate" gives management flexibility of how to enforce this requirement.

The seventh of the hygienic practice requirements is that clothing or other personal belongings must not be stored in areas where components, dietary supplements, or contact surfaces are exposed, or where contact surfaces are washed.

The eighth in this series of requirements speaks about not eating food, chewing gum, drinking beverages, or using tobacco products in areas where components, dietary supplements, or contact surfaces are exposed or where they are washed. This gives firms flexibility to determine where employees may or may not eat, chew gum, drink, or use tobacco products.

The ninth and final portion of §111.10 is something of a nonspecific catchall, requiring taking any other precautions to protect against the contamination of components, dietary supplements, or contact surfaces with microorganisms, filth, or any other extraneous materials, including perspiration, hair, cosmetics, tobacco, and medicines applied to the skin.

It is important to keep in mind that these principles of personal hygiene must apply to *any* and *all* individuals (not only company employees) that enter areas where components, products, or contact surfaces could become contaminated.

This includes visitors, maintenance people, consultants, construction crews, service or calibration personnel, etc. To control this, while not specifically mandated, many firms take steps to limit the access to specific areas to be sure the individuals who enter do, in fact, comply.

THE ROLE OF TOP MANAGEMENT

Although the regulations are silent on this, it is well known that what the "boss" wants, says, and does sets the tone and example for all employees. It is, therefore, a good business practice for the key executives to make it clear that quality is the lifeblood of the company, and that the GMPs are both required by law and valued by management as an excellent basis for helping ensure quality. This helps set the attitudes of the supervisors and the hourly workers. Without full and enthusiastic support from the top, the quality system is unlikely to perform well. Moreover, top management must make the necessary financial commitments to support the quality endeavor. This is true regardless of the size of the firm, and applies to sole proprietors of small firms as well as to CEOs of "giant" establishments.

The fact that the FDA holds the top management of regulated firms accountable is generally well understood. Top executives are responsible for *all* activities in the firm, including the assurance of quality. They may delegate the day-to-day duties involved to subordinates, but the top executives must be sure that all functions are, in fact, being well handled. During FDA inspections, the investigators are instructed to document the chain of command, not only the organizational details, but also the names of executives responsible for various functions. If unsuitable conditions are found, both the company *and* the accountable executives can be in trouble. Warning Letters are always addressed to the highest-ranking executive in the firm. When regulatory action is required, in severe instances, criminal charges can be brought against the accountable executives. The courts have ruled that the CEO (whatever title the top executive may have) cannot be excused simply because an underling failed to perform well. This is called strict vicarious liability and means that even a corporate executive who has no direct involvement in or knowledge of conditions that result in a criminal violation can be held criminally accountable since that person at least had the opportunity of informing himself or herself of the violative conditions. The U.S. Supreme Court has taken the position that executives have a positive duty to implement measures to ensure that GMP violations will not occur.

In summary, both for good business reasons and to avoid the possibility of serious legal problems, top management should take all appropriate steps to be sure that the GMP regulations are well understood and are properly and fully complied with in the company.

QUALITY CONTROL PERSONNEL
As required by §111.12(b), the person who is responsible for quality control operations must be identified, and each person who is identified to perform quality control operations must be qualified to do so and have separate responsibilities related to performing such operations, distinct from those that the person has when *not* performing quality control. Also, subparagraph (c) goes on to say that each person engaged in performing any quality control operations must have the education, training, or experience to perform the assigned functions.

This topic is discussed further in the chapter on quality control.

SANITATION SUPERVISORS
One or more employees must be assigned to supervise overall sanitation, according to §111.15(k). The usual requirement for qualification by education, training, or experience to develop and supervise sanitation procedures applies.

Again, this gives the firm flexibility in determining the appropriate combination of qualification attributes, and such supervisors should be selected and trained accordingly.

SUGGESTED READINGS
Gallup D, Beauchemin K, Gillis M. Competency-based training program design. PDA J 1999; 53(5):240–245.
Morris W. Personnel training: a growing compliance concern. PDA Lett 2006; 42(4):14–18, 24–25, 38.
O'Keefe DF Jr., Shapiro MH. Personal criminal liability under the Federal Food, Drug and Cosmetic Act—the Dotterweich Doctrine. Food Drug Cosmet Law J 1975; 30:5.

Redway K, Knights B. Hand Drying: Studies of the Hygiene and Efficiency of Different Hand Drying Methods. London: School of Biosciences, University of Westminster, 1998.

Sethi SP, Katz R. The expanding scope of personal criminal liability for corporate executives—some implications of United States *v*. Park. Food Drug Cosmet Law J 1977; 32:544.

Snyder JE. Management responsibility for the quality system: a practical understanding for the CEO in FDA-regulated industries. J GXP Compliance 1999; 3(3):53–59.

Spiegel K. Handwashing and sanitizers. Food Quality, February/March, 2006:73–82.

U.S. Supreme Court, United States v. Dotterweich, 320 U.S. 277, 1943, and United States v. Park, 421 U.S. 658, 1975.

Vesper JL. Training for the Healthcare Industries: Tools and Techniques to Improve Performance. Boca Raton, FL: CRC Press LLC, 1993.

4 Physical plant and grounds

It is obvious that the physical plant and grounds do indeed play a prominent role in product quality. One must do what is feasible to eliminate the possible entry of contaminants in the first place, and further, to take positive steps to avoid spreading dirt, dust, and other forms of contamination from area to area within the facility. While firms building a completely new facility have the distinct advantage of being able to plan the size, layout, and construction details, since constructing a new facility is both costly and time-consuming, what is more usual is to take whatever steps as appropriate to renovate, upgrade, and refurbish existing plants, and to also do what is needed to keep the facility in a good state of maintenance.

SPACE REQUIREMENTS

In §111.20(a), it is stated that any physical plant used to manufacture, package, label, or hold dietary supplement products must be of *suitable* size, construction, and design to facilitate maintenance, cleaning, and sanitizing, and must have *adequate* space for the orderly placement of equipment, as well as for the storage of materials needed for maintenance, cleaning, and sanitizing operations to reduce the potential for mix-ups or contamination.

Note the words "suitable" and "adequate," which give management some flexibility, but of course can lead to disagreements with the FDA.

Thus, firms should conduct a review of space available and how it is being used. Modifications, or enlarging the facility may possibly be required to comply with this portion of the regulations, and it is prudent (although not specifically mandated) to prepare and retain a report on this and the decisions reached. It is also advisable to periodically review the space issues to keep abreast of changes that may occur, such as the addition of new products, and the increase or decrease in the production volume of certain products with the passage of time. The insidious hazard is that sales volumes tend to change gradually, and operating personnel learn to cope with the increases when they occur, and do not necessarily recognize the resulting overcrowding that could result in noncompliance. Changes in space utilization can, of course, also impact the utilization of electrical power, steam, water, compressed air, drains, etc. A complete understanding of all the issues is necessary, not merely a narrow focus.

The usual concern in modifying an existing facility is a financial one, that is, that any required changes will be unaffordable, but this is not necessarily true. With careful evaluation and planning, it is often possible to hold the costs of upgrading facilities within reason.

In §111.20(b), it is specified that adequate space must be provided for the orderly placement of equipment and supplies for *maintenance*, and *cleaning* and *sanitizing operations* to prevent contamination and mix-ups, including contamination from chemicals, microorganisms, filth, or other extraneous materials.

Section 111.20(c) calls for having *separate* defined areas of adequate size in the physical plant, *or other* suitable systems (such as computerized inventory controls or automated systems of separation) to prevent contamination and/ or mix-ups of components, packaging, labels, and/or dietary supplements when they are received, stored, or rejected, or while they are awaiting a material review and disposition decision, or reprocessing, separating them *as necessary*. This also calls for separating products and their raw and packaging materials of different types of dietary supplements and other foods, cosmetics, and pharmaceutical products. This applies while performing laboratory analyses, the storage of laboratory supplies and samples, during cleaning and sanitizing contact surfaces, and in warehousing components or dietary supplements. This means that components, products, etc., in the warehouse do *not* necessarily need to be in "separate defined areas" *if* there is a suitable system in place to avoid contamination and mix-ups. It also means that different materials can be processed in the same room concurrently as long as adequate controls are in place. This gives firms considerable flexibility, but again it is prudent to conduct a study to see if these requirements are being met and if not to take corrective action.

PHYSICAL DESIGN AND CONSTRUCTION

The plant must be designed and constructed in such a way as to prevent contamination of components, dietary supplements, or contact surfaces, according to §111.20(d). This requirement, too, is not specific and is subject to interpretation.

WALLS, FLOORS, AND CEILINGS

Walls, floors, and ceilings must be capable of being adequately cleaned, and must be kept in good repair, as stated in §111.20(d)(1)(i). In some parts of the plant, as discussed below, this may imply a need for walls, floors, and ceilings with smooth and hard surfaces that are easily cleaned and are not damaged by the cleaning chemicals and process. The response to Comment 101, in the preamble to the regulations, gives firms considerable latitude regarding the specific *types* of walls, floors, and ceilings that are acceptable, since it recognizes that the appropriate materials of construction *can* be varied in different parts of the plant. For example, an area where labels are stored could have different requirements than an area where wet processing is done since wet environments can create numerous issues such as harboring bacteria, molds, and other unwanted pests that could lead to contamination. Moreover, office areas present different requirements than manufacturing areas.

The GMPs are silent on *how* walls, floors, and ceilings should be made to comply with §111.2(d)(1), so the following comments are *suggestions*, not requirements.

Concrete or cinder block walls, or plaster or even gypsum wallboard walls, coated with appropriate high-performance interior architectural wall coatings (HIPACs) offer good service at a reasonable cost. The HIPACs are tough, extra-durable coating systems that are applied as a seamless film and cure to a hard easily washable finish that can withstand heavy-duty cleaning agents. There are of course many additional methods for constructing walls, including tile, glazed block or glazed facing tile, preformed fiberglass reinforced polyester (FRP) panels mounted in anchored floor channels, precast tilt-up concrete wall panels with a sealer on the surface, and others.

It is a good idea to eliminate ledges and window sills if feasible, but if unavoidable they should slope downward at about a 45° angle to prevent accumulation of dust and dirt.

Floors in dietary supplement plants tend to pose problems in part because of human traffic, forklift trucks, and moving heavy equipment over them, plus they require frequent cleaning since dust, dirt, and debris of course fall to the floor. Often the base of the floor is poured concrete, but concrete tends to be porous, and the pores can harbor microorganisms. Moreover, concrete surfaces tend to dust because of the abuse from traffic. To counter these difficulties in critical areas it is common practice to use toppings or coatings of which many types and grades exist, such as various kinds of epoxy systems (with or without aggregates), urethanes, vinyl esters, polyureas, methyl methacrylates, and others. The selection of the optimum topping or coating requires careful consideration, in part because in some areas the floor needs to be resistant to chemicals that may be spilled as well as to harsh cleaning and sanitizing products used. In general, it is best to discuss the selection and installation technique with the manufacturer of the flooring material, rather than with the flooring contractor since the manufacturer often tends to be better informed about the product.

It is advisable to round the juncture between the wall and the floor with a cove having a radius of at least 1 in. Coved junctures help eliminate the crack that would otherwise be present, and also help facilitate cleaning and maintenance. A freestanding concrete curb is sometimes used, both to achieve the cove and to act as a bumper to prevent materials-handling equipment from damaging the wall. Where a topping is used on a concrete slab, a preformed cove piece is usually placed first, then the topping run right up to it to avoid a joint.

Depressions and low areas in floors should be eliminated to the extent possible, to prevent the accumulation of dirt and moisture. If floor drains are used, a slope toward the drain(s) of about one-quarter inch per linear foot is recommended.

Three types of joints are commonly used in concrete floor slabs, *isolation* joints to separate the floor from walls and columns to permit horizontal movement upon drying of the slab, and vertical movement of the walls and columns due to differential settling rates, *control* joints to permit horizontal movement of sections of the slab because of shrinkage during drying as well as thermal expansion and contraction, and *construction* joints used when it is not feasible to pour the entire slab at one time. The joints need to be taken into consideration in the application of toppings or coatings, particularly regarding expansion and contraction that can result in heaved or cracked floors.

In addition to coatings and toppings over concrete, there are also several other types of floors used in some dietary supplement plants, including ceramic tile, and others. The important considerations are to provide a surface easy to clean, free from cracks, crevices, and pores that can be well maintained. Where only light and dry work is done, such as in packaging areas, if forklift trucks or other heavy equipment would not be involved, even vinyl tile could be considered. In office areas, carpet can be appropriate. In warehouse areas plain concrete with a surface-sealer, may suffice. Each area of the plant is likely to have differing requirements, so there is no "one size fits all." The details of the optimum floor in each section should be specifically tailored to the conditions in that area. This helps the sanitation effort and can also result in savings in repair and maintenance costs over the life of the building.

Ceilings present still another set of problems in that they must be easy to clean, smooth and flat, moisture resistant, and permanently devoid of cracks and fissures that could accumulate dirt and give harborage to microorganisms. The underside of concrete roof slabs can form adequate ceilings, as can plaster on metal lathe, although plaster tends to be porous. In multistory buildings, the underside of a concrete slab on one level can serve as the ceiling for the level below. Painted ceilings should be avoided, particularly when moisture is involved, in that peeling is likely to occur, which can cause contamination.

If moisture or steam is present in the area, vents should be installed to prevent moisture from accumulating on the ceiling.

There continues to be an ongoing debate over the advantages and disadvantages of using suspended ceilings, that is, large panels in T-shaped tracks hung on wires. The advantages include fast and simple installation, the ability to cover and conceal pipes, conduits, ducts, etc., the ease of getting to overhead utilities for maintenance, and the flat, smooth, attractive, easy-to-clean surface that result. Light fixtures are also easy to install in such ceilings, and some brands of panels are made of fiberglass, ceramics, or have sealed plastic covers making them waterproof and washable. The disadvantages are that dust and dirt can collect behind the panels, and insects or rodents may live in the space behind the panels. Some suspended ceilings have the panels gasketed to avoid these problems, but that adds to the cost and diminishes the ease of maintenance.

FIXTURES, DUCTS, AND PIPES

According to §111.20(d)(1)(ii), fixtures, ducts, and pipes must not contaminate components, dietary supplements, or contact surfaces by dripping or other leakage, or condensate.

While not specifically mentioned in the regulations, *any* overhead structural details could also possibly contaminate by collecting dust and dirt over time, which could fall into products or onto contact surfaces, or could be runs for insects or rodents. Therefore, if I-beams, H-beams, channels, or other overhead structural elements have been used, it is prudent to either box them in or to completely fill the webs with a suitable inert material such as polyurethane foam. Similarly, gaps and voids between structural members and walls or ceilings can be filled. Piping or conduit directly above exposed product or contact surfaces should be avoided to reduce the possibility of contamination.

The basic concept is to eliminate unnecessary surfaces to be cleaned, as well as to eliminate hard-to-clean surfaces. Such elimination can be accomplished in part by running electrical conduit, small pipes, etc., when feasible, in or under floor slabs. Another way is to use single-sealed enclosures or raceways to include several items.

One approach to minimize the chances of cross-contamination *between* areas where different manufacturing steps are conducted is to isolate such areas through the use of interior partition walls. While this does "chop up" the layout and adds to problems in materials handling and supervision and also requires the use of hallways or corridors, isolation of individual operations is an effective way of reducing cross-contamination.

Although not mentioned in the GMPs, the plant layout preferably should be planned in such a way as to eliminate the necessity and opportunity for employees and visitors to pass through critical areas where product, components, or contact surfaces are exposed. Observation windows can be provided to enable seeing activities within such areas without the necessity of actually entering.

Offices, cafeterias, and laboratories can be located on exterior walls with windows to the outdoors, but separated from manufacturing and packaging areas. The location of entrance and exit doors is worthy of thoughtful consideration to avoid vandalism or burglary from outsiders and pilferage and theft from insiders. Careful attention to security systems, and ways of monitoring and controlling who comes into and who leaves the plant should also be considered. Outside truck drivers are frequent and necessary visitors to most plants, and often have free time while there, awaiting loading or unloading operations. It is worth considering providing a small lounge for truckers adjacent to the dock area to make it unnecessary for them to leave a restricted area.

VENTILATION OR ENVIRONMENTAL CONTROL EQUIPMENT

§111.20(d)(1)(iii) requires having adequate ventilation or environmental control equipment, such as airflow systems, including filters, fans, and other air-blowing equipment, that minimizes odors and vapors (including steam and noxious fumes) in areas where they might cause contamination. Similarly, §111.20(d)(1)(iv) requires equipment that controls temperature and humidity when such equipment is necessary to ensure the quality of the dietary supplement. These sections, plus §111.20(d)(2), are the only ones that speak of ventilation and temperature and/or humidity control. §111.20(d)(2) states that when used, fans and related equipment must be located and operated in a manner that minimizes the potential for microorganisms and particulate matter to contaminate components, dietary supplements, or contact surfaces. This includes exhaust equipment and venting equipment.

FDA agrees that controls for temperature and humidity are only *required when necessary* to ensure the quality of a dietary supplement, but it is important to keep in mind that bacteria multiply rapidly in a warm environment, while molds thrive in a moist environment. So care must be taken to not ignore these factors.

This falls under the topic generally called heating, ventilating, and air-conditioning (HVAC). It is somewhat surprising that these terms are not used in the regulations nor is the subject further elaborated upon since such factors are significant both for the comfort of personnel and also in many instances for protecting product quality. HVAC installations and maintenance tend to be vital but costly, and involve separate detailed technology. The design, installation, and operation of even the most basic HVAC systems require specialized knowledge and careful planning and execution, in part because of the impact on indoor air quality.

HVAC systems usually involve at least partial recirculation of air, which can result in particulate matter and microorganisms being transferred from one area of the building to other areas. Therefore, HVAC can be an important factor in controlling airborne contamination. Such systems usually involve bringing in some fresh air from outdoors, which gets mixed with recirculated air, and is filtered, heated, or cooled, and often humidified or dehumidified. The degree of filtration is critical, and is dependent in part on the types and grades of filters used. Pressure differences between areas of the building are critical to the directional flow of air, and are usually controlled according to the cleanliness requirements of specific areas, in a cascade effect with the most critical ("cleanest") areas being at the highest pressure. The number of air changes per hour in different sections of the plant can also be important.

The physical location of the fresh air inlets can also be of importance in controlling airborne contamination. Dust and other forms of potential contamination may enter the air from outdoors, or may be picked up in different areas of the plant, in particular wherever dust is generated in operations. Humans also contribute skin scales, as well as droplets and microorganisms from the nasal passages and from speaking. Floor drains can also create aerosols that may involve unwanted organisms. For these reasons, the design and maintenance of all forms of HVAC deserve a high priority.

HVAC systems use air monitoring devices (AMDs), sensors, and controls, the adjustment and maintenance of which are of importance. Although not mandated, it is prudent to periodically check air velocities and the number of air changes per hour as well as the pressure differentials between rooms or areas of the facility, and to also monitor the temperatures and relative humidity achieved. Filter integrity also needs to be checked from time to time and filters changed on a predetermined schedule. It is also advisable to at least occasionally check the nonviable particle counts in various areas, and also do environmental monitoring on a periodic basis to assure there are no ongoing potential problems with airborne microbiological contamination.

In small plants without sophisticated HVAC, even the simple use of fans and similar equipment needs careful consideration to avoid airborne contamination.

AISLES AND WORKING SPACES

§111.20(d)(1)(v) calls for aisles or working spaces between equipment and walls that are sufficiently unobstructed and of adequate width to permit all persons to perform their duties and to protect against contamination by clothing or personal contact.

The concept here is that if equipment is too close to walls, people working there might press against and contaminate the equipment. Therefore, adequate space is needed, although "adequate" is not defined.

LIGHTING

Adequate light is required by §111.20(e) in all areas where components are examined, processed, or held, and in all areas where contact surfaces are cleaned. Similarly, adequate light is required in hand-washing areas, dressing and locker rooms, and in bathrooms. There are no standards explaining specifically how much light is required in each area, so this becomes a matter of judgment.

Good illumination is essential and cannot be overemphasized for dietary supplement manufacturing facilities. This contributes to production efficiency, quality, sanitation, and safety, but too often is not given sufficient attention. The needed lighting power density obviously varies from one part of the plant to another, depending on the nature of the work performed in each area. For example, tasks involving inspection require more light than do warehousing operations. Proper lighting also increases workplace attractiveness, enhancing morale.

Factors to consider in selecting or retrofitting lighting include freedom from glare, contrast (i.e., color differences between details and the background), and the positioning and direction of the sources of light. Energy-wasting incandescent lamps are being phased out in favor of more efficient sources, including solid-state light-emitting diodes (LEDs) now available in many different shapes and sizes, and improved high-intensity fluorescent fixtures now starting to be used in some instances to replace high-intensity discharge (HID) light sources (such as high-pressure sodium lamps, mercury vapor, and metal halide lamps) long used for lighting warehouses and other high-bay areas.

Tubular fluorescents remain in wide use, including the T8 32-W lamps that in some instances are replacing the T12 40-W lamps. Also, the highly energy-efficient European technology T5 lamps are gaining popularity for industrial illumination. The "T" designations indicate that the lamps are tubular in shape, while the "8" means 1-in. diameter, the "12" means 1½-in. diameter, and the "5" is a metric dimension.

Lighting fixtures are now usually referred to as "luminaries," and consist of the entire lighting unit, including the lamps and the internal and exterior parts designed to hold the lamps and to connect to the electric power supply.

Some facilities make good use of natural daylight through windows and skylights, both for energy savings and to improve the quality of the work environment.

Exterior lighting, at doorways, walkways, and parking lots, also needs consideration for the convenience and safety of employees and visitors. An effort should also be made to minimize attracting flying insects, particularly near doors.

Since lighting is critical and tends to involve complex technology, it is advisable to use the services of someone knowledgeable in this to conduct a survey of needs and to make appropriate recommendations for corrective action if needed.

Luminaries yield less light as dust and dirt accumulate on them, and lamps depreciate in their output with age, so it is advisable to establish and follow a routine maintenance schedule for lighting equipment.

§111.20(f) requires the use of safety-type light bulbs, fixtures, skylights, or other glass or glass-like materials whenever these are suspended over an area where breakage might result in shattering and thereby could cause contamination.

Several manufacturers of lighting equipment supply light bulbs, fluorescent tubes, and skylights, and lighting fixtures, with a clear plastic coating that would prevent shattering in the case of breakage. These must be used in the places specified to help contain glass fragments in the event of breakage.

BUILDING CODES

In addition to the GMPs, it is of course essential to follow all applicable building codes, mandates of the Americans with Disabilities Act (ADA), and regulations of the Occupational Safety and Health Administration (OSHA), as well as other governmental agencies, to assure compliance in the design, construction, and maintenance of physical plants.

BULK FERMENTATION VESSELS

§111.20(g) requires protection against contamination in bulk fermentation vessels by using protective coverings when appropriate, putting such vessels in areas where harborages for pests over and around the vessels are eliminated, and placing the vessels where they can be checked regularly to be sure there are no pests, pest infestation, filth, or extraneous materials. This section also requires the use of skimming equipment when needed.

SCREENING OR OTHER PROTECTION AGAINST PESTS

§111.20(h) requires the use of adequate screening or other protection against pests, where necessary. This is also inferred in §111.15(d), which is a sanitation requirement to not allow animals or pests in the plant.

To help prevent rodents from entering the building at loading docks, it is prudent to use a metal strip barrier at least 12-in. wide together with a dock overhang of at least 12 in. Instead of having stairs for personnel to get onto the dock from the ground level it is advisable to use a built-in rung-type ladder that can be used by people but not by rodents.

KEEPING THE PLANT CLEAN AND SANITARY

In §111.15(b)(1), it is stated that the physical plant must be kept in a clean and sanitary condition. This is a rather general requirement, but of obvious importance albeit without specific details as to how this should be accomplished and the standards that need to be met. This is because of the diversity in kinds, types, and size of plants in the dietary supplement industry, and gives management the flexibility for, and requirement of deciding *how* to meet these needs. Therefore, a specific plan and strategy should be developed for each facility to achieve effective sanitation. This should include determining the necessary frequency of cleaning and sanitizing floors, walls, ceilings, drains, restrooms, cafeterias or other eating facilities, locker rooms, etc.; when these steps should be done; and by whom and specifically how to be done. Thus, a carefully planned written cleaning and sanitation program is needed. Since these steps must be carried out by personnel, they need to be trained and well supervised, and they should understand both the practical and the regulatory reasons *why* it is important that this function be handed well.

It is advisable, although not mandated, to color code mops, brooms, buckets, and other cleaning implements to identify the specific areas in which they should be used. This helps minimize the possibility of causing cross-contamination.

The cleaning procedures and equipment, as well as the cleaning compounds and sanitizing agents used in each area of the plant should be specified in detail, together with how they are to be used. Generally, these differ in areas where only dry items are handled from those involving wet processes. The major vendors of cleaning supplies often can be helpful in selecting the optimum techniques for cleaning (and sanitizing, where appropriate) each area of the facility.

It is, of course, necessary to select the appropriate *time* for cleaning, but since this is often done at night, after the close of business, it is particularly important that the workers know just what is expected of them, and that they be well supervised. It is also important that the established procedures be carefully followed, and done in the correct sequence. Some firms employ outside contractors to do the cleaning, but these need to have specific directions, training, proper monitoring, and supervision.

Good housekeeping during operations helps minimize the need for cleaning, and also helps create a working environment where cleanliness is emphasized and appreciated. Thus, all supervisors, not just the cleaning supervisors, should stress housekeeping steps and should set the example for other workers by their own habits.

MAINTENANCE OF THE PLANT

§111.15(b)(2) requires that the physical plant be kept well maintained. This implies the need to promptly make repairs as the need arises. Cracks or holes in walls, ceilings, and floors need to be fixed, broken or cracked glass replaced, roof leaks promptly corrected, leaking pipes repaired, lighting equipment kept in good condition, and many other such details adequately handled. Supervisors should know to alert whoever is responsible for maintenance whenever such needs are seen.

CLEANING COMPOUNDS AND SANITIZING AGENTS

§111.15(c) speaks to requirements for cleaning compounds, sanitizing agents, pesticides, and other toxic materials. Subparagraph (c)(1) requires that such compounds be free from microorganisms of public health significance and that they are safe and adequate under conditions of use. The FDA, in the preamble to the regulations, has indicated that a company may accept a supplier's guarantee or certification that such items are not contaminated with unwanted microorganisms and are safe and adequate under conditions of use. Therefore, manufacturers of dietary supplements should insist on obtaining such continuing guarantees or certifications, and should keep them on file to be available to show to the FDA, should an issue ever arise. It is not required that the firm using such products conduct these types of tests as long as they have the required documents.

STORAGE OF TOXIC SUPPLIES

§111.15(c)(2) requires not using nor holding toxic materials in a physical plant in which components, dietary supplements, or contact surfaces are manufactured or exposed, *unless* those materials are necessary to maintain clean and sanitary conditions, or for use in laboratory testing procedures, or for maintaining or operating the physical plant or equipment, or for use in the plant's operations.

In other words, toxic materials *other than those actually needed* must not be kept in the plant.

Similarly, subparagraph (c)(3) requires the *identification* and *holding* of cleaning compounds, sanitizing agents, pesticides, pesticide chemicals, and other toxic materials in a manner that protects against contamination of components, dietary supplements, or contact surfaces. The term "identification" implies that all the containers of such materials should be clearly labeled, including the name of the compound and instructions for its proper use, while the proper *holding* of these materials implies the wisdom of keeping them in a locked room or cabinet to avoid the possibility of an unauthorized person having access to them.

SUPERVISION OF SANITATION

§111.15(k) requires having one or more people assigned to supervise overall sanitation, each of whom must be qualified by education, training, or experience to develop and supervise sanitation procedures.

PEST CONTROL

Pest control in the plant is covered in §111.15(d).

Subparagraph (d)(1) states that animals or pests must not be allowed in any area of the plant, except that guard or guide dogs can be allowed into *certain* areas if their presence will not result in contamination.

Subparagraph (d)(2) requires taking effective measures to exclude pests from the physical plant and to protect against contamination by pests.

This can be difficult to accomplish in some instances since some ingredients used in making dietary supplements are attractive to certain insects and rodents, and indeed inbound shipments of some of these materials may already have such pests in them, thereby introducing them into the plant. Therefore, it is important to make it difficult if not impossible for pests to enter the facility in the first place, and to do whatever is feasible to make conditions unfavorable to any pest that may get in. This implies carefully monitoring arriving shipments, plus good housekeeping and eliminating where possible any sources of food, water, and nesting that may exist. Garbage, trash, and refuse should be promptly removed and kept in tightly covered and frequently cleaned containers to deny access to pests. Places where personnel eat, and toilet facilities, should be kept scrupulously clean. Dark corners should be minimized, and possible harborages monitored. All holes or openings where pipes or conduit enter or leave through exterior walls should be sealed to prevent them from becoming ways for insects or rodents to get in. Dock levelers are also potential entry points. Doors should be kept closed, and gaps at door bottoms eliminated.

Pests may also enter the plant in employees' clothing, or in whatever personal belongings they bring into the facility.

Floor drains should be cleaned on a regular schedule.

Loading docks should preferably have an air-lock arrangement with large doors through which trucks can enter after which those doors are closed, with smaller doors from the dock area into the plant through which hand trucks or forklift trucks can load or unload the trucks. If these conditions are not feasible, plastic strip door covers, or air curtains are helpful.

Insect light traps of the electrocution type are helpful in the control of flying insects that may get into the plant, although these must be regularly cleaned and maintained. Such traps should be indoors, near possible entry points, but not visible from outside since if the blue light can be seen it merely attracts more insects to enter the facility.

INSECTICIDES, FUMIGANTS, FUNGICIDES, AND RODENTICIDES

§111.15(d)(3) states that insecticides, fumigants, fungicides, and rodenticides must not be used *unless* proper precautions are taken to protect them from causing contamination.

Many firms use the services of professional pest management firms to handle pest control in and around the plant. If such services are used, it is important to establish that they fully understand the GMPs as related to this topic. It is also advisable to keep records of when they were in the plant, what they found, what action they took, what pesticides were used, and where they were used.

It is further advisable to have one member of the DS firm specifically charged with the overall responsibility for pest control. This might be the facility manager or someone from his or her staff, someone from quality control, or whoever is the most appropriate in the organizational structure. But, whether the person actually handles the pest control operations or supervises others who do so (including outside contract services), this individual should be thoroughly knowledgeable of the pest control situation and should keep management informed of its status.

WATER SUPPLY

An important consideration in manufacturing is the quality of the water that is used both as a component and for other purposes such as cleaning the plant and equipment.

This is covered in §111.15(e), which requires water that is safe and sanitary, and at suitable temperatures, and under pressure as needed for all purposes where the water does *not* become a component. However, where water *is* used as a DS component, or contacts components, dietary supplements, or contact surfaces, it must as a minimum comply with applicable federal, state, and local requirements, and must not contaminate the dietary supplement. In the United States, the applicable specifications are covered by the National Primary Drinking Water (NPDW) Regulations found in 40 CFR 141, administered by the Environmental Protection Agency (EPA).

This portion of the DS GMPs resulted in many comments to the FDA on the proposed regulations, in part because the requirements for water used in conventional foods merely requires that it be safe and of adequate sanitary quality, whereas for drug products the water must comply with the U. S. Pharmacopoeia (USP) specifications for Purified Water. The FDA considered these comments, and concluded with the wording that is in the final rule.

The FDA made it clear that water meeting the USP standards would certainly be acceptable, but it is not mandated. Since the GMPs are *minimum* requirements, firms may of course use water that exceeds these requirements, and many do use Purified Water USP processed either by deionization, reverse osmosis, or distillation. However, such units require good maintenance, and careful attention must be paid to storage and distribution systems.

Another discussion centered around what overseas manufacturers that export dietary supplements for sale in the United States must do about water quality, since the requirements for drinking water do vary from country to country. The outcome of this was the requirement to meet all of their *own* national or local water safety requirements, *plus* their water must not contaminate the products.

There was also discussion on why the final DS GMPs call for two different water standards, one for water used in cleaning walls and floors, and another when it may become part of a DS product either directly or indirectly. In this respect, it is important to note that for water that may become a component of a DS it is required to have and keep *documentation* of its quality under §111.23(c). This is *not* required for water used to clean floors or in employee bathrooms.

It is advisable to obtain a certificate of analysis of the water if it is supplied by the municipality to be sure it complies with the applicable requirements, but it is also necessary to ensure that nothing happens to the water to contaminate it *after* it enters the plant. Municipal water may be of proper quality when it leaves the water treatment facility, yet because of old or defective piping between there and the DS plant, or within the plant, the water at the point of use may not necessarily be of adequate purity. Therefore, in addition to certification from the municipal source (if that is how the incoming water is obtained), it is also important to periodically test the water *in* the plant, particularly at the points farthest from the point of entry into the plant (which would indicate any impact that the plant piping might exert on the water quality). Records of such testing must be retained.

While not specifically mentioned in the GMPs, it is also advisable to keep up-to-date diagrams of the plant's plumbing system, covering both water and sewage piping. It is also advisable to have all piping clearly identified by color-coding, stenciling, or by labels applied at appropriate intervals. A voluntary code exists for this, known as ASME/ANSI A.13.1, although any system of identification marking of pipes and valves can be useful.

Water from a private source must comply with the state and local requirements, but it might possibly also require treatment such as filtration and/or chlorination to comply with §111.15(e)(2). It is also important to keep in mind that the impurities in water, either from municipal sources or private sources, often vary from season to season, so it is prudent to have water tested at various times of year.

PLUMBING

According to §111.15, the plumbing in the plant must be adequate in size and design and adequately maintained to

1. carry sufficient amounts of water to required locations throughout the plant;
2. properly convey sewage and liquid disposable waste from the plant;
3. avoid being a source of contamination to components, dietary supplements, water supplies, or creating an unsanitary condition;
4. provide adequate floor drainage in all areas where floors are subject to flooding-type cleaning or where normal operations release or discharge water or other liquid waste on the floor;
5. not allow backflow from or cross-connection between piping systems that carry water used for manufacturing dietary supplements, for cleaning contact surfaces, or for use in bathroom or hand-washing facilities.

These requirements are all essentially self-explanatory, except the portions on floor drains and backflow prevention. The following remarks are suggestions, not regulatory mandates.

Areas where the floors are necessarily frequently wet should be provided with floor drains properly located and installed. An acceptable rule of thumb is to have one 4-in. drain for about each 400 square feet of floor area, and to have the floor slope toward the drain at about one-quarter inch per lineal foot. However, if lesser volumes of water are likely to be encountered, the spacing may be greater and the slope more gentle.

PHYSICAL PLANT AND GROUNDS

Where floors are washed down frequently, trench drains may be preferable. These are available in pre-built modular pre-sloped sections made of polymer-concrete or stainless steel with appropriate gratings, or they can be poured-in-place concrete.

The concept is to provide adequate drainage to avoid stagnant puddles without the necessity of pushing water to the drains with a squeegee. Floor drains need to be equipped with traps, and infrequently used drains should have threaded plugs, to prevent sewer gases entering the plant.

Drainage lines should be large enough to allow rapid removal of water. To keep out rodents, each floor drain should be equipped with a sturdy grill-type cover and such covers should be kept in place. It is preferable to use drain fixtures that are adjustable in height to compensate for any variances in the floor surface. Floor drains should be kept free from buildup of solid material, and kept clean.

The requirement about backflow or cross-connections has to do with unintended reversal of flow of nonpotable water or other substances into the water distribution system that could result in contamination. Backflow can happen when the supply pressure of the water is exceeded by the pressure of downstream liquids. Various scenarios can cause this, for example when the incoming water pressure dips below normal because of water main breaks, fire fighting, etc. Cross-connections are unintended temporary or permanent piping arrangements that could allow such contamination to occur. There are several different types of "backflow preventers" that either eliminate the possibility of a cross-connection or provide a mechanical barrier to backflow. Among these are simple air gaps, which physically separate the end of a water outlet from the tank or other vessel into which water is being supplied, making it impossible for anything in that vessel to get sucked into the water piping system. Other backflow preventers are more complex arrangements involving check valves.

The mechanical types of backflow preventers have moving parts that are subject to corrosion or wear and therefore should be periodically checked to be certain they are still operating properly.

The requirement in §111.15(g) is for adequate disposal of sewage, either through a sewage piping system or another satisfactory means. National and local codes usually detail what must be done in this regard.

RESTROOMS

Restrooms for employees and visitors (referred to in the GMPs as "bathrooms") are covered in §111.15(h). The regulations leave the details and features of these up to management to decide, although these are worthy of careful attention in that such facilities affect both employee morale and personal sanitation habits.

It is usual (but not required) to have separate men's and women's facilities, but generally located adjacent to each other to simplify plumbing. The number of toilets is usually specified in local codes. Restrooms need to be properly ventilated to avoid objectionable odors reaching other nearby areas. Floors, walls, and ceilings should be easily cleanable, and free from cracks and crevices.

Restrooms should be kept scrupulously clean, well lighted, and attractive. Doors should be the self-closing type, and should not open onto areas where dietary supplements, components, or contact surfaces could be exposed to contamination.

HAND-WASHING FACILITIES

The GMPs, in §111.15(i), require hand-washing facilities designed, built, and maintained to ensure that the hands of employees and visitors are not a source of contamination. These must be conveniently located to encourage and facilitate use by employees, and must furnish running water at a suitable temperature. Arrangements should be made to be certain that hand-washing facilities are frequently monitored to assure that the needed supplies are always readily available. However, firms are given flexibility in specifically how these goals should be achieved.

TRASH DISPOSAL

According to §111.15(j), trash must be conveyed, stored, and disposed of in ways that minimize the development of odors and the potential for trash to attract, harbor, or become a breeding place for pests. Moreover, the methods used must protect against contamination of components, dietary supplements, any contact surface, water supplies, and grounds surrounding the physical plant.

The issue of odor from trash relates to issues of microbial contamination, such as decomposition, decay, and the proliferation of microorganisms some of which may be harmful. Moreover, such odors can attract rodents and other pests Therefore, the proper handling and disposal of trash is a proper GMP concern.

SITE AND GROUNDS ISSUES

The grounds bordering the physical plant must be kept in good and neat condition. This helps eliminate possible sources of contamination, as stated in §111.15(a). This includes properly storing equipment outdoors, removing litter and waste, and cutting grass and weeds within the immediate vicinity of the plant to avoid attracting and harboring pests. Similarly, roads, yards, and parking lots must be maintained to help prevent contamination.

Wet areas should be drained if they might contribute to contamination due to seepage, filth, or other extraneous materials, or if they could provide a breeding place for pests.

While not specified, a reasonable rule of thumb would be to consider an area of about 100 ft from each side of the actual building to be the "grounds of the physical plant."

If one were selecting a site on which to construct a new plant in which dietary supplement products were to be manufactured, it would be prudent to attempt to minimize environmental issues, both those that might impact on operations of the plant and those having to do with how the plant may affect the environment.

Unfortunately, some sites currently in use for housing DS manufacturing *may be* adjacent to sources of air pollution or bad odors, pollutants, insects or rodents, or even annoying noise, so it is necessary to accommodate to existing conditions. The FDA would not judge the DS plant's location or suitability solely on the facility's proximity to possible sources of contamination, but more on the adequacy of the measures the firm takes to prevent such possible contamination from adversely affecting the product quality of the dietary supplements being manufactured. The task of protecting products from such contamination is challenging but not insurmountable.

In a similar way, *if* the grounds bordering the firm's property, but not under the firm's control, are *not* well maintained, it is necessary under §111.15(a)(5) to use care by means of inspection, extermination, or other ways, to exclude pests, dirt, filth, or other extraneous materials that could be sources of contamination.

§111.15(a)(4) requires operating systems for waste treatment and disposal so that they do *not* constitute a source of contamination.

SUGGESTED READINGS

Cramer MM. Food plant sanitation: design, maintenance and good manufacturing practices. Boca Raton, FL: CRC Press, Taylor & Francis Group, 2006.

Imholte TJ (revised by Imholte-Tauscher TK). Engineering for Food Safety and Sanitation. 2nd ed. Medfield, MA: Technical Institute of Food Safety, 1999.

Lydersen BK, D'Elia NA, Nelson KL. Bioprocess Engineering: Systems, Equipment and Facilities. New York, NY: John Wiley & Sons, Inc., 1994.

Signore A, Jacobs T, eds. Good Design Practices for GMP Pharmaceutical Facilities. New York, NY: Informa Healthcare, 2005.

5 Equipment and utensils

BASIC REQUIREMENTS

According to §111.27(a), it is required to use equipment and utensils that are of appropriate design, construction, and workmanship to enable them to be suitable for their intended use and to be adequately cleaned and properly maintained. This includes (but is not limited to) equipment used to hold or convey; equipment used to measure; equipment using compressed air or gas; equipment used to carry out processes in closed pipes and vessels; and equipment used in automated, mechanical, or electronic systems.

These terms are deliberately somewhat vague, leaving it to the discretion of the firm as to what is "appropriate," "suitable," and "adequate," but FDA feels these descriptions have been used for such a long time in food and pharmaceutical manufacturing that they are generally understood by the industry, by equipment vendors, and by those who enforce the regulations. It is advisable to be prepared to defend the determination that items do meet the required criteria, in event of being challenged on this point.

Equipment must be designed and constructed to withstand the environment in which it is used, including the action of components or dietary supplement products, and, if applicable, cleaning and sanitizing agents, according to §111.27(a)(3)(iv).

The regulations do *not* require formal validation of equipment and utensils.

MEETING GMP REQUIREMENTS FOR EQUIPMENT AND UTENSILS

Firms are afforded the flexibility to select and use equipment that best satisfies their particular needs, as long as it meets the criteria described above. Each firm is responsible for being sure that all equipment used in their manufacturing process will produce quality products in accordance with GMP. Firms are free to modify standard equipment designs to best suit their process. Plants, processes, and products vary so widely that it would be impractical to spell out in detail just what designs and constructions must be used. However, the FDA does take seriously the fact that all equipment must meet the stated criteria, and firms have been cited for GMP violations when equipment was considered to be inadequate, insanitary, or in poor repair.

The *suitability* of any given item of equipment depends in part on its capacity as compared to what is needed, in part on the design and construction to enable it to properly perform its function, and in part on the condition and maintenance status of the item. *Suitability* also implies that the equipment is capable of easily being cleaned and inspected, and sanitized if needed. The "cleanability" is dependent in part on both the nature of the material being handled and the likelihood of possible physical and/or microbiological contamination. Again, this is a judgment decision left to the firm, but it is advisable to be in a position to defend such decisions if questions arise.

EQUIPMENT AND UTENSILS

To enhance cleanability, equipment should have as few component parts as feasible; should have rounded (as opposed to sharp) corners; should have all contact surfaces smooth and nonporous, devoid of pits, cracks, crevices, and deep scratches that could harbor filth or microorganisms; should be free from areas that would be difficult to clean; and should be easy to disassemble and reassemble (if this is needed) by using simple tools. Exposed threads should be minimized to the extent feasible. Product contact surfaces in and on equipment should be easily accessible for cleaning and inspection. Exterior noncontact surfaces should be kept as simple and smooth as possible, uncluttered, self-draining, and without places that could trap product, water, dirt, insects, or microorganisms (Fig. 5.1).

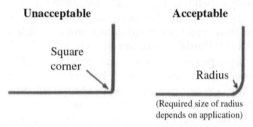

Figure 5.1 Acceptable versus unacceptable internal angles of food equipment. *Source*: From FDA, 2000.

According to §111.27(a)(4), equipment and utensils must have seams that are smoothly bonded or maintained to minimize accumulation of dirt, filth, organic material, particles, or other extraneous materials or contaminants (Fig. 5.2).

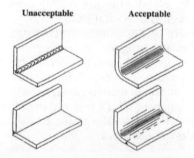

Figure 5.2 Corner welds in food equipment. *Source*: From FDA, 2000.

As also mentioned in chapter 7, §111.27(a)(2)(v) mandates that equipment must be properly maintained, for many reasons, one of which is to protect components and dietary supplement products from being contaminated by any source.

"FDA-APPROVED" EQUIPMENT
The FDA neither approves nor prohibits specific items of equipment and does not maintain a list of approved items. Occasionally, equipment vendors claim they supply "FDA-approved" items, but this is not correct since no such approvals exist.

VOLUNTARY STANDARDS FOR SANITARY EQUIPMENT

To assist both equipment manufacturers and the using firms, for many years voluntary standards and codes have been published regarding the sanitary design, construction, and installation of machinery and equipment used in the industries regulated by the FDA. These standards are not intended to limit or stifle creative engineering, but rather innovation is encouraged and provision is made for the consideration of new and advanced materials and designs to achieve the objectives of cleanability, easy inspection, and product protection. These standards, the use of which although *not* legally required, can be very helpful to dietary supplement manufacturers when purchasing equipment.

The first such standards evolved in the 1920s for the sanitary design of fittings used in milk pipelines. These were published as a joint effort of three industry associations, and became known as the "3-A standards," the name that is still used. These are now continuously updated, and available for purchase online, either as individual standards or as compilations, from 3-A Sanitary Standards, Inc. Among the many types of equipment covered are storage tanks, metal tubing and fittings, heat exchangers, flow meters, mixers of various types, pneumatic and mechanical conveyors for dry materials, several types of pumps, many types of sanitary valves, and other kinds of equipment useful to firms manufacturing DS products. Since 1956, the 3-A symbol has been used to identify equipment that has been verified to meet the 3-A standards for design and fabrication.

Other voluntary standards for sanitary equipment include ISO-14159 Hygiene Requirements for the Design of Machinery, standards published by NSF International (formerly called the National Sanitation Foundation) and by the BioProcess Equipment Committee of the American Society of Mechanical Engineers. In addition, several such standards have been developed by the European Committee for Standardization (CEN) working with the European Hygiene Equipment Design Group (EHEDG).

The U.S. Department of Agriculture (USDA) Agricultural Marketing Service (AMS) has a voluntary program to inspect and certify equipment and utensils used to process livestock and poultry products. Many of the items covered by this program can also be useful in the manufacture of dietary supplements. Therefore, although there is neither any direct connection between this and the dietary supplement GMPs nor any mandatory requirement to use such items, equipment certified as acceptable to the USDA is made to rigid standards of sanitary design and construction.

AVOIDANCE OF CONTAMINATION DURING EQUIPMENT USE

Section 111.27(2) states that equipment and utensils, when used, must not result in contamination with lubricants, fuel, coolants, metal or glass fragments, filth or any other extraneous material, contaminated water, or any other contaminants.

It is possible that leakage or drippage can occur when equipment is in use, so this must be considered and eliminated. For example, piping should not be placed immediately above where product is open to the atmosphere where such contamination could occur, or directly above contact surfaces, since piping joints might develop leaks or condensation or dirt might accumulate and fall into product or onto surfaces. Similarly, elevated motors and drives might drip lubricant, and therefore their placement can be critical. Covers or lids should be placed on vessels to avoid inadvertent contamination. Hinges on covers should

EQUIPMENT AND UTENSILS

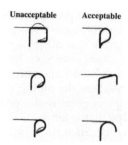

Figure 5.3 Top rim of food equipment. *Source*: From FDA, 2000.

be eliminated since they tend to have crevices that could harbor dirt and microorganisms. All such possibilities should be considered and risks of contamination minimized or eliminated. Common sense is the rule in seeking out potential problems of this sort and taking corrective action if needed.

Although the condition of contact surfaces is of importance for protecting product from contamination, noncontact surfaces should also be given careful attention. For example, mounts for machinery typically have adjustments for height, which involve threaded parts, but exposed threads can harbor microorganisms. Ledges and voids on noncontact surfaces can harbor insects, microorganisms, dirt and filth, and stagnant water (Fig. 5.3).

Tanks and kettles containing liquids or semiliquids should be so designed as to be completely self-draining so that no residue remains when the vessel is empty.

Similarly, piping and tubing carrying product or components should be sloped to be self-draining, with a suggested slope of at least one-eighth inch per lineal foot (except that slope is less important for sanitary tubing that is disassembled at frequent intervals).

Piping and tubing should be well supported to avoid sagging that could prohibit good drainage. If welded piping or tubing is used, orbital welding (as opposed to manual welding) is generally preferred, and the welds must be of high quality with no pockets or voids where product might accumulate. It is advisable to inspect the interior of welded piping or tubing with a borescope to confirm that the welds are adequate and well formed. If gaskets are used, they should be selected and installed to provide the desired seal with cleanability but without contributing contamination. Take-apart types of stainless steel sanitary tubing are also in wide use for ease of cleaning and excellence in sanitation. The general standards for sanitary tubing and piping are promulgated by ASTM International, previously known as the American Society for Testing and Materials, which include their standard A269 (Seamless and Welded Austenitic Stainless Steel Tubing for General Service) and A312 (Seamless and Welded Austenitic Stainless Steel Pipe). These standards are useful, but not mentioned in nor mandated by the GMPs.

Valves and fittings should be selected and installed with "drainability" and "cleanability" in mind. Many styles of valves allow some entrapment of liquid, so it is important to avoid this possible source of contamination. Examples of satisfactory valves include diaphragm types, which are excellent since they have no recesses to harbor contaminants, are easy to clean, and do not trap residue in the valve body. Full-port plug-type and ball-type valves also serve well. In general, it is advisable to use valves and fittings meeting the 3-A standards, but this does not imply that others are necessarily unsatisfactory.

Deadlegs in piping are unused portions where the liquid in the system can become stagnant, giving a place where microorganisms may proliferate. These may exist when changes are made in the piping, for example, when a section is removed at a tee, with a short piece of unused pipe remaining. The FDA has said that any section of unused pipe that is longer than six times the diameter of the pipe is considered to be a deadleg. Such situations should be avoided since they can cause contamination.

Many motors driving pumps, agitators, conveyors, sifters, and other equipment may have places where dirt and organisms can accumulate. The housing around the working parts of a motor is called the motor's *enclosure*, of which there are two general types, open and totally enclosed. Open motors are ventilated through openings allowing passage of air into the interiors for ventilation and cooling, whereas totally enclosed types are built to prevent direct passage of air from outside to inside the frame but employ other methods of keeping the motors from overheating. In critical areas, it is advisable to use totally enclosed motors to minimize the likelihood of contamination from dust or other debris accumulating and existing within the motors. Where frequent washdown of the motors is anticipated, special models of motors and drives designed to withstand such service are available, with the exterior surfaces typically either epoxy painted or made of stainless steel. Detailed specifications for the many varieties of electric motor frames and enclosures are published by the National Electrical Manufactures Association (NEMA) and by the International Electrotechnical Commission (IEC).

Drives and power transmission equipment also need to be considered from a sanitation point of view. Drives are usually either by belts or by chains, with belts being simpler of the two, being easily cleanable, and requiring minimal maintenance. Polymeric belts are available that tend to resist deterioration from cleaning and sanitizing agents, although their sprockets (if made from ordinary carbon steel) may rust from frequent washdowns. Chain drives, which are available in either plain carbon steel or stainless steel and other corrosion-resistant alloys, unfortunately involve multiple parts that can harbor microorganisms and therefore require frequent thorough cleaning. Chain drives also require lubrication so that care must be taken to avoid contamination from this source.

For handling bulk solid materials (including powders, granules, and other dry forms of materials), sifters, conveyors, feeders, and mixers of many types are available. Such equipment is also often used in other industries where sanitation is not an important consideration, but a few suppliers of such items do design and build special models that meet sanitary standards, made of easily cleaned corrosion-resistant nontoxic materials.

Similarly, a wide range of types and styles of sanitary mixers exists, including portable electric or air-driven propeller mixers, in-line mixers, homogenizers and colloid mills, ribbon blenders, V-type and double-cone mixers, change-can mixers, dough-type mixers, and many others.

In some process steps in manufacturing dietary supplements, a vessel may be equipped with a shaft for an internal agitator. The impeller shaft is typically connected to a motor through a gear drive. To avoid contamination through the point where the shaft enters the vessel, it is usual to use a seal or packing of some type. The impeller shaft may enter from the top, side, or bottom of the vessel. The selection and maintenance of the seal for shaft entry points is another item that must be carefully considered.

Liquids and semisolids in dietary supplement manufacturing are frequently transported from one location to another by means of pumps. The possibility of contaminating the product stream during pumping exists, either through chemical reaction between the product being pumped and the material of construction of the pump, or through the interior surfaces or recesses of the pump, which may not be entirely free from foreign matter, or through unwanted substances that may enter from the outside due to cracks, crevices, or other forms of leaks. Therefore, pumps need to be carefully selected, installed, and maintained. Fortunately, centrifugal, positive rotary, positive displacement, progressive cavity, rotary lobe, diaphragm, air piston, and other types of pumps exist, built to the sanitary design specifications of the voluntary standards organizations described above.

Selecting the optimum pump for any given operation involves consideration of several factors, but pump vendors generally have the expertise to be helpful.

The important concept is that it is necessary to consider many aspects of hygiene in assessing the suitability of equipment, but there are helpful sources of information available to assist in doing so.

LUBRICANTS

The science and practice of lubrication, friction reduction, and wear is called tribology and is of importance in the design and operation of most moving machinery and equipment. As mentioned above, items of equipment used in the manufacture of dietary supplements require lubrication frequently, and there is a potential for contamination of products, components, or contact surfaces from this source. This is specifically mentioned in §111.27(a)(2)(i).

Minor leaks of oil or grease are not uncommon and not always even avoidable. Therefore, care must be taken both in the location and in the maintenance of lubricated items to minimize risk of contamination from lubricants. However, the potential problems that could result from this can be mitigated through the use of *food-grade* lubricants, which are nontoxic and essentially innocuous. Until 1988, such lubricants were certified by the USDA's Food Safety and Inspection Service, but this is now done by NSF International under the standard NSF-116. Food-grade lubricants are designated by the term "H-1" for use where there might be incidental contact with product, components, or contact surfaces; "H-2" where there would be no possibility of such contact; and "H-3" for edible oils. A large range of viscosity grades of food-grade oils and greases is now available, and these have performance characteristics equivalent to nonfood grades.

There is also an international standard ISO-21469 titled "Safety of Machinery—Lubricants and Incidental Product Contact."

The GMPs do not require the use of food-grade lubricants, but they do require that contamination from lubricants must be avoided.

FDA recognizes that certain lubricants are an integral part of encapsulation of gelatin-enrobed products and other dosage forms, but since these do not result in contamination, they are satisfactory under §111.27(a)(2).

COOLANTS AND HEAT TRANSFER FLUIDS

In a similar way, §111.27(a)(2)(iii) mentions the requirement to avoid contamination from coolants, which would be included in the larger category of heat transfer fluids, which involve both heating and cooling. NSF International also certifies food-grade heat transfer fluids under the designation "HT-1."

COMPRESSED AIR

The requirement for compressed air and other gases introduced into or onto a component, dietary supplement, or a contact surface, or used to clean any contact surface must be treated in such a way as to avoid contamination. This is stated in §111.27(a)(7) and is also implied in §111.27(a)(1)(iii) and §111.27(a)(2)(vii).

Most facilities use compressed air for many purposes, and this is a potential source of contamination. Air compressors usually result in certain amounts of both oil and water entering into the compressed air. So-called oil-free compressors are a step in the right direction, but even these are less than perfect and the vague term "oil-free" can be misleading since it refers to only the compression chamber and not necessarily to the air delivered to the point of use. Therefore, it is advisable to use well-designed desiccant or refrigeration-type dryers as well as filters in the system, particularly for critical areas where compressed air is employed. Filters rated at 0.45 μm or better at the point of use reduce both particulate content and the bioburden of the compressed air, thereby significantly minimizing the probability of contamination from this source.

The location of the air intake to the compressor can be a significant factor in the quality of the compressed air since the compressor tends to concentrate whatever particles or other contaminants may be present in the ambient air. Intakes for air compressors should have adequate and well-maintained filters.

Contaminants, purity classes, and test methods for compressed air can be found in the ISO-8573 standard. The required compressed air quality of course differs depending on the situation.

It is unwise to use compressed air to blow dust off of machinery since this merely resuspends the dust into the immediate environment where it will settle again. It is also unsafe to use compressed air to blow dust off of people or clothing since particles in a jet of compressed air can be hazardous to eyes or can abrade skin.

METAL OR GLASS FRAGMENTS

§111.27(a)(2)(iv) also addresses the need to use equipment of appropriate design and construction so that its use will not result in contamination from metal or glass fragments. This is essentially self-explanatory except that in rare instances it may be possible for such fragments to exist despite best efforts to prevent it. For example, screws, bolts, rivets, pieces of screen from sifters, or other such items may become loose and enter the product or may come in components. This requires frequent and thorough monitoring of equipment not only by maintenance people but also by the equipment operators. Moreover, it is prudent where appropriate to use magnetic separators capable of capturing and removing iron objects or small ferrous debris. The use of metal detectors can also alert the presence of nonferrous metal fragment contaminants, which can then be removed either manually or by means of an automatic mechanism.

INSTALLATION AND MAINTENANCE OF EQUIPMENT

§111.27(a)(3)(i) dictates that all equipment and utensils must be installed and maintained to facilitate cleaning the equipment, utensils, and all adjacent spaces.

Although not specified in the GMPs, a useful rule of thumb is to provide at least 36 in. of space around each piece of equipment to allow for cleaning, sanitation, and maintenance, and to keep the equipment at least 8 in. off the floor and at least 18 in. from the ceiling. If the distance off the floor is not feasible, an alternative is to seal the equipment directly onto the floor so that cleaning under it is not required. However, the space required to accomplish cleaning, sanitizing, and maintenance varies according to the design and function of the equipment, and therefore judgment is required in evaluating each instance.

It is up to the firm to decide what space must be given to each piece of equipment to comply with this requirement.

MATERIALS OF CONSTRUCTION FOR EQUIPMENT AND UTENSILS

§111.27(a)(3)(ii) states that equipment and utensils that contact components or dietary supplements must be *corrosion resistant*, while subparagraph (iii) says these material must be *nontoxic* and subparagraph (iv) calls for equipment and utensils to be constructed to withstand the environments in which they are used.

In essence, from a practical point of view, this limits metallic materials largely to stainless steel, as well as a few other special alloys, and titanium. Titanium is strong, nontoxic, and corrosion resistant and therefore satisfactory, but costly.

Aluminum finds limited application in dietary supplement equipment because it is both mechanically soft and vulnerable to corrosion by some common cleaning and sanitizing compounds. Plain carbon steel is used to some extent, but it is severely limited by the tendency to rust under moist conditions. Copper and brass also have corrosion concerns, plus the fact that copper ions can have an adverse effect on the stability of many ingredients often used in dietary supplements.

The most commonly used materials therefore are the stainless steels, of which many grades exist but only a few are widely used in dietary supplement manufacturing. The main grades are categorized both by their composition and by their microstructure. The grades are called martensitic, ferritic, and austenitic. Of these, the austenitic stainless steels constitute the most frequently used materials for this industry. Other grades are used for architectural, ornamental, and other applications.

These austenitic grades nominally contain about 18% chromium and 8% nickel, by weight, with the balance being mostly iron, and thus are sometimes referred to as "18 and 8 or 18/8 stainless." These are strong, relatively easy to fabricate, and have good corrosion resistance to most environments, although the term "stainless" is something of a misnomer in that these alloys are subject to corrosion under certain conditions.

The chromium present in stainless steel tends to react with oxygen in the air, forming a very thin invisible film or layer of a complex chromium oxide on the surface of the alloy, which yields protection from corrosion. This surface layer is called a "passive film," the formation of which occurs spontaneously but can be chemically enhanced by a process called "passivation." When this film is intact, the alloy is said to be "passive," as opposed to being "active" when the film is absent or discontinuous. The passive film tends to be self-healing if disrupted, particularly in oxidizing environments. A number of types of chemical treatments for passivation are known. However, since stainless steel does, by its very nature, automatically form the needed passive film, in many instances there is no need to enhance the passive film. However, the chemical approaches to passivation, when needed, are explained in detail in ASTM A380 "Standard Practice for Cleaning, Descaling, and Passivation of Stainless Steel Parts, Equipment, and Systems," and ASTM A967, "Standard Specification for Chemical Passivation Treatments for Stainless Steel Parts." It is important to note that from a practical point of view, passivation has significant implications, but the GMPs are silent about this, and therefore such considerations are not mandated.

The types of stainless steels are usually identified in the United States by either of two numbering systems, the most common being that of the American Iron and Steel Institute (AISI) and the other being the Unified Numbering System (UNS), which is managed jointly by the Society of Automotive Engineers (SAE) and ASTM International. In some instances, suppliers of stainless steel identify the grades by their own proprietary trade names.

The AISI numbering system uses four digits. Those with "1" as the first digit are plain carbon steels, "2" are nickel-containing steels, "3" are nickel-chromium–containing steels, "4" are nickel-chromium-molybdenum–containing steels, and "5" are chromium-containing steels. The two austenitic 18/8 stainless steels mostly used in dietary supplement manufacturing equipment are types 304 and 316. Type 316 contains molybdenum (which 304 does not), which yields increased resistance to corrosion in certain specific environments, and in particular helps inhibit pitting caused by chlorine and chlorides, but it is appreciably more costly than 304. Although plain carbon steels can be hardened by heat treatment, 304 and 316 cannot, but they can be hardened by cold working. Both 304 (UNS S30400) and 316 (UNS S31600) are usually sold in their softer or "annealed" condition.

Both 304 and 316 types are available with extra-low carbon content, which helps prevent the precipitation of chromium carbide caused by heat at and near welds, the so-called heat-affected zone (HAZ). The formation of chromium carbide reduces the amount of chromium in the alloy, and therefore lowers the intergranular corrosion resistance of the alloy. These low-carbon grades are designated 304L (UNS S30403) and 316L (UNS 31603). These are often used for the increased corrosion resistance they provide, particularly where welding is to be employed in the fabrication of the equipment.

Another alloy, trade named AL-6XN (UNS N08367), is also useful for equipment for the dietary supplement industry. This is sometimes called a "super-austenitic" grade since it has low carbon and high levels of nickel and chromium, leading to improved corrosion resistance.

Another consideration for equipment made of stainless steel plates or sheets is the surface finish. The finishes are designated by numbers, 2B, 3, 4, 7, 8, and electropolished. Number 2B is also called bright cold rolled and is neither ground nor polished. Numbers 3, 4, 7, and 8 finishes are formed by mechanical abrasive action on the 2B plates and sheets. The higher the number, the finer the polish, and the more costly it is.

For equipment intended to be used in dietary supplement manufacturing, No. 4 is the most commonly specified and is generally entirely suitable. The finer the polish, the smoother the surface is, and the fewer scratches and other imperfections, which aids in cleaning and yields fewer harborages for microorganisms. In general, smoother finishes are also more conducive to ease of cleaning. The No. 4 finish continues to be acceptable for manufacturing equipment in the dietary supplement industry.

A bright mirror-like finish can be obtained through the use of an electrochemical process called electropolishing, which is essentially the reverse of electroplating, where small amounts of the surface metal are removed. This is quite attractive, and helps with ease of cleaning, but tends to be costly.

The measurement of surface texture (smoothness or roughness) is done through the use of an instrument called a profilometer, which uses a stylus to measure the vertical deviation of the peaks to the troughs in the surface irregularities, in microinches or micrometers, by a procedure spelled out in ANSI/ASME Standard B.46.1. The results are expressed as the "roughness average" or Ra.

In addition to metallic materials of construction, a number of kinds of plastics are also useful for utensils, hoses, piping, conveyor systems, chains, and other applications.

Most plastics are lightweight and lower in cost than metals. Heat-sealing or welding is often possible. However, their tensile strength is usually lower than that of metals, and many physically increase in size through water absorption. Some tend to be somewhat porous, which can lead to absorption of product constituents and may also cause problems due to harboring microorganisms. Plastic piping and hose can be extremely useful in certain situations. Polymeric or elastomer materials must be selected with care since some may leach unwanted chemicals used as plasticizers and fillers.

The 3-A Sanitary Standard 20-20, titled "Multi-use Materials Used as Product Contact Surfaces," 20-23 "Standards for Multi-use Plastic Materials," and 18-03 "Multi-use Rubber and Rubber-like Materials" are helpful sources of information on this topic, although these standards were written primarily for dairy applications.

COLD STORAGE EQUIPMENT

In some instances, components or dietary supplement products need to be held at low temperatures, requiring the use of refrigerators or freezers. Where such cold storage equipment is used, §111.27(a)(5)(i) requires that each such item must be fitted with an indicating thermometer, temperature-measuring device, or temperature-recording device that indicates and records (or allows recording by hand) the accurate temperature within the device, while §111.27(a)(5)(ii) requires having an automated device for regulating temperature or an automated alarm system to indicate a significant temperature change in a manual operation.

EQUIPMENT LOGS

As also mentioned in chapters 6 and 7, there is a requirement specified in §111.35(b)(2) to document, either in the batch records or in separate logs, the history of each item of equipment, including the dates of use, maintenance, cleaning, and if appropriate also sanitation.

AUTOMATED, MECHANICAL, OR ELECTRONIC EQUIPMENT

As mandated by §111.30(a) and (b), for any automated, mechanical, or electronic equipment used to manufacture, package, label, or hold a dietary supplement, it is required to design or select equipment to ensure that the established dietary supplement specifications are consistently met. This must be taken into consideration when selecting equipment at the time of purchase or in employing old equipment for a new use. Under both circumstances, it is necessary to ensure that the equipment is capable of operating satisfactorily within the operating limits required, that is, the equipment must be capable of doing what is intended. This would include (but not be limited to) considerations of how the equipment would be affected by existing environmental conditions as well as whether the equipment can be maintained properly to ensure satisfactory operation. Again, the terms "capable" and "satisfactory" are deliberately vague to give a degree of flexibility to the firm in applying them to particular circumstances.

Such equipment must be routinely calibrated, inspected, or checked to ensure proper performance, according to §111.30(c), which also requires quality control personnel to periodically review these calibrations, inspections, or checks. Written records of these calibrations, inspections, and checks are required.

It is required by §111.30(d) to establish and use appropriate controls for this kind of equipment (including software for a computer-controlled process) to ensure that any changes are approved by quality control personnel and are instituted only by authorized personnel. Further, §111.30(e) requires establishment and use of appropriate controls of the equipment to ensure it functions as intended. Quality control personnel must approve these controls.

It is required to make and keep backup files of current software programs (and any outdated software needed to retrieve records, if the current software cannot do so) and of data entered into computer systems that are used to manufacture, package label, or hold dietary supplements, according to §111.35(b)(5). Such backup files may be hard copy, diskettes, tapes, microfilm, or compact disks. This ensures that data can be retrieved if the primary software should develop problems.

There is also a requirement to make and keep documentation of the controls used to ensure that equipment functions in accordance with its intended use. This is stated in §111.35(b)(6). This requirement has been interpreted by some as requiring formal validation of the equipment design, selection, and capability, but this is *not* the FDA's intention. Under §111.30(e), firms are required to ensure that equipment operates in accordance with its intended use, that is, the equipment must be suitable and the production process must consistently deliver the expected results. That is the intent of §111.35(b)(6), to make and keep documentation of the controls used to ensure that the equipment does function as intended. Examples of such controls would include temperature settings, fill rates, and blending times that must be set, checked, and adjusted as necessary.

SUGGESTED READINGS

Anon. Stainless Steels. Materials Park, OH: ASM International (American Society for Metals), 1995.

Bricher JL. Effective sanitary design. In: Equipped for excellence: a blueprint for success. Food Safety Magazine, December 2004–January 2005.

Graham D. Equipment for sanitary design. Food Safety Magazine, August–September 2009.

Graham D. Sanitary Design of Food Plants and Equipment. Boca Raton, FL: Chapman & Hall, 1998.

Imholte TJ. Engineering for Food Safety and Sanitation. Medfield, MA: Technical Institute of Food Safety, 2006.

Lange BH. GMP manufacturing and qualification. Pharmaceutical Engineering, January–February 1997:18–24.

Pack TM, Cunfer EA. Equipment selection for the pharmaceutical industry. Proceedings Interphex, USA, July 19, 1988:149–156.
Peckner D, Bernstein IM, eds. Handbook of Stainless Steels. New York: McGraw-Hill Book Company, 1977.
Saravacos GD, Kostaropoulos AE. Handbook of Food Processing Equipment. New York: Springer, 2003.
Valentas KJ. Handbook of Food Engineering Practice. Boca Raton, FL: CRC Press, 1997.

6 Cleaning and sanitation

Both for pragmatic business reasons and for regulatory compliance, it is essential to take proactive steps to ensure clean and sanitary conditions in manufacturing and warehousing areas as well as with production machinery and equipment.

It is prudent for management to establish an effective and comprehensive cleaning and sanitation program. This requires both appropriate training and supervision to ensure the program is properly carried out. It is advisable to consult with a reputable supplier of cleaning and sanitation supplies, and to similarly consult with a competent pest control professional, in developing such a program. It is one of management's responsibilities to see to it that this program is adequately established, handled, and documented.

In the cleaning and sanitation program, it is advisable to define and justify the decisions on the procedures and agents used and on the means of verification that these accomplish the desired goals.

CLEANING AND SANITATION LOGS

It is required by §111.35(b)(iii) to have written procedures for cleaning and sanitation (as necessary) of all equipment, utensils, and any other contact surfaces that are used to manufacture, package, label, or hold components or dietary supplements. It is also required to document, either in individual logs or in the batch records, the date of cleaning and sanitizing equipment, according to §111.35(b)(2) and §111.260(c).

CLEANING THE PHYSICAL PLANT

The facility must be maintained in a clean and sanitary condition, according to §111.15(b), as discussed in chapter 4, whereas §111.16 requires establishing and following written procedures for cleaning the physical plant and for pest control. Cleaning operations must be conducted in a consistent manner, regardless of who conducts such operations or when they are conducted. Records of these procedures must be made and kept, according to §111.23(b).

Heavy accumulations of dirt on floors can be removed with brooms and brushes or by portable or central vacuum cleaners. Mopping or conventional wash-down procedures are useful in eliminating lighter accumulations. Larger floor areas can be effectively cleaned through the use of walk-behind or riding-type floor scrubbing machines.

Heating, ventilation, and air-conditioning (HVAC) equipment requires regular periodic attention, particularly the air handling units (AHUs), which contain blowers and heating and cooling elements. It is of importance to ensure filters and coils remain clean, to avoid spreading particulate contamination.

Interior walls and ceilings must be kept reasonably clean, as must overhead structural elements, piping, conduits, and ducts. Brushes on long poles and extension tubes on industrial vacuum cleaners are useful in these applications.

Floor drains, although essential, offer challenges in that they tend to provide favorable growth conditions for microorganisms due to being dark and wet, plus drains typically contain nutrient material. However, drains tend to be quite difficult to clean. It is crucial to use the right tools and the appropriate drain sanitizing agents coupled with frequent rinsing using copious volumes of low-pressure water. Care should be taken, however, to avoid forming aerosols of droplets when cleaning drains, which could spread contamination.

Employee eating areas, locker rooms, bathrooms, and hand washing/sanitizing facilities are other plant areas requiring particular attention.

Receiving areas, docks, and warehousing areas for storage and for physical distribution must also be maintained in a clean and sanitary condition.

One facet of facilitating cleanliness in the interior of the plant is to ensure that the *exterior* surrounding premises are also well maintained, as discussed in chapter 4. This helps minimize the likelihood of the entrance of pests. Moreover, it helps minimize possible sources of contamination either getting into the building through open doors or on the shoes of employees and visitors.

CLEANING PROCESSING EQUIPMENT AND MACHINERY

§111.27(d) requires cleaning and sanitizing (as necessary) all equipment, utensils, and any other contact surfaces used to manufacture, package, label, or hold components or dietary supplements. §111.360 also requires that all manufacturing operations be conducted in accordance with adequate sanitation principles.

Equipment and utensils must be taken apart, as necessary, for thorough cleaning and sanitizing, as required by §111.27(d)(1).

§111.27(d)(4) requires cleaning *other* surfaces that do *not* come into direct contact with components or dietary supplements as often as necessary to protect against contamination.

Types of Soil

The chemical and physical properties, and the water solubility, of the soil to be removed are important factors to consider in each cleaning task. In this context, "soil" refers to unwanted matter on contact surfaces. Not only the *type* of soil but also the nature of the surfaces involved influence the optimum cleaning techniques as well as the selection of cleaning compounds best for use. Soils can, in general, be classified as being either organic (fats, oils, proteins, carbohydrates, and mixtures of these) or inorganic (salts and scales). Both organic and inorganic materials may be present in the soil, where the greasy component serves as a kind of adhesive binding the soil to the surface that needs to be cleaned. In general, neutral or alkaline cleaners are best used to remove organic soils, whereas acid cleaners are more efficient against inorganic soils. Occasionally, two separate cleaning cycles are needed, in sequence, one using an alkaline cleaning agent and the other an acid cleaning compound.

Water Used for Cleaning

Water is a key constituent of most cleaning operations. Cleaning compounds and sanitizers are usually dissolved in water. Water is also used for flushing and rinsing. Water used in cleaning should be essentially free from suspended matter and should be of potable quality. The water used for cleaning often contains mineral impurities that tend to reduce the effectiveness of cleaning and sanitizing solutions, sometimes resulting in the formation of thin films or deposits on surfaces, which can be difficult to remove and are in themselves a form of unwanted soil. Water "hardness" results from the presence of salts of calcium and magnesium, usually expressed in terms of parts per million (ppm) of the salts present. The hardness depends to a large degree on the origin of the water, that is, from wells where it has trickled through many feet of earth picking up mineral impurities, or from lakes, ponds, rivers, or reservoirs where there usually is less opportunity to pick up salts of calcium and magnesium and is therefore "softer." Water hardness also tends to differ significantly from one geographic area to another. The minimum specifications for water used on contact surfaces are stated in §111.15(e)(1) and (2).

CIP Vs. COP Cleaning Methods

Cleaning can be accomplished by either clean-in-place (CIP) or clean-out-of-place (COP) methods. CIP systems use controlled circulation and spraying of cleaning solutions (followed by sanitizing solutions) without the need for disassembly. This relies on the chemical and physical actions of the solutions, their temperature, and the time period of surface contact. Special attention is required for the design and installation of equipment and pipelines when CIP methods are used. The technique almost completely removes the human element, thus reducing labor costs and increasing uniformity and reliability. The resultant cleaning tends to be more thorough and complete than that by manual methods, and is accomplished faster with less production downtime. However, the equipment to be cleaned by CIP must be properly designed to facilitate intimate contact of the cleaning and sanitizing solutions with the surfaces, and the contact surfaces need to be self-draining. Although widely used in the food and dairy industries, and to a growing extent also the pharmaceutical industry, the production volumes typical of the dietary supplement industry usually do not justify the high cost of CIP installations. Therefore, COP is much more common in this industry.

COP is used for all equipment not deliberately designed and installed for in-place cleaning. This involves manually cleaning machine parts, fittings, utensils, take-down piping, and similar items. It is usual to provide a wash area for the smaller items, whereas tanks, kettles, and large exterior surfaces of machinery are cleaned by hand, where they are, using brushes and other implements.

When an area is specifically devoted to COP cleaning, it is advisable (if feasible) to either wall-off the area, or at least curb it from adjacent areas, to avoid spreading water. The wash area should have adequate floor drains. Racks made of stainless steel pipe, or structural members such as angles or channels with the legs pointing downward, are useful for air-drying cleaned items. Whether or not a wash area is provided, it is usual to have a large sink for hand-cleaning machine parts, small utensils, etc., preferably made of stainless steel, and often with two compartments.

Frequency of Cleaning

In preparing the cleaning and sanitizing program, management needs to decide and list the frequency of cleaning (and sanitization, where needed) of each area of the plant and of each major item of equipment. This schedule differs from

CLEANING AND SANITATION

company to company depending on many variables, including (but not limited to) the products handled, the quantities produced, the processes used, the kinds of equipment and machinery involved, and whether frequent small runs or semicontinuous longer runs of product are the norm. Some operations require cleaning at the end of each shift or day, some at the completion of each run, and some after specific periods of time have elapsed. Small items of equipment, such as utensils and spatulas, should be cleaned after each use.

It is advisable to set limits on the time between the end of use of equipment and when the cleaning must be done. Leaving equipment uncleaned for substantial periods of time can cause difficulties such as soil drying and hardening on surfaces, making effective cleaning more difficult.

Cleaning Procedures

The usual first step is thorough flushing or soaking of the item(s) in potable water, not so cold as to set greasy deposits, nor so hot as to cause chemical changes within the soil, which could cause the soil to bond more firmly to the surfaces. The object of this is to remove as much of the soil as possible at this stage. Agitating, recirculation, or brushing during this step can be useful.

Following this, the next step is the application of a solution of the appropriate cleaning compound, following the directions on the label. Care should be taken to be certain that the proper dilutions are used, as opposed to simply "eyeballing" the amounts of cleaning agent and water. Using too much dilution can cause the resulting cleaning solution to be ineffective, whereas using excess amounts of the compound can be costly and potentially unsafe. Overuse of cleaning chemicals can be harmful and can result in leaving a film or residue on the surfaces being cleaned.

The application of the cleaning solution can be accomplished in a number of ways, such as by sprays, by soaking, by recirculation, by hand using a cloth or sponge, or with the aid of foam or gel-type additives to increase adhesion of the cleaning agent to the surfaces.

After visible deposits have been removed, the soil-laden cleaning solution is drained away, and the surfaces thoroughly rinsed with potable water to remove all traces of the remaining cleaning compound. The surfaces are then usually air dried.

Brushes

There are several manufacturers of brushes specifically for these tasks, employing different kinds of bristles, including but not limited to nylon, which tends to be superior to animal hairs or vegetable fibers as being stiff but not brittle and chemically resistant to cleaning solutions. Brushes are available in many sizes and shapes, and it is advisable to select the optimum brush for each task. An oversized brush may not clean well because the bristles may be forced to lay flat as opposed to standing upright. Since it is the *tips* of the bristles that do much of the cleaning, it is essential that the tips contact the surfaces to be cleaned. Brushes should be replaced when the bristles no longer return to their original shape after use. They should be cleaned, sanitized, and dried after each period of use. Many vendors color code the brush handles, indicating where each should be used, or the brushes may be numbered, with the brush color or number to be used indicated in each specific cleaning SOP.

Parts Washers

Parts washers are often used, which have rotary motor-driven brushes to contact both interior and exterior surfaces. They often provide for the recirculation of hot cleaning solutions through perforated pipes or enclosed stainless steel or plastic baskets. Cabinet-type parts washers also exist for automatically cleaning portable utensils, removable machinery parts, disassembled valves and pipe fittings, and similar items. Such washers are not unlike household washers in principle.

Steel Wool

Regular steel wool, or steel brushes, should not be used on stainless steel surfaces, since they tend to leave traces of iron, which can later rust and cause streaking. Stainless steel wool is commercially available, and is superior for such purposes, providing care is taken to avoid scratching product-contact surfaces. Stainless steel "sponges" are also available, made from flattened round wire to avoid the tendency to scratch surfaces.

High-Pressure Cleaning

Removal of surface deposits can be effectively accomplished through the use of high-pressure, low-volume blasting with water or cleaning solutions. Handheld spray guns or permanently mounted nozzles direct the stream onto the surface to be cleaned. Such units do not use much water, which yields cost advantages. This tends to be more reliable than scrubbing with brushes and, being rapid, makes efficient use of labor. Hard-to-reach places can often be effectively cleaned this way. However, care needs to be taken to avoid spreading contamination to other equipment or plant areas during the use of high-pressure water or solutions.

CLEANING COMPOUNDS

Cleaning compounds must be adequate for their intended use and safe under their conditions of use, according to §111.27(6). The choice of the optimum cleaning product for each occasion depends on the nature of the surface to be cleaned and the type of soil involved. This implies the necessity to stock various cleaning compounds. The cleaning and sanitizing SOPs should clearly state *which* cleaning compound is to be used for each task, how it is to be prepared for use, and how it is to be applied and rinsed.

Most cleaning compounds are used as solutions in water. Through their inherent detergency, such compounds lower the surface tension, which increases the solution's ability to penetrate, dislodge, emulsify, and suspend the soil. The emulsifying action is particularly important where fats and oils constitute a major portion of the soil.

Commercially available cleaning compounds are fairly complex mixtures of chemicals, formulated to achieve the intended results. These are generally identified by trade names and should be selected depending on the surface to be cleaned, the nature of the soil, the characteristics of the water used, the specific method of application, and the rinsability.

Cleaning compounds should be noncorrosive to equipment, stable in storage, and preferably nontoxic and relatively nonirritating to skin, although some types may not meet these criteria if not used properly or in the correct dilutions. Under §115(c)(1), cleaning compounds and sanitizing agents must be free from microorganisms of public health significance and must be safe and adequate under the conditions of use. The FDA will accept a supplier's guarantee or certification that a given cleaning compound or sanitizing agent meets these requirements.

Although some effort has been made to reduce the selection of cleaning compounds to a rational basis, it still remains largely empirical based on trial and error. Moreover, it is usually impractical to stock a wide variety of cleaning compounds, tailoring each to a specific situation, requiring reasonable compromises. Cleaning compound vendors and sanitation consultants can be of assistance in determining which cleaning agents should be used for each task.

Cleaning compounds should not be judged on the amount or persistence of the foam they produce, since there is no direct correlation between foam and cleaning power, although bubbles can help prolong the contact time and can help entrap and hold loosened particles of soil. Moreover, some types of soils impact the amount of foam produced either positively or negatively. For certain types of cleaning, such as in spray and circulation systems, excess foam can be problematic, requiring the addition of defoaming agents.

Cleaning compounds may produce solutions that are either acidic or alkaline. Most are alkaline, having a pH above 7 in solution, and these tend to be based on formulations containing sodium carbonate (soda ash), phosphates such as trisodium phosphate or tetrasodium pyrophosphate, silicates, or hydroxides such as sodium hydroxide (caustic soda or lye). Such products are usually combined with surfactants (wetting agents). Alkaline cleaning compounds neutralize and disperse acid soils and saponify fatty soils, helping solubilize them.

Acid cleaning compounds, having a pH below 7, are used for specific applications involving alkaline soils. Sometimes strong mineral acids are called for, such as nitric, hydrochloric (muriatic), phosphoric, and sulfuric acids. More commonly, weaker acids are used in formulating acid-type cleaners, such as citric, tartaric, lactic, formic, and acetic acids.

VERIFICATION OF CLEANING EFFECTIVENESS

The FDA does *not* require formal validation of cleaning methods used in dietary supplement manufacturing operations, since the written procedures required by §111.25(c) are considered to be sufficient. However, although not mandated, it is at least prudent to *verify* that cleaning regimens are adequate and are successfully accomplished when properly followed. It is left to the discretion of the firm as to whether and how to verify cleaning regimens.

The question in essence becomes how much clean is *sufficiently* clean, since the term "clean" is subjective and open to interpretation.

Although not required by the regulations, some firms do occasionally take swab samples and/or rinse samples from the cleaned equipment to confirm that their established cleaning methods are truly effective. This can be established by checking the samples for residue content by any of several analytical methods, such as the total organic carbon (TOC) technique. High-performance liquid chromatography (HPLC) is also useful for such testing.

Probably the most widely used technique for such verification is simply visual inspection. Several studies have shown that there is justification in using the visible-residue limit (VRL) method. It has been established that a trained observer can adequately detect whether cleaning methods are, indeed, effective.

There are also techniques and tools available for monitoring surface contamination through the use of what is known as ATP bioluminescence. The term "bioluminescence" comes from the Greek "bios" meaning living and the Latin "lumen" meaning light. "ATP" is an abbreviation referring to the chemical, adenosine triphosphate, which exists in all organic matter. Detection of the presence or absence of ATP on surfaces is a useful tool to determine the effectiveness of cleaning. ATP can be made to transfer energy to an enzyme called *luciferase, which* causes a reaction that emits light. The intensity of the light is proportional to the amount of ATP present. This minute amount of light can be amplified electronically to give a direct reading of the amount of soil on the surface being tested. There are small handheld instruments commercially available, based on this technology. Such instruments are used to some extent in the food and pharmaceutical industries, and, of course, they can also be used in the dietary supplement manufacturing industry to quantitatively verify the effectiveness of cleaning procedures.

EQUIPMENT SANITIZING

The term "sanitize" as used in the GMPs is defined as "to adequately treat cleaned equipment, containers, utensils, or any other cleaned contact surface by a process that is effective in destroying vegetative cells of microorganisms of public health significance, and in substantially reducing numbers of other microorganisms, but without adversely affecting the product or its safety for the consumer." This definition does *not* require any specific quantitative reduction in the number of microorganisms remaining on sanitized surfaces.

The regulations allow the firm discretion to decide when sanitizing treatments are necessary and to only sanitize when needed. In other words, it is up to the firm whether to both clean *and* sanitize, or just to clean without sanitizing, on the basis of the risks associated with the materials and process used. However, it is advisable to be prepared to justify such decisions in the event of being challenged.

In §111.27(d)(2), it is stated that it is required to ensure that all contact surfaces used for manufacturing or holding low-moisture components or dietary supplements are in a dry and sanitary condition when in use. When such surfaces are wet-cleaned, they must be sanitized (when necessary) and thoroughly dried before subsequent use.

Under §111.27(d)(3), if *wet processing* is used during manufacturing, it is required to clean and sanitize all contact surfaces, as necessary, to protect against the introduction of microorganisms into components or dietary supplements. When cleaning and sanitizing are necessary, it is required to clean and sanitize all contact surfaces *before use* and after any interruption during which the contact surface may have become contaminated. In continuous production, or consecutive operations involving different batches of the same product, it is required to clean and sanitize the contact surfaces as necessary.

Similarly, §111.27(d)(4) states that it is required to clean surfaces that do *not* come into contact with components or dietary supplements as often as necessary to protect against contamination.

Single-service items (such as utensils intended for one-time use, paper cups, and paper towels) must be stored in appropriate containers, and handled, dispensed, used, and disposed of, in a manner that protects against contamination of components, dietary supplements, or any contact surface. This is stated in §111.27(d)(5).

Cleaned and sanitized portable equipment and utensils that have contact surfaces must be stored in a location and manner that protects them from contamination, under §111.27(7).

SANITIZERS

Sanitizers are chemical agents used to reduce microbiological contamination on surfaces to an acceptable level. These do not necessarily leave the treated surfaces completely sterile (i.e., free from *all* living organisms). For practical consideration in dietary supplement manufacturing and packaging, sanitizers are useful in avoiding possible microbial contamination that could lead to adulteration. Most cleaning operations are relatively inefficient in destroying the microorganisms present on surfaces. Therefore, in many instances, both cleaning *and* sanitizing are employed.

As matter of nomenclature, the term *disinfectant* refers to a chemical agent used to destroy microorganisms on inanimate objects and surfaces, whereas an *antiseptic* is used for similar purposes but on living tissues. The suffix *-cide* indicates that the agent effectively kills organisms, whereas *stat* indicates that the agent merely inhibits or greatly reduces the reproduction of organisms. For example, a bactericide eliminates living bacteria, whereas a bacteriostat does not necessarily kill existing organisms but does arrest their further growth.

Sanitizers should be bactericidal, not merely bacteriostatic, and should also be effective against a wide spectrum of the organisms likely to be encountered. They should also be relatively nontoxic and nonirritating to humans.

A variety of commercially available sanitizers meet these criteria. In general, sanitizers can be classified into four groups: compounds that release chlorine, iodine and iodophors, quaternary ammonium compounds, and acid-type sanitizers. In addition, bromine compounds, phenolics, and certain aldehydes are potentially useful as sanitizers, but are seldom used in dietary supplement establishments for a variety of reasons.

Under §111.15(c)(1), sanitizing agents must be free from microorganisms of public health significance and be safe and adequate under the conditions of use.

Also, §111.27(d)(6) requires that cleaning and sanitizing agents be properly stored.

Chlorine Sanitizers

The chlorine-releasing sanitizers are usually inorganic hypochlorites, such as potassium, sodium, and calcium hypochlorite. These are relatively inexpensive, readily available, and effective against a broad spectrum of microorganisms. Negative factors are that they tend to deteriorate on storage, particularly when exposed to light or elevated temperatures. They also tend to break down in the presence of traces of iron or copper in solution. To prolong their shelf life, it is advisable to store these items in cool, dark places. Prolonged contact with hypochlorites on certain grades of stainless steel may result in pitting corrosion.

The most commonly used hypochlorite sanitizer is sodium hypochlorite, the common household bleach. A number of commercial sanitizers are based on sodium hypochlorite. It is important to heed the label directions on dilutions. Hypochlorites must never be used together with acid products or with ammonia, as doing so results in the release of a toxic gas.

Calcium hypochlorite is a dry powder which must be dissolved in water. This dry form is considerably more stable than the liquid products, although it is somewhat more costly.

Another chlorine sanitizer is chlorine dioxide (sometimes called chlorine peroxide). This is a gas, but is readily soluble in water. The gas form is explosive, making transportation difficult, so it is sometimes generated by users on-site.

There are also other commercially available organic chlorine sanitizers, sold under a variety of trade names. These include chlorinated isocyanurates, dichloro dimethyl hydantoin, N-sodium-N-chloro-p-toluenesulfonamide, and others.

Quaternary Ammonium Compounds

The quaternary ammonium compounds, commonly referred to as "quats" or "QACs," were developed during World War II as germicides and antiseptics, and are still widely used. They are odorless and tasteless, exhibit low toxicity and low skin irritation, and have surfactant properties giving them wetting and penetrating properties. They are stable in storage and noncorrosive to most materials. They have bactericidal activity at low concentrations, but are somewhat selective in the organisms they can control. Water hardness and the presence of iron ions tend to reduce their activity.

Iodophors

The term "iodophors" stems from the Latin, meaning "iodine carrier." These are loosely bound complexes of iodine with surfactants. They have rapid killing power for a broad spectrum of organisms. They are useful in hard water, less affected by organic matter than are the chlorine-based sanitizers, and are also less subject to becoming inactivated by anionic surfactants than are the quaternary ammonium sanitizers. They typically have a mild "iodine-like" odor and are usually amber in color. They must not be used at elevated temperatures, as they tend to break down releasing iodine vapor. These compounds have enhanced activity in acid solutions and, therefore, are often formulated with phosphoric acid to a pH around 2.5 to 3.5.

Acid Sanitizers

Acids, usually phosphoric acid, can be formulated with certain anionic surfactants to form useful sanitizers. Although somewhat slow in action, acid sanitizers are effective against a wide range of bacteria as well as against yeasts and molds. They do not corrode nor discolor stainless steel, and when used as directed, a water rinse following application is generally not required. They are effective in either hard or soft water, and their sanitizing action is not decreased by the presence of organic matter. Several trade named sanitizers are based on this concept.

Ozone

Although ozone has been used for years as a sanitizing agent, its acceptance has recently accelerated. It is a highly reactive form of oxygen, having three oxygen atoms as opposed to the usual two. Consequently, it is unstable and is prone to combine with other molecules. It is a safe, powerful sanitizing agent, often used to sanitize water, but is also used on nonporous surfaces. It leaves no chemical residues. Ozone can be easily generated on-site, with commercially available ozone-generating equipment.

Thermal Sanitizing

The use of heat, for example hot water, has long been recognized as an effective way of sanitizing. It is a slow process, requiring time for the surface to reach an effective temperature, and then a cooldown period. But, heat is relatively inexpensive, easy to use, and readily available.

SANITIZER USE CONSIDERATIONS

For sanitizers to function properly, they must be applied to *clean* surfaces. Therefore, they are used *after* the cleaning step has been completed.

Some (but not all) sanitizers require a final rinse with water.

The labels of commercially available sanitizers give explicit directions regarding dilutions, methods of application, contact time, and whether a postrinse is needed. It is important that these directions be followed. In the United States, sanitizers are required to be registered with the Environmental Protection Agency (EPA), and an important step in acquiring such registration is that the manufacturer of the sanitizer must submit the proposed labeling together with effectiveness data, toxicity data, stability data, and other pertinent information. Users are required by law to follow the use directions on the labels.

BIOFILMS

The slimy, slippery films often encountered on surfaces that are wet for prolonged periods of time are known as "biofilms." It is now believed that a large percentage of microbial life exists as biofilms, as opposed to separate individual organisms. Biofilms are what makes wet rocks slippery, and they also account for the slime found in drains. The plaque on teeth is a form of biofilm, and medical devices implanted in humans frequently are troubled with biofilms.

Biofilms start forming by a few individual organisms lodging in cracks, crevices, or indentations in a surface. As these organisms reproduce, a colony forms, at first as a thin film. The microbes give off a sticky, slimy material that both "glues" the colony together and helps capture additional microbes and nutrients. Other microbes joining the colony may be of the same or different species. Therefore, any given biofilm may be heterogeneous containing many different species. Over time, the colony forms layer on layer, held together by a polymeric matrix produced by the living organisms.

In the early stages, biofilms can be removed rather easily, by scrubbing, scraping, or killing the organisms with a sanitizer. However, as time goes by and the films grow in depth, they become extremely difficult to eliminate. Sanitizers kill only the outer layers of the film, which quickly regrow.

The existence of biofilms on contact surfaces can create a number of quality problems.

Biofilms can lead to microbially induced corrosion of metals and the degradation of plastics. They are often present in water pipes and in drains.

OUTSOURCING CLEANING AND SANITATION

As discussed in chapter 31, outsourcing specific operations is allowable. However, GMP compliance is still required, and it becomes the responsibility of the firm named on the product labels to ensure such compliance.

If cleaning and/or sanitation are outsourced to a contracting firm, that organization must be fully aware of the GMP aspects of their work and must see to it that those obligations are properly met. This makes the selection of the contract service an important consideration, and further makes good communication and supervision of their activities essential.

Much of the cleaning and sanitation is usually done after normal working hours. Thus, the contracting firm must understand and follow the relevant established SOPs, and they must be regularly provided with proper tools in good operating condition as well as with the required supplies.

SUGGESTED READINGS

Adair DL, Carroll L. Clean as a whistle while you work. Food Quality Magazine, July–August 2002.
Brandt J. The case for ozone in sanitation. Food Quality Magazine, December–January 2009:32–36.
Cords BR, Dychdala GR. Sanitizers: halogens, surface-active agents, and peroxides. In: Davidson PM, Branen AL, eds. Antimicrobials in Foods. New York: Marcel Dekker, 1993:36–52.
Eilers JR. You can't sanitize a dirty surface. Food Processing, December 1990:144–146.
Ghannoum M. Microbial Biofilms. Herndon, VA: American Society for Microbiology Press, 2004.
Gibson RT. Recommended procedures for cleaning plant surfaces and equipment. Plant Eng 1979; 33(23):133–135.
Jarvis LM. Communal living: biofilms. C&E News, June 9, 2008:15–23.
Marriott N. Principles of Food Sanitation. 5th ed. New York: Springer, 2006.
Mermelstein NH. Cleaning and sanitizing. Food Technol 1998; 52(3):82–84.
Redemann R. Basic elements of effective food plant cleaning and sanitizing. Food Safety Magazine, April–May 2005.
Schaule G, Flemming H. Pathogenic microorganisms in water system biofilms. Ultrapure Water, April 1997:21–28.
Traver T. Biofilms: a threat to food safety. Food Technology, February 2009:46–52.
Troller JA, Taylor S. Sanitation in Food Processing. 2nd ed. Maryland Heights, MO: Academic Press Elsevier, 1993.
Valigra L. ATP bioluminescence moves mainstream. Food Quality Magazine, June–July 2010:30–32.
Verstraeten N. Living on a surface: swarming and biofilm formation. Trends Microbiol 2008; 16(10):496–506.
Zottola EA. Microbial attachment and biofilm formation: scientific status summary. Food Technol 1994; 48(7):107–114.

7 Maintenance and GMP

THE CRITICAL IMPORTANCE OF MAINTENANCE
The upkeep of, and making necessary repairs to, facilities and equipment is closely linked to product quality. Studies of FDA Establishment Inspection Reports, Warning Letters, and the reasons for recalls make it clear that a definite relationship does exist between quality and maintenance.

Shortcomings in maintenance can result not only in regulatory actions but also in customer dissatisfaction from resulting defects, leading to consequent lost sales plus the possibility of liability litigation. Equipment failure can result in costly downtime, which, in turn, can lead to low finished goods inventories, which can cause difficulty in filling orders. Equipment and buildings are vital assets, and for this reason, they need to be carefully and properly cared for.

Although the regulations are silent on maintenance details, it is expected that each company will have a well-designed and implemented plan. Reactive maintenance, which is making repairs only when equipment actually breaks down, is highly unwise.

In some firms, there is a tendency for top management to look on maintenance as a necessary evil, an overhead expense required to keep production functioning. In such firms, when funds are in short supply, maintenance appears to be a cost that can be reduced, at least temporarily, by reducing the maintenance staff and deferring all but urgent maintenance tasks. However, this approach ignores the well-established interrelationship between maintenance and product quality, and often results in unfortunate outcomes. Management at all levels should recognize this and give continuous support and a high priority to ensuring an adequate number of qualified maintenance personnel supported by ongoing training as well as keeping an appropriate inventory of needed spare parts. These are not places where cost cutting is necessarily appropriate. Moreover, proper maintenance usually ends up costing less in the long run than more reactive repair approaches.

REGULATORY ASPECTS OF MAINTENANCE
The mandate for proper equipment maintenance is stated in §111.25(c) and §111.27(a), whereas §111.15(b)(2) covers the need to keep the physical plant in good repair. Similarly, the requirement to maintain instruments and controls (which includes but is not limited to laboratory instruments) is stated in §111.27(a)(6)(ii). These portions of the regulations require maintenance, but without specific details, leaving how to achieve these goals to each company since facilities, equipment, and instruments differ from firm to firm. However, written procedures for such activities are required by §111.35(b)(1), and the procedures must be followed.

EQUIPMENT LOGS
It is a GMP requirement to keep records of the history of each piece of equipment, the date of installation, the date of each use, and the date and time

of all maintenance and repairs. These can be kept either as individual equipment logs or with the batch record, according to §111.35(b)(2) and §111.260(c).

EQUIPMENT MAINTENANCE PROGRAM MANUAL
Although the requirement for equipment maintenance is clear, the GMPs are silent on the specific *ways* of conducting such maintenance, leaving this for each individual firm to decide. A good approach to this is to establish a written maintenance program, outlining the specific equipment covered, the maintenance personnel roles and responsibilities (job descriptions), spare parts policy, and the documentation required. The maintenance program should also include how unplanned emergency repairs should be handled. It is advisable to have specific SOPs detailing what needs to be done for each item of equipment. Such SOPs serve for use by the maintenance technicians and as training tools.

If a formal work order system is used, provision should be made for that in the maintenance program manual.

If the policy of the firm establishes a predetermined frequency for performing specific tasks on each item of equipment, such scheduling should be included in the program. Preventive maintenance plans should be detailed and should include how decisions were reached in preparing this portion of the manual, such as the significance of historical data and consideration of the experience with similar equipment as well as recommendations from the equipment's manufacturer (if available).

MAINTENANCE ORGANIZATION
The GMPs are also silent regarding how and by whom the maintenance function should be structured, but it is prudent to spell out these personnel policies in the manual, with specific roles and responsibilities documented. This should include establishing who has the authority to prioritize maintenance projects and who can change the maintenance plans and systems.

CLEANLINESS IN MAINTENANCE
Individuals involved in maintenance operations should understand the importance of using special care to avoid the possibility of contaminating products. The work of maintenance technicians often necessarily involves "dirty" functions, and therefore steps must be taken to protect products and contact surfaces. Although all employees must understand the basic GMP requirements, these are of particular importance for those involved with maintenance functions. In addition to company workers, outside maintenance personnel invited to help solve equipment problems may not be fully aware of the requirement to exercise care in this regard and therefore they should be informed of these issues and closely supervised.

EQUIPMENT IDENTIFICATION
It is advisable to identify each major piece of equipment with a clearly visible unique identification number. This facilitates keeping logs and communicating actions required both with regard to maintenance and with regard to cleaning and sanitizing activities.

PREVENTIVE MAINTENANCE

As the name implies, preventive maintenance (PM) is a proactive approach to *prevent* equipment problems and breakdowns as opposed to fixing problems only when they occur. PM is based on using routine scheduled actions such as equipment inspections, adjustments, regular and proper lubrication of moving parts as appropriate, replacing worn parts before they fail, and similar steps to avoid unexpected malfunctions. The actual details of any PM program depend on the specific equipment and machinery in use and therefore vary from company to company.

There are several variations on the PM theme, one of which is termed *predictive* maintenance (PdM), which involves trying to establish when failure is likely to occur if specific steps are not taken. An example of a PdM "tool" is vibration monitoring for rotating or reciprocating parts. This helps detect imbalance, bearing wear, gear problems, etc., in rotating equipment such as mixers, fans, blowers, pumps and compressors, cavitation in pumps, and other such problems. This is an old technique, recently significantly improved by miniaturization of vibration sensors and data loggers.

Predictive tools also include analysis of lubricants, temperature tracking, infrared thermographic scanning, and many others, that aid in the complex tasks of foreseeing possible failures to allow for timely replacement of parts. In using PdM, it is up to the firm to decide which of these tools is appropriate in specific circumstances. Some firms reduce the overall costs associated with PdM through prioritizing their equipment, applying PdM only for the most critical items.

Another type of PM is *reliability-centered maintenance* (RCM) designed to ensure that a given piece of equipment continues to properly perform its intended functions. This was originally developed for use in aircraft to help prevent serious accidents. This involves studying probable failure modes (the causes of failures) to establish where and when PM tasks are needed to minimize the risks and consequences of unexpected failure of equipment. This requires a highly complex and technically sophisticated process, where the analysis required takes considerable time and therefore tends to be costly. In most instances, in the dietary supplement manufacturing industry the costs involved in establishing and operating an RCM system cannot be justified.

Still another type of maintenance program is known as *total productive maintenance* (TPM). This requires involvement of operators of equipment working in tandem with the maintenance technicians in making minor adjustments, being sure the equipment is kept clean and properly lubricated, and alerting maintenance personnel when an impending problem is suspected. The people using given pieces of equipment can be quite helpful in diagnosing and preventing maintenance problems.

Although there are various ways of achieving PM, none of which are mentioned in the regulations, it is prudent to choose and apply whichever approaches best apply to the equipment and the availability of properly trained maintenance technicians. However, none of the forms of PM (or combinations of the forms) is ever perfect, so even under the best of situations, equipment failures *will* sometimes occur. When this happens, *corrective* maintenance (CM) must be applied. CM refers to the steps taken *after* a problem has occurred to fix the situation to restore the equipment to operable condition.

COMPUTERIZED MAINTENANCE MANAGEMENT SYSTEMS

A computerized maintenance management system (CMMS) is a computer software system designed to facilitate PM through scheduling activities, keeping track of spare parts inventories, recording the history of the maintenance of each item of equipment, aiding in decision-making related to maintenance activities, and other useful information management activities related to the maintenance functions. There are many types of such programs commercially available to help plan and track the status of maintenance activities, minimizing or eliminating much of the paperwork otherwise needed.

A well-designed CMMS can be very useful in maintenance planning and scheduling, helping eliminate downtime and rework, prolonging the life of assets, and in cost reduction.

BUILDING MAINTENANCE

In addition to equipment considerations, keeping buildings and building services in good repair is also an important consideration. Older structures may have cracks or holes in walls or floors where dirt or water may pose problems. Windows and doors also need to be maintained in good condition, and drains must function well to avoid accumulation of solid debris or stagnant water. If dropped ceilings are used, care must be exercised to avoid the possibility of insects or rodents nesting above them. Chipped or peeling paint must be corrected.

Building services, such as heating, ventilation, and air-conditioning (HVAC) systems, must be kept in good operating condition, as must the lighting, electrical, and plumbing systems. If dust collectors or air compressors are used, they too need routine attention. Roofing should be periodically inspected, and repairs made as necessary.

Although these items may seem obvious, FDA inspections do sometimes cite substandard building maintenance, making it advisable to periodically audit the condition of the facility and take corrective action if needed. As previously mentioned, the requirement to maintain the physical plant in good repair to prevent contamination is specified in §111.15(b)(2).

OUTSOURCING AND OUT-TASKING MAINTENANCE

Although it is clear that proper maintenance must have a high priority, most dietary supplement manufacturers probably would not consider maintenance to be one of their core competencies. Therefore, in some circumstances, it may be prudent to consider using an outside contract service to handle the maintenance function. This can be a complex and difficult management decision, based on many items to be taken into consideration, such as the product line and annual volume of goods produced, the nature of the manufacturing processes, the type of equipment involved, whether it is difficult to identify and attract qualified maintenance personnel in the geographic location of the manufacturing facility, the availability and reputation of contract maintenance services with GMP knowledge in the area, and many other factors. However, contract maintenance is becoming more prevalent, and in some circumstances can result in cost savings and improved production reliability and quality.

Even the largest firms usually do outsource some infrequent but major maintenance projects, such as replacing a roof, resurfacing a parking lot, making significant changes in or additions to the building, upgrading the HVAC system, etc. These occasional short-term uses of contracted services fall into what is often called "out-tasking." Similarly, if a firm decides to make a long-term arrangement to handle a specific need, such as cleaning and sanitizing or pest control, this may be considered to be out-tasking as opposed to "outsourcing." It is a matter of semantics how the terms are applied.

Outsourcing and out-tasking of maintenance and other functions are permissible under the GMPs, although it remains the responsibility of the firm that introduces the products into commerce to ensure that all applicable regulations are properly met. Therefore, careful monitoring and supervision of outside contract services are essential.

SUGGESTED READINGS
Anon. ISPE Good Practice Guide: Maintenance. Tampa, FL: International Society for Pharmaceutical Engineering, 2009.
Bagadia K. Computerized Maintenance Management Systems Made Easy: How to Evaluate, Select, and Manage CMMS. New York: McGraw-Hill Professional, 2006.
Levitt J. Complete Guide to Predictive and Preventive Maintenance. New York: Industrial Press, 2002.
Levitt J. Handbook of Maintenance Management. New York: Industrial Press, 1997.
Mead WJ. Maintenance: its interrelationship with drug quality. Pharm Eng 1987; 7(3):29–33.
Mead WJ. Proper maintenance: a key to product quality. Pharm Technol 1985; 9(10):42–43.
Moubray J. Reliability-Centered Maintenance. 2nd ed. New York: Industrial Press, 1997.

8 Calibration

THE REQUIREMENT FOR CALIBRATION

The instruments and controls used in manufacturing and in laboratory testing must perform properly to achieve consistent quality. To ensure reliability, some types of mechanical or electronic equipment, especially measuring devices and instruments, require periodic *calibration*. This consists of a comparison of their actual performance to that of a known standard, coupled with appropriate adjustments if needed. This concept is basic in the science of measurement, known as *metrology*. Therefore, calibration is *required* for specified items, that is, those used either in the testing of a component or a dietary supplement, or in the manufacturing of a dietary supplement, as stated in §111.27(b).

Calibration must be done *before* the first use of such instruments and controls, and then routinely at appropriate intervals of time. Often, vendors precalibrate such devices, but in such instances it is prudent to have the supplier take steps to ensure the devices remain properly calibrated *after* installation and to leave certification that this has been satisfactorily completed.

ACCURACY AND PRECISION

The term *accuracy* refers to how close the measurements are to the true value, whereas *precision* is an expression of the degree of reproducibility of the measurements. An illustration of these meanings is to consider a number of shots or arrows fired at a target, where hitting the bull's eye is "accurate," while missing it is "inaccurate," and where if the grouping of the shots on the target is small and compact, they are "precise," while if widely spread they are "imprecise."

RECORDS AND WRITTEN PROCEDURES

Written procedures must be established and followed under §111.25(a) for calibrating instruments and controls used in manufacturing or testing a component or dietary supplement, whereas §111.25(b) requires such procedures for calibrating, inspecting, and checking automated, mechanical, and electronic equipment.

Records are required documenting calibration, according to §111.35(a).

§111.35(b)(1) requires having written procedures for calibrating instruments and controls used in manufacturing or testing a dietary supplement and for calibrating, inspecting, and checking automated, mechanical, and electronic equipment.

Documentation is required each time a calibration is performed on instruments and controls used in manufacturing or testing, according to §111.35(b)(3).

The *type* of information that must be documented is stated in §111.35(b)(3)(i) through (b)(3)(vii), including the identity of the instrument or control, the date of the calibration, the reference standard used including the certification of its accuracy, the calibration method used including appropriate limits for accuracy and precision, the calibration reading or readings found, information on any recalibration method used, and the initials of the person who performed the calibration and any recalibration.

Records of calibrations, inspections, and checks of automated, mechanical, and electronic equipment are required by §111.35(b)(4).

Calibrations may be performed either on-site (at the plant where the manufacturing is done) or off-site, and in either case they can be done by qualified services or contract agencies.

CALIBRATION PROGRAMS

Although not specifically called for in the regulations, many firms establish a written program establishing specific directions for calibrating each type of instrument or control, the frequencies for such calibrations, tolerances for the results, records of the calibrations, and the procedures for handling out-of-tolerance deviations when they occur, since the possible impact on products produced since the last calibration must be assessed and appropriate decisions reached and documented.

Such programs are quite useful in helping to ensure that calibrations are carried out in an agreed-upon and consistent way, whether done by in-house personnel or by an outside contract service.

CALIBRATION STATUS LABELS

Although not specifically mentioned in the regulations, it is usual to identify all instruments and controls that require calibration with an identification label showing the current calibration status, typically the date of the last calibration and the due date of the next calibration. Each such piece of equipment is usually given a unique identification number, which appears on the label and in the appropriate records.

These labels should be made as tamper proof as feasible, and they should be so placed as to be readily visible. The significance and meaning of the calibration labels should be clearly understood by all of the appropriate employees. Some firms also use similar labels on items that do *not* require calibration, saying "No calibration required" or something equivalent, to avoid confusion.

CALIBRATION FREQUENCY

The FDA does not prescribe in detail how frequently calibration must be done, but it must be sufficiently often to ensure that instruments and controls are operating within correct parameters. According to §111.27(b)(2) and (3), aside from the first use, calibration must be done at the frequency specified in writing by the manufacturer of the instrument or control, or at routine intervals, or as otherwise necessary to ensure the accuracy and precision of the instrument or control.

Calibration takes time, and it often involves equipment downtime, which can interfere with production schedules. However, several sources offer calibration scheduling software, which helps adherence to necessary calibration intervals while minimizing costly downtime.

Many types of instruments and controls tend to drift out of calibration over time. Therefore, the frequency of calibration depends on the type and stability of the instrument or control, on the frequency of use, on temperature and humidity in the area where the equipment is used as well as on many other variables. Therefore, in addition to recommendations from the maker of the item, the calibration intervals are usually based on maintaining the desired level of accuracy and reliability, coupled with historical experience over time. There are helpful statistical techniques that can be used in establishing realistic frequency of calibration for each instrument or control. However, improper functioning of instruments and controls can lead to costly recalls and to regulatory action, so this topic is worthy of careful concern. This is also one of the details that often comes into focus during FDA inspections.

TRACEABILITY

Calibration is in effect a process of comparing the functioning of an instrument or control to a standard and, where feasible, adjusting the instrument or control to agree with the standard. Therefore, the *authenticity* of the standard is of significant importance.

National standards typically are kept or tracked by well-recognized government metrology laboratories such as in the United States, the National Institute of Standards and Technology (NIST). These national standards are then copied by using state-of-the-art equipment and techniques as *primary* standards from which instrument and control manufacturers obtain *secondary* working standards. The term *traceability* refers to the ability to relate measurements to these standards through an unbroken chain. This topic is not mentioned in the GMPs but is of importance in that the accuracy of measuring instruments depends on their inherent capabilities and also on the reliability of the traceable standard.

THE ROLE OF QUALITY CONTROL IN CALIBRATION

According to §111.30(c), quality control personnel must periodically review the calibrations, inspections, and checks of automated, mechanical or electronic equipment, and in §111.117, they must review and approve all processes for calibrating instruments and controls, must periodically review all records for calibration of instruments and controls, and must also periodically review all records for calibrations, inspections, and checks of automated, mechanical, or electronic equipment as well as reviewing and approving controls to ensure that such automated, mechanical equipment functions in accordance with its intended use.

It is important to note that quality control personnel's role is not limited to circumstances where there has been a calibration failure, nor that they need to be directly involved in the calibration processes, nor they need review each calibration as it is conducted. Rather, their task is to review the records *periodically* (e.g., every 6 or 12 months) to uncover *trends* in the performance of the equipment that could adversely impact product quality. Spotting such trends enables quality control personnel to recommend actions to correct the trend.

SCALES AND BALANCES

Among the most commonly used items of equipment for measurement in manufacturing operations are scales, and in laboratory operations are analytical balances. Errors in these could obviously impact quality, and some types do tend to occasionally develop nonlinear errors. Therefore, in addition to periodic *calibration*, it is usual and strongly recommended (but not specifically mandated) to do frequent checks of scales and balances. Some firms do this weekly, daily, or even more frequently, and maintain records of having done so.

Typically, such checks are done with appropriate "standard" weights, which can be purchased from various metrology firms and from vendors of scales and balances. These weights are usually certified as being traceable to NIST primary standards. Such weights must be handled and stored with care to preserve their accuracy.

It is advisable that each scale used in production be labeled with the acceptable range of use, and the weight checks performed at the midpoint and near the upper and lower edges of the range, that is, three-point weight check.

Some may contend that such frequent three-point checks of scales are not really needed, since the electronic scales most often used in manufacturing now are not nearly as subject to the same vulnerabilities as were the older mechanical types. Others contend that a frequent one-point check is adequate. However, the regulations are silent on this issue, leaving the matter to the judgment in each firm.

ANALYTICAL INSTRUMENT CALIBRATION

In addition to equipment, instruments, and controls used in manufacturing, the GMPs also require calibration of equipment and instruments used in laboratories to examine and test components and incoming packaging materials, in-process production, finished batches, finished dietary supplement products, returned goods, reprocessed goods, etc. This requirement for the calibration of laboratory equipment is applicable both for in-house laboratories and when outside laboratories are used.

Each instrument or control should have a separate and detailed SOP on how and when the calibration is to be done, with detailed records of the laboratory calibration program. The FDA considers laboratory operations, including calibration, as being extremely important, and consequently, this tends to be a major item of focus during establishment inspections.

Considerable technical literature exists on the methods and techniques of calibrating laboratory instruments and systems, with which most analytical chemists are familiar. Instrument vendors can often be helpful in supplying further information, and methods and suggestions appear in compendia such as the USP, which can be useful albeit not mandated by the regulations.

It is useful to maintain a listing of laboratory instruments and equipment requiring periodic calibration, together with files describing the history of the calibration and maintenance of each such item and with logs recording the daily use of such equipment.

Some manufacturers of analytical balances provide "autocalibration" features on their equipment. Although quite useful and may lengthen the frequency with which analytical balance calibration is needed, these autocalibration methods should not be exclusively relied upon. Periodic performance checks should be conducted to verify that the balances are still performing accurately and satisfactorily as measured against accredited standard weights. Moreover, care should be taken to keep analytical balances level and protected from vibration, excessive humidity, and air currents that could impact readings.

It is worth noting that since the dietary supplement GMPs are relatively new, much of the existing literature on calibration is from related FDA-enforced industries, particularly the pharmaceutical industry. However, the concepts and details of calibration needs, methods, and techniques are similar regardless of the end products involved.

SUGGESTED READINGS

Anon. Calibration Control Systems for the Biomedical and Pharmaceutical Industry, Recommended Practices 6, National Conference of Standards Laboratories (NCSL), Boulder, CO, 1999, and ANSI/NCSL Z540.3-2006 Calibration Laboratories and Measurement and Test Equipment.

Anon. NIST Handbook 44: Specifications, Tolerances, and Other Technical Requirements for Weighing and Measuring Devices. Gaithersburg, MD: National Institute of Standards and Technology, 2008.

Anon. GAMP Good Practice Guide on Calibration Management. Tampa, FL: International Society for Pharmaceutical Engineering (ISPE), 2002.

Bremmer RE. A calibration approach to improve measurement accuracy. PDA J Pharm Sci Technol 1982; 36(5):193–195.

Campbell P. An Introduction to Measurement and Calibration. New York: Industrial Press, 1995.

Curtis M. Handbook of Dimensional Measurement. New York: Industrial Press, 2007.

Dills DR. Establishing a calibration program for the FDA-regulated industry. J Validation (IVT) 2000; 7(1):4–17.

Evans DM, Erickson J. Establishing calibration intervals, traceability, and documentation. BioPharm 1998; June:74–75.

Hauck WW, Koch W. Making sense of trueness, precision, accuracy, and uncertainty. Pharm Forum 2008; 34(3):837–842.

Nielsen CL. Calibration and maintenance. Pharmaceutical Formulation and Quality, Part I, March–April 1999:46–47; Part II, May–June 1999:44–45; Part III, July–August 1999:29–30; Part IV, September–October 1999:41–42.

Palumbo R. Gage calibration is an investment in quality production. Quality Digest, February 1997:60–61.

Payne GC. Calibration: why it is important. Quality Progress, September 2005:66, 69.

Wehman T. Equipment calibration. J Validation Technol 1995; 1(2):62–66.

9 Production and process controls

THE BASIS FOR PRODUCTION AND PROCESS CONTROLS

The basis of the GMPs is that quality must be built into the product at every stage of its manufacturing, packaging, and labeling and that it is inadequate and unacceptable to simply test the final product to determine that it is satisfactory. To reliably achieve the desired outcome of consistent excellence in quality and safety, certain *controls* are necessary. This requirement is stated in §111.55.

Dietary supplement manufacturing consists of a series of interconnected processes, where the term "process" refers to specific steps or tasks that must be carried out to accomplish a stated goal. For example, in producing tablets, a typical scenario is to weigh each of the dry ingredients, place them into a blender and mix them for a given period of time, then add a liquid to the dry mix and blend it under specified methods and time to form a wet granulation, then dry the granulation in a cabinet dryer or a fluid bed dryer to the proper final moisture content, then place the dried granulation into a tablet press to compress it to the desired form and to achieve the proper thickness, hardness, and dissolution properties, and then, in some instances, apply a coating by a specified process. Each of these steps must be properly controlled to achieve the critical quality attributes. No matter what form of product is made, there will be a series of steps, or *processes* involved, each of which must be conducted properly.

Processes are also involved in other business operations, not just manufacturing. For example, there are processes for managing an organization and for managing finances and other resources. The firm's many processes interact in various ways, and each process needs to be identified, planned, managed, and monitored appropriately to achieve the desired outcome.

Processes need to be *robust*, meaning that they must be capable of performing satisfactorily even under difficult or stressful conditions.

PROCEDURES VS. PROCESSES

A *procedure* is the method or way that each step in a process is carried out. Therefore, each process consists of an interrelated *set* of procedures. As the word implies, a procedure is the way to proceed. To help ensure that each procedure is done as intended, instructions as to how to conduct a procedure are reduced to a written document called a *Standard Operating Procedure* or SOP, discussed in chapter 25. These typically detail the preparation for, conduct of, and completion of a specific task.

DESIGN REQUIREMENTS FOR PRODUCTION AND PROCESS CONTROLS

While §111.55 requires implementing production and process controls, §111.60 outlines the design requirements for such a system. Subparagraph (a) says that the system must be designed to ensure that products are manufactured, packaged, labeled, and held in a manner that will ensure that the master manufacturing record (MMR) is properly followed.

Subparagraph (b) states that the system must include *all* of the requirements of subparts E through L of Part 111 and that they must be reviewed and approved by quality control personnel.

The items requiring attention are covered in the following chapters:

- Specifications—chapter 10
- Sampling—chapter 11
- Deviations and Corrective Actions—chapter 12
- Incoming Components, Packaging Materials, Labels—chapter 13
- Master Manufacturing Record—chapter 14
- Batch Production Record—chapter 15
- Manufacturing Operations—chapter 16
- Packaging and Labeling Operations—chapter 17
- Quality Control Responsibilities—chapter 18
- Laboratory Operations—chapter 19

In addition, as discussed in chapter 29, it is important to have supply chain integrity, using only qualified vendors, coupled with identity testing of each incoming component. Other factors of significance in helping ensure the processes used remain in control include having well-designed and properly constructed and maintained facilities, utilities, and equipment; suitably qualified, trained, and supervised personnel; appropriate in-process testing; and steps to prevent contamination from any source. Excellence in record-keeping and documentation is also essential, as discussed in chapter 25.

Formal validation is *not* required, but rather, it is mandated that all of the controls necessary to ensure the quality of the dietary supplements are in place. This includes the fact that it is necessary to meet all of the specifications established under §111.70(e) and the fact that in-process controls established under §111.75(b) must also be properly monitored.

Since production and process controls are the main mechanisms to ensure the identity, purity, quality, strength, and composition of the products manufactured, it is critical that they cover *all* the stages from receipt and acceptance of components, packaging and labeling materials, through the release for distribution. It is also important to identify the points in the manufacturing process where control is necessary to prevent adulteration.

Designing such controls involves careful decision-making, based on thoroughly *understanding* the manufacturing processes used. Product and process understanding clearly helps provide more consistent quality. The activities and studies resulting in such understanding should be documented, although this is not specifically required by the regulations. It is often helpful to use flow diagrams, or written descriptions of each of the steps, explaining the operations. This also helps formulate procedures and write the applicable SOPs.

Well-designed controls help build quality into the products, with extensive written procedures, and allow less burdensome testing of the finished products. The representative sampling and testing at the end essentially becomes a check on whether the overall manufacturing process is truly under control.

The goals of a properly designed and executed production and process control system are to achieve product quality that is consistent from batch to batch and unit to unit, with freedom from contamination, and to ensure the identity, strength, purity, and composition that the products purport or are represented to have.

PROCESS IMPROVEMENT

Even well-designed processes need to be periodically reviewed, based on performance and manufacturing experience. Changes to improve established processes may become warranted, and improvements made. If so, the proposed changes should be carefully evaluated by all concerned, and if it is decided that changes should be made according to a well-justified and documented rationale, with the agreement with appropriate quality control personnel, the change should be implemented. This may possibly also require modifications in other processes.

The topic of continuous improvement is discussed in chapter 28, and change control in chapter 26.

KNOWLEDGE MANAGEMENT

Although not a new topic, in recent years there has been increased interest in knowledge management (KM) as related to quality control. The International Conference on Harmonization (ICH) guideline Q10 states that KM is a "systematic approach to acquiring, analyzing, storing, and disseminating information related to products, processes and components." The FDA is quite aware of this and encourages firms to consider applying the KM principles in the improvement of production and process control systems, albeit there are no requirements to do so.

KM approaches take advantage of experience acquired by the firm and/or individuals within the firm as well as information obtained from the literature, from consultants, from technical and trade associations, from attending conferences and seminars, and from many other sources. There are various ways and techniques of deliberately collecting, storing, and making use of accumulated experience that helps avoid "reinventing the wheel." These can have a positive impact on decision-making related to implementing and improving production and process control systems.

The information explosion in recent years, brought about in part by the Internet, has greatly increased the availability of useful knowledge. Moreover, as employees move from one company to another, their accumulated knowledge often goes with them. Conversely, new employees may bring with them valuable experience gained elsewhere. The object of KM is to capture such knowledge and make use of it in furthering the firm's activities.

VARIABILITY

A factor worthy of consideration is that variation occurs in essentially all manufacturing processes. Variation, in the form of minor changes in raw materials, processing times and temperatures, and other such operating characteristics, does impact quality.

Variation is common in essentially everything and is almost always present to some degree. Consider, for example, any group of people and note the variation from one person to another in height, weight, age, intelligence, education, and a host of other characteristics. No two people are identical due to this variability.

If body temperature of each individual in a group were to be accurately measured, there would be noticeable differences, and if the temperature were to be retaken later, even more differences would be detected. Similarly, most manufactured items also vary to some extent despite efforts to exercise strict control. Such variation may, or may not, be significant, but is ever present.

Some of the variables in any manufacturing system occur in the ingredients used, in the training and skills and workmanship of employees, in plant environmental conditions such as temperature, humidity, and lighting, in laboratory testing, in sampling, in instrument calibration, and a myriad of other factors. It is usually worthwhile to attempt to identify and reduce variability in each process and each product.

Variation in quality characteristics is generally classified as due either to *common* causes that are inherently part of the process, or to *special* causes that arise due to specific circumstances. For example, in making tablets, there may be a very slight variation in tablet weight from the various punch stations, which is normal, expected, and acceptable. That would be "common cause" variation. However, there may also occasionally be significant variation in tablet weight due to poor mixing in preparing the granulation. That, as a one-time event, would be considered to be "special cause" variation, for which corrective action is needed.

With proper identification and data analysis, it is often possible to identify and eliminate the special causes of variation. As a general rule, the magnitude and impact of common cause variation are slight, whereas those of special cause variation are much greater.

A process that has only common cause variation affecting the outcomes is termed a *stable* process, where the variation remains essentially constant over time. A process that has outcomes affected by both common causes and special causes is called an *unstable* process.

Making *any* changes in a stable process usually results in increased variability, so efforts to "adjust" any system tend to be counterproductive. Therefore, it is generally inadvisable to try to "fine-tune" a more-or-less stable situation, such as making intuitive but actually unnecessary adjustments to a machine.

It is prudent to monitor process variability and, where feasible, identify and reduce the sources. Doing so helps yield stable processes, which tend to maximize productivity while minimizing costs. Process understanding involves identification and management of the sources of variability. Consistent product quality is influenced by variability. There are clear advantages to reducing variability when it is possible to do so.

HUMAN ERRORS

As important as it is to control variability, the reduction of mistakes is an even more urgent concern. Human errors can and do occur all too frequently at all stages in manufacturing. Predicting and controlling mistakes are a definite necessity, when possible.

Inadequate training is sometimes the cause of mistakes, so an obvious solution is to resort to more and improved training. However, there is evidence that suggests that retraining tends to be only marginally effective since most errors result from causes *other* than failure to understand the task. Instead, a large percentage of errors result from failure to remember precisely what must be done and how it must be accomplished, by taking unjustified shortcuts, from missing a step in the required procedure, from memory lapses due to fatigue or illness, from lack of availability or access to needed items, from inappropriate behavior, from faulty decision-making, and a host of other such causes not directly related to training per se.

To minimize mistakes in carrying out procedures, it is prudent to strive to design the procedures to be as error tolerant as possible. Errors are almost always unintentional. Carefully and clearly written detailed SOPs, readily accessible to the appropriate persons, are a key factor in reducing mistakes. Good supervision is of course another important factor. For complex procedures, it is sometimes advisable to provide a checklist of the steps to be followed, somewhat like the preflight checklists often used by aircraft pilots.

It is advisable to track errors to see where and why they most frequently occur. This can point to steps that can be taken to prevent recurrence of these same mistakes.

It is also advisable to strive to inculcate a culture of excellence, where everyone cooperates in attempting to recognize and prevent errors, not only in their own spheres, but also in helping others and in eliminating error-prone conditions.

SUGGESTED READINGS

Abel J. Rethinking process control for pharmaceutical manufacturing. Pharmaceutical Technology, November 2002.
Amy CK, Strong A. Teaching the concepts and tools of variation using body temperature. Qual Prog, March 1995:168.
Anderson LH. Controlling process variation is key to manufacturing success. Qual Prog 1990; 23(8):91–93.
Davis W. Using corrective action to make matters worse. Qual Prog 2000; 33(10):56–61.
Dekker S. The Field Guide to Understanding Human Error. Farnham: Ashgate, 2006.
Hickey AJ, Ganderton D. Pharmaceutical Process Engineering. 2nd ed. Informa Healthcare, New York, 2009.
Jolner BL, Gauand MA. Variation, management, and W. Edwards Deming. Qual Prog 1990; 23(12):29–37.
Moran RD, Nolan TW. Process improvement. Qual Prog, September 1987:62–68.
Nally JD, Karaim MD. Production and process controls. In: Bunn G, Nally JD, eds. Good Manufacturing Practices for Pharmaceuticals. 6th ed. New York: Informa Healthcare, 2006.
Nolan TW, Provost LP. Understanding variation. Qual Prog 1990; 23(5):70–78.
Peters GA, Peters BJ. Human Error: Causes and Control. Boca Raton, FL: CRC Press, 2006.
Snee RD. Creating robust work processes. Qual Prog 1993; 26(2):37–41.
Snee RD. Process variation—enemy and opportunity. Qual Prog, December 2006:73–75.

10 Specifications

DEFINITION OF SPECIFICATIONS

A *specification* is a detailed set of appropriate acceptance criteria, which a finished dietary supplement, component, in-process batch, labels, and other packaging materials must meet. Specifications typically include a list of required characteristics and properties, together with tests and test methods, and also acceptable numerical limits or ranges on the test outcomes to which the item needs to conform. Specifications are necessary guidelines for those who are in the role of decision-makers as to what is and what is not acceptable.

To consistently produce products of the same uniform quality, it is obviously necessary to first define precisely what the manufacturer wants the quality of the finished products *to be*, and then also to provide a "road map" to achieving this through establishing the desired quality of all of the raw materials and in-process steps to be used as well as the quality of the finished and packaged products ready for distribution. Specifications are a necessary and critical part of a total control strategy.

The dietary supplement GMPs (unlike the GMPs for pharmaceutical and medical device products) leave it up to the *manufacturer* to decide the parameters of the quality of the finished products (as long as they are not legally adulterated). This gives a considerable degree of flexibility to each manufacturer. In other words, a firm may elect to sell only products that are excellent or products that are less so. Such decisions impact the cost of goods and the selling price, as well as the continued acceptance by the consuming public, which in turn will usually affect the volume of sales and the profits gained by the firm. Therefore, establishing the *level* of product excellence desired is an important business decision for every manufacturer. The GMPs are not about that kind of decision, but instead are related to consistently making products that meet the *manufacturer's* own predetermined specifications.

The FDA's definition of *quality* means that the product consistently meets the manufacturer's own (not FDA's) established specifications for identity, purity, strength, composition, and limits on contamination to ensure adulteration is prevented, and has been manufactured, packaged, labeled, and held under conditions to prevent adulteration (which includes but is not limited to full compliance with the GMP regulations).

Therefore, in establishing specifications, it is important to understand what FDA means by the terms "identity," "purity," "strength," and "composition." Although in Comment 57 of the preamble to the 2007 final regulations FDA declined to define these terms since the ways they are used by the industry vary, some clarification was provided, saying in effect that identity refers to consistency with the description in the master manufacturing record (MMR), purity does not necessarily imply freedom from impurities but instead means the portion or percentage of a dietary supplement that represents the intended product, strength relates to the *concentration* meaning the amount of an ingredient per serving or other specified unit of measure, and composition refers to the specified mix of product and product-related substances in a dietary supplement. Using these definitions, each manufacturer has the flexibility to define the "quality" of products made by choosing the specifications for those products and for the materials used in manufacturing them.

Most manufacturers do have and have had formal specifications established. However, with the advent of the GMPs, it is now prudent to do a thorough review of this topic to be sure that all required specifications are in place and are well documented. There is no mandated format for this, but each specification should be both complete and clearly comprehensible. Each should be dated, reviewed, approved, and signed by the appropriate persons within the organization, including quality control personnel as required by §111.70(c)(3). Specifications must be properly kept up-to-date through an adequate change control program to ensure that only the current version of each is in use.

It is advisable to be sure that specifications are set neither too "tight" nor too "loose." Each firm is required to diligently *follow* their own established specifications, so including details not truly necessary or difficult to consistently achieve may overly complicate operations. However, adequate and well-written specifications are clearly the key to consistent quality. Moreover, good specifications for purchased items are quite useful in dealing with suppliers, and in fact it is often useful to get vendors' input in preparing such specifications to ensure their ability to consistently meet the specifications once finalized.

REQUIRED SPECIFICATIONS

According to §111.70(a), one must establish a specification for *any* point, step, or stage in the manufacturing process where control is necessary. Very similar wording is also contained in §111.210(h)(1), as a requirement of the MMR. This is obviously somewhat subjective since differing opinions might arise as to what is truly required. However, this "catchall" mandate does provide the manufacturer with the flexibility (and responsibility) to determine what specifications *are* needed that are appropriate to the circumstances. Specific requirements are called for in subparagraphs (b) through (g) of §111.70, whereas (a) covers additional items that the *manufacturer* considers necessary, which are not otherwise specifically stated. Moreover, the GMPs do not prevent firms from establishing additional specifications that are not specifically required by the regulations, such as those related to appearance or aesthetic matters.

IDENTIFICATION OF COMPONENTS

The definition of a *component* in the GMPs is any substance intended for use in the manufacture of a dietary supplement, including those that may not appear in the finished product. This includes *dietary ingredients* as defined in the FD&C Act, and also *other* materials such as excipients (i.e., inactive ingredients). The term "component" can be confusing in that some firms also use this word to mean items such as packaging materials, using the term "ingredient" or "raw material" in place of component. However, since the dietary supplement GMPs use the term component in a specific way, attention needs to be paid to this definition.

It is required by §111.70(b)(1) to establish an *identity* specification for *each component* used in the manufacture of a dietary supplement. This is true whether the firm purchases components from an outside vendor or manufactures the component(s), or has the component(s) manufactured under a contract arrangement. In other words, 100% of the components used, whether dietary ingredients or not, must undergo identity testing and therefore require identity specifications. However, for those components that are *not* dietary ingredients (and only those), the identity testing can be done by the vendor and included in a Certificate of Analysis (C of A) under certain prescribed conditions mentioned below.

The identity testing is critically important to avoid disasters such as those from the substitution of diethylene glycol for glycerin or the mix-up between *Digitalis lanata* and *plantain*, which have occurred in the past. Clearly, the identification issue is more difficult with botanical ingredients. Similarly, blood products, tissue products, cartilage, and other such items present still different identification issues. New analytical technology and modifications to existing methods are continually being developed. However, good and indisputable identification *is* both possible and required.

OTHER REQUIRED COMPONENT SPECIFICATIONS

In addition to identification, §111.70(b)(2) and (3) require specifications for components that will ensure that the dietary supplements manufactured from them meet their specifications for purity, strength, and composition and also to set limits on the types of contamination that may adulterate or lead to adulteration of the finished batches of dietary supplements. There are many sources of information that can be useful in preparing such specifications for components, including various pharmacopoeia, compendia, recommendations from industry trade associations, and suggestions from vendors of the items. For dietary supplements, compliance with monographs in official compendia is voluntary (unless claimed on the label), but can be quite helpful background in writing component specifications, determining ranges and limits, and supplying analytical methods.

Flexibility (and responsibility) is given to manufacturers in §111.70(b)(2) by permitting the firm to decide *which* component specifications (other than identity) are or are not needed to ensure that the *finished products* meet their specifications.

Regarding setting limits on contaminants, it is not feasible to consider every possible type of contaminant, and not all ingredients are subject to the same types of contaminants. Therefore, the FDA expects manufacturers to be aware of the types of contamination *likely* or *certain* to contaminate a given component that could lead to adulteration, and includes these (with limits and appropriate test methods) in the specifications.

SPECIFICATIONS FOR EXCIPIENTS

Excipients are components *other than* dietary ingredients, which are vitally important in manufacturing finished products. These inactive ingredients include such items as fillers, lubricants, disintegrants, binders, colors, suspending or dispersing agents, flavors and sweeteners, and other such processing aids, required to make the final formulation. For example, in making tablets or capsules, such ingredients may impact uniformity and consistency, appearance, taste, disintegration, and stability, among other essential features.

Excipients have uses not only in dietary supplements, but also in foods, pharmaceuticals, and cosmetics. In recent years, much emphasis has been put on the quality of excipients, particularly those used in drug products, by the industry association of manufacturers and end-users of excipients, the International Pharmaceutical Excipients Council (IPEC). This organization has published excipient GMPs, has worked with compendia such as the *United States Pharmacopoeia–National Formulary* (and others overseas) in establishing and harmonizing excipient monographs, has established Good Distribution Practices for excipients, has developed a document for guidance in establishing quality agreements between buyers and suppliers of excipients, and has taken many other helpful steps. However, their emphasis has been on excipients used in *pharmaceutical* products, not dietary supplements, although there is an obvious overlap.

In addition to chemical specifications, including limits on impurities and contaminants, the *physical* characteristics of excipients are critical to their applications. Functionality is a key attribute, which is not always easy to determine by laboratory testing, yet such characteristics are critical to how an excipient performs in a formulation. These physical criteria may include such variables as particle shape and size distribution, crystal structure, flowability, bulk density, and many other parameters. Thus, the supplier's "grade" must be specified, which is then typically verified by the vendors' certificates of analysis.

A typical example is microcrystalline cellulose (widely used in many solid dosage products), which is available from various suppliers, each of which offers several grades having significant differences in functionality, necessitating the specification to cite the vendor's trade name and grade to help ensure consistently receiving the optimum component.

It is important to note that in §111.35(d) of the *proposed* GMPs, the regulatory status of excipients in dietary supplements was mentioned, but the FDA deleted this portion in the final regulations, saying in effect in Comment 237 of the preamble that *other* regulatory requirements cover the proper use of ingredients in dietary supplements, but that it is the responsibility of the manufacturer to ensure that ingredients used are lawful. Further, the response to Comment 238 points out that *dietary ingredients* as defined in section 201(ff)(1) of the Act are *not* food additives, nor required to be generally recognized as safe (GRAS) substances, but *other* ingredients used (i.e., excipients) *are* subject to FDA's food additive authority. The regulations for determination that a substance is GRAS are found in Parts 182, 184, and 186 of Title 21 of the Code of Federal Regulations. However, the FDA has further stated in response to Comment 239 that excipients used in drug products are not necessarily satisfactory for use in dietary supplements since the use of drugs may be intermittent or short term whereas the intake during routine and long-term use of dietary supplements could lead to unsafe levels of some excipients, and therefore it is advisable to consider the requirements outlined in §170.30.

IN-PROCESS SPECIFICATIONS

It is required by §111.70(c)(1) that in-process specifications be established for any point, step, or stage in the MMR where control *is necessary* to help ensure that the finished product speculations are met. This is similarly mentioned in §111.210(h)(3). This might include, for example, information about the viscosity or pH that must be achieved during the manufacture of a batch of a liquid product. Once such in-process specifications are established, of course they must also be met, as stated in §111.75(b). Where necessary, limits must be set on the types of contamination that may adulterate or lead to adulteration of the finished batch, which clarifies that it is important to also establish limits on the contamination that may occur from the manufacturing facility and/or the equipment used.

The regulations do not establish what the specific requirements are for in-process monitoring, leaving it to the *manufacturer* to determine what is necessary in each instance to ensure that the finished product specifications are met. Moreover, such monitoring is intended to detect any deviations or unanticipated occurrences that might result in a failure to meet specifications as described in chapter 12.

Under §111.70(c)(2), the firm must provide documentation for *why* meeting the in-process specifications, together with meeting the component specifications, helps ensure that the finished batch of product also meets specifications. Quality control personnel must review and approve this documentation under §111.70(c)(3) and §111.105(e). This requirement is based on the fact that meeting in-process specifications *alone* does not necessarily ensure meeting the finished product specifications. There is obvious interplay between component specifications, in-process specifications, and finished batch specifications, and this requirement to document the basis why this will help ensure finished product quality essentially forces consideration of the various facets of control.

Although not mandated, it is prudent to set in-process specifications "tighter" than the finished product specifications to help ensure that normal variability does not cause the finished batches to fall outside the specifications.

PACKAGING MATERIAL SPECIFICATIONS

It is required under §111.70(d) that packaging material specifications be established. Any packaging that comes into physical contact with the product must be safe and suitable for its intended use and must not be reactive or absorptive or otherwise affect the safety or quality of the product. This refers to primary packaging items meaning those that do actually contact the product (often plastic or glass bottles or jars), caps and cap liners, and blister packs. The integrity of the packaging material tends to be a more significant issue with liquid or semiliquid products than with solid dosage forms, such as tablets, capsules, and powders and granules, but it is important regardless of the physical nature of the product contained.

With plastic containers, there is the possibility of extractables and/or leachables that might impact product quality or stability or lead to adulteration over time. These "foreign" materials may originate in the manufacturing of the containers, such as plasticizers, mold release agents, colorants, antistatic agents, and others. It is prudent to know about the possible existence of these and to include maximum allowances of them in the specifications to ensure that interactions with the products will not be a problem.

Various plastics as used in packaging differ in moisture and oxygen permeability. Even rigid plastic bottles have some permeability. These factors need to be considered in selecting and specifying the materials used in bottles, jars, caps, and blister packs. Literature exists for guidance, plus ASTM Committee D10 Division III has published a number of useful standards.

Care should be taken in writing specifications that the dimensions for bottles and their closures match to ensure that the fit will be good and that the optimum cap liner is selected and specified. Cap liners are of importance in that they act as barriers to moisture and oxygen, which in time can degrade dietary supplement products.

Although most dietary supplement products are not required by regulations to be packaged with tamper-evident and/or child-resistant closures, in some circumstances these may be a wise business choice to use even if not mandated. This is at least worth considering when developing specifications for packaging materials. In packaging products containing iron, attention should be given to the requirement of the Consumer Products Safety Commission (CPSC) regulation in 16 CFR 1700.14 as well as FDA's required warning statement in 21 CFR 101.17(e).

As an example of the type of information that should be contained in packaging material specifications, for glass or rigid plastic bottles and jars, it is typical to include not only the type or grade of glass (e.g., USP Type I, II, or III, etc.) or plastic resin (e.g., PE, PET, PVC, P/P, PETE, etc.) but aso color, size, weight, dimensional design including screw threads or provision for snap caps or other types of closures, defining a "lot" of bottles (especially since bottles and jars are usually produced continuously in very large quantities, so what constitutes "a lot" is often an arbitrary decision based on convenience rather than uniqueness), test methods or standards to be used, establishing how the empty containers will be packed for shipment and the outer shipping containers marked, establishing what constitutes noncomformities and internal cleanliness, and establishing sampling plans with acceptable quality limits.

Similar lists of items for other types of packaging can of course be devised, albeit what is included in such specifications is at the discretion and judgment of the firm and is not detailed in the GMPs.

LABEL SPECIFICATIONS

Under Section 201 of the FD&C Act, there is a significant difference between the terms "label" and "labeling." Label refers to a display of written, printed, or graphic matter on the immediate container and also on the outside container or wrapper if it is used as a retail package. Labeling refers not only to labels but also to *other* written, printed, or graphic matter accompanying the article such as folders, inserts, outserts, or booklets.

Physical specifications for labels are required by §111.70(d). Although it is up to the manufacturer to establish such specifications, for labels, these would typically include size, printing stock, colors, etc. Although not covered by the GMPs, it is also important to be certain that the labels *say* both what is required by the appropriate regulations and that the claims and other copy are properly included as intended by the manufacturer.

All too often, printing errors on labels and mix-ups as to which labels are applied to packages result in recalls. Therefore, label control systems should be implemented even beyond what is mentioned in §111.70 and §111.120, although such systems are not specifically mandated. Label controls need to be developed according to the needs of each facility. This would usually include a Standard Operating Procedure on an artwork approval process, using only reliable suppliers of labels (not selected on price alone, as vendors' internal controls tend to add to their costs), obsolete printing plates and obsolete labels should be promptly destroyed to avoid the possibility of misuse, and label storage areas should be secure and well controlled.

Some label printers use a system called "gang printing," where more than one kind of label is printed using large sheets and after die-cutting the various labels are hand separated. This system is fraught with the danger of label mix-ups, so even if economical, should not be used, although this is not stated in the GMPs. Roll labels are much safer than cut labels, even if gang printing is not used, although errors can still occur with roll labels if proper controls are not employed.

Some firms sell the exact same product to various accounts, each with their own label. In doing this, it is common practice to produce many unlabeled packages of product, store the unlabeled goods, and then apply the appropriate labels to a portion of the stock as needed. This system is often called "bright stock" or "brite stock." This term is derived from the canned food industry where it is fairly common to package food into cans, then retort the cans, and *then* apply the labels, "bright" (or "brite") referring to the unlabeled metal cans. Although this is an economical way of producing small quantities of goods for particular accounts, it can obviously easily cause the wrong label to be applied to specific containers. Therefore, the use of bright stock should be avoided if feasible, but if this approach must be used, careful controls should be instituted to avoid possible product mix-ups.

One way of helping avoid label and product mix-ups is to use different colors, sizes, shapes, or graphic designs on the labels for different products, making it easy for operators on the packaging lines to be alert for problems that could lead to recalls. On more automated equipment, barcodes and other machine-vision techniques can help assure the use of the correct labels.

Label printers are not required to follow GMPs and may or may not understand the detailed requirements for quality control. That is why careful selection of label suppliers is important. It is wise to audit label vendors and work closely with them to try to minimize errors. Label problems constitute one of the most significant causes of product recalls.

These concepts are not only limited to labels per se but also apply to individual cartons, preprinted containers, tubes, and insert or outsert vendors.

FINISHED PRODUCT SPECIFICATIONS

It is required by §111.70(e) to establish finished product specifications for the identity, purity, strength, and composition of the batches of finished product, *including* limits on the types of contamination that may adulterate or lead to adulteration of the product. Although limits on contaminants were established for the components used, it is important to readdress this issue for the finished product to ensure that the manufacturing process has not adversely affected such levels or has contributed an additional source of contaminants.

CONTRACT PACKAGERS AND LABELERS

If the product comes in from a manufacturer for packaging and/or labeling (and for final distribution as opposed to being returned to the manufacturer), under §111.70(f) specifications must be established to ensure that the product received is adequately identified and is consistent with the purchase order for the goods.

Contract packagers and labelers must establish specifications for the dietary supplement products they handle, but may rely on the content of the product they receive from the manufacturer. The contractor may obtain information regarding this from an invoice, certificate, or other form of verification of what the product consists of, so that the contactor has adequate information and to ensure that the product is consistent with the purchase order. However, the contractor who handles some portion of the production of finished dietary supplements must be aware that the GMP requirements do apply. The specifications established by the contractor must provide assurance that the product received is adequately identified and is consistent with the purchase order. In other words, this section applies to the product that has left the control of the firm that manufactured the batch. However, if the product is *returned* to the manufacturer (as opposed to going directly into distribution), §111.70(f) does not apply to the contractor.

ENSURING THAT THE PROPER PACKAGING AND LABEL ARE USED

As part of the finished product specifications, under §111.70(g), it is required to establish specifications for the operations of packaging and labeling to be certain that the correct packaging is used and the correct label has been applied. In other words, separate specifications must exist to ensure the use of the specified packaging, and the specified label, as required by the MMR. This concept also appears in §111.415(g), which requires examining representative samples of each batch of packaged and labeled product to ensure that the requirements of §111.70(g) have been met.

MICROBIOLOGICAL TESTING

During the development of the regulations, it was proposed to require testing of *all* finished products to ensure freedom from microbiological contamination, but it was decided this is not always required, leaving it up to the manufacturer to determine if and when such testing is necessary, depending on many possible variables. Therefore, there is no mandatory specification requiring such testing, although obviously certain types and amounts of microbiological contamination could lead to adulteration, sothis issue requires vigilance.

RESPONSIBILITY FOR DETERMINING SPECIFICATIONS ARE MET

It is required by §111.73 to determine that all of the specifications established under §111.70 are actually met. The methods and criteria for doing so are detailed in §111.75. This is a significant step in the focus on building quality into dietary supplements throughout the entire production and control system.

METHODS OF DETERMINING WHETHER SPECIFICATIONS ARE MET
Component Identity

As mentioned above, it is critically important to know the identity of ingredients, particularly those classified as *dietary* ingredients is defined in section 201(ff) of the Act. Therefore, §111.70 (b)(1) requires establishing an identity specification for *each* component and §111.75(a) requires that before the use of any *dietary ingredients* it is necessary to conduct *at least one* appropriate and reliable test or examination to identify that component. The FDA correctly believes that misidentification of dietary ingredients poses a risk to public health.

For many dietary ingredients, identity testing is fairly easy, using methods outlined in various compendia, and in some instances using well-established instrumental techniques such as near-infrared (NIR) or the recent improvements in thin layer chromatography (TLC). But, as mentioned above, *some* components (including but not limited to herbals and botanicals) pose more difficult issues. For components of agricultural origin, the optimum testing is usually based on whether the component is in the form of whole plant, or whether it is cut, powdered, or extracted. When sufficient morphological characteristics are present, these and organoleptic evaluation (sight, odor, taste, smell, etc.) can often be used. Microscopy is also useful, but depends on the skills of a trained microscopist aided by good reference samples. Chemical analysis for marker compounds can also be useful, for which many validated analytical methods exist (many based on chromatographic techniques) and more are currently under development.

Dietary ingredients from other sources, such as blood products, organ tissue, cartilage, etc., may present similar problems in identity testing.

Most manufacturers use only a limited number of dietary ingredients, so they become quite familiar over time with the appropriate identity testing methods. Some ingredient vendors can offer helpful suggestions regarding identification techniques. Various trade associations have also been extremely helpful with this, some offering seminars and workshops on the topic. Moreover, identification is not really working with *total unknowns* since it is more or less a matter of saying "*is* this or is it *not* the item we ordered," which narrows the field considerably. However, the manufacturer must establish and use identification tests for *all* dietary ingredients.

Certificates of Analysis

For all components, both dietary ingredients and nondietary ingredients (excipients) firms may either conduct appropriate tests or examination to ensure all components meet their specifications (aside from the mandatory identity testing for dietary ingredients), or rely on certificates of analysis from qualified suppliers. A C of A is a document provided by a supplier of a component prior to or upon receipt of the component, which certifies certain prescribed characteristics and attributes of the component, making it unnecessary for the dietary supplement manufacturer to repeat conducting such testing unless it is desired to do so. It is advisable to have the original manufacturer furnish the C of A, as opposed to a subsequent distributor.

It is of critical importance that certificates of analysis be truthful and accurate. To help ensure this, the FDA requires that suppliers be properly qualified, to establish their reliability, if certificates of analysis are to be accepted in place of the firm doing the testing. The details of *how* the firm qualified the supplier must be documented and retained, and such documentation must be reviewed and approved by the firm's quality control personnel. The details of how to verify the reliability of supplier's certificates are a bit vague, but involve confirmation of the supplier's tests or examinations. One way of accomplishing this would be to test samples of given component batches (either in-house, or by sending them to an outside contact laboratory) and comparing the results with the C of A. Another, or additional, way is to audit the supplier, looking into their quality systems, laboratory, training, etc. Many in the dietary supplement manufacturing industry hoped during the development of the GMPs that dietary component suppliers would be required to comply with GMPs, but FDA declined that suggestion. The minimum criteria for qualifying suppliers are stated in §111.75(a)(2)(ii), and are further discussed in chapter 29. The details of how the firm qualified each supplier must be documented and retained, and such documentation must be reviewed and approved by quality control personnel.

Meeting In-Process Specifications

The requirement to establish in-process specifications is stated in §111.70(c). Also, in §111.75(b) is the requirement to follow these specifications, that is, to monitor in-process points, steps, or stages where control is necessary to ensure the quality of the finished batch. In addition, this portion of the regulations requires detecting any deviation or unanticipated occurrence that might result in failure to meet the specifications. See also the discussion of this in chapter 12.

Meeting Finished Product Specifications

In the proposed regulations, considerable emphasis was put on testing finished batches. However, after considering the comments received and in view of the stringent requirements to meet both the component specifications and the in-process specifications, the FDA decided that it is not necessarily required after all to do full testing of every batch of finished product. The GMPs were designed to ensure that the manufacturer establishes what the product is to be, and then spells out specifications for the ingredients, in-process controls, and the finished product. If all of the manufacturing processes are well controlled, a burdensome amount of finished batch testing is not needed.

Therefore, the manufacturer has the flexibility to either test *every* batch or to test only a *subset* of such batches, provided that the subset is chosen using a sound statistical sampling plan. Moreover, provision is made to allow for testing with one or more (not necessarily all) of the finished product specifications to ensure that the product meets all of the specifications for identity, purity, strength, composition, and for the limits on contamination that may adulterate or lead to the adulteration of the product. This concept appears in §111.75(c)(1). Under this, the firm has the flexibility to select one or more of the established finished batch specifications and conduct the appropriate tests or examinations to verify that the production and process control system is producing a dietary supplement that meets *all* of the product specifications.

In §111.75(a)(2)(D), it is required to periodically reconfirm the supplier's certificates of analysis, but no definition is given as to how frequently this must be done.

In §111.75(a)(2)(B), it is stated that a C of A includes a description of the tests or examination method(s) used, the established limits for the results, and the actual results of each test. In fact, most CoAs cite the name and a concluding statement interpreting the outcome. For herbal, botanical, or other ingredients of natural origin, certificates of analysis often give more detail on the origin and morphology of the item.

The term "appropriate tests or examinations" implies that such tests must be *scientifically valid*. The FDA considers test methods published by the Association of Official Analytical Chemists (AOAC), the various compendia, and the FDA itself, in scientific journals, textbooks, proprietary research, etc., to *be* scientifically valid. This is further mentioned in §111.75(h). However, such tests and examinations must include at least one of the five kinds of testing mentioned in this portion: gross organoleptic analysis, macroscopic analysis, microscopic analysis, chemical analysis, or other scientifically valid methods. This effectively covers the gamut of available test methods. This is discussed further in chapter 19.

It is required to provide adequate documentation for the basis of determining that such reduced testing will ensure that the finished batch does meet *all* of the product specifications, under §111.75(c)(3), and quality control personnel must review and approve this documentation under §111.75(c)(4).

There may be circumstances where it is *not* feasible to verify that one or more of the product specifications show that the process control system is producing a product that satisfactorily meets the complete specifications. It is then possible to *exempt* one or more of the product specifications from the verification requirements as called for in §111.75(c)(1), *providing* the firm provides adequate documentation of the inability to conduct the specific verification(s) in that there is no scientifically valid method of testing or examining the exempted specification at the finished batch stage, but that the component and in-process testing does indicate compliance with the specification(s) and that periodic testing of finished batches confirms this. Moreover, the documentation must show why meeting the remainder of the specifications ensures that the batch does meet all of the product specifications, aside from that or those exempted. Quality control personnel must review and approve such documentation.

Requirement for Visual Examination Prior to Packaging or Labeling

As mentioned above, §111.70(f) requires that a product *received* from a supplier for packaging and/or labeling (for distribution, as opposed to being returned to the supplier and has left the control of the firm that made the batch) must be properly identified upon receipt. In other words, specifications must be established to ensure that the product received is adequately identified and is consistent with the purchase order or the invoice or other documents from the supplier to be certain that it *is* the ordered product. In connection with this, §111.75(e) requires that before the packaging or labeling operation is begun, the product must be visually examined and the documentation from the supplier examined. This is to avoid the possibility of a mix-up.

Visual Identification of Incoming Packaging Materials and Labels

As a precaution, §111.75(f) requires, prior to use and as a minimum, conducting a visual examination of containers and their caps or other closures, and of labels, as well as reviewing the purchase order or the supplier's invoice or other appropriate documents, to ensure those items conform to their specifications. This ties with §111.160, which, among other requirements, calls for quarantining incoming packaging materials and labels until they have been tested and released for use. See further discussion of this in chapter 13.

Visual Examination of Finished Packaged and Labeled Product
In §111.75(g), it is stated that as a minimum, it is required to conduct a visual examination of finished packaged and labeled dietary supplement products to determine whether the specified packaging and labeling were used as specified in the MMR. This is to ensure that the proper packaging materials and labels were used. See also chapter 17 for more details.

Some firms use bar codes or other types of marks that can be automatically read on the packaging lines by electronic devices or machine vision to ensure the proper packaging materials and labels have been used.

REQUIREMENTS IF SPECIFICATIONS ARE NOT MET
As required in §111.77, if established specifications are not met, quality control personnel must reject the items involved, *unless* these personnel approve a corrective treatment or adjustment or reprocessing that ensures the quality of the finished product and compliance with the MMR. Moreover, compliance with §111.123(b) (discussed elsewhere in this book) is required.

This requirement for rejection is made specific for component specifications in §111.70(b), which if not met, quality control personnel must reject the component(s) involved, which must not be used in the manufacturing of dietary supplements. Similarly, packaging materials and/or labels that fail to meet the specifications in §111.70(f) must be rejected and not used. However, it is permissible to correct existing problems causing rejections through treatment or reprocessing, if this is approved by quality control personnel.

SPECIFICATION ARCHIVES
Although not mentioned in the GMPs, some firms keep files of all previous quality specifications, properly numbered and dated, in order to refer back to revisions and previous editions, including information on when and why such revisions were made.

Corrections or changes to specifications must be properly handled to ensure that only the proper and current version is in use. Photocopies must be accounted for and controlled to be certain of this.

STANDARDS VS. SPECIFICATIONS
A *standard* is a document issued, developed, and distributed by a recognized national or international organization, applicable to many companies or organizations affected, to help simplify interchangeability. For example, screw threads, the width of railroad tracks, many electric and electronic parts, the dimensions of sanitary tubing, etc., are covered by agreed-upon, transparent, fair, and generally available standards, greatly facilitating the use of items and parts made by different firms in any given industry. Most of such standards are considered to be "voluntary consensus standards," meaning that although quite useful, they do not have the force of law and are issued by such standards-setting organizations (SDOs) as ASTM International (formerly the American Society for Testing and Materials), ANSI (American National

Standards Institute), ASME (American Society of Mechanical Engineers), Bioprocessing Equipment Standards, 3A-SSI (3A Sanitary Standards, Inc.), GS1 (an international not-for-profit association involved with standards for barcodes and radio frequency identification systems or RFID), and many other similar organizations. Such standards are established through established detailed procedures, where "consensus" implies general agreement (but not necessarily unanimity) among the interested parties.

The ANSI is a nonprofit membership organization, which helps oversee the creation and publishing of voluntary consensus standards, working in conjunction with and accrediting SDOs, in the interest of facilitating both the global economy and U.S. competitiveness in it. Often, standards include both the ANSI name and the SDO name, for example, "ANSI ASQ Z1.4."

Other types of standards include those issued by government agencies, which may be mandated by law, as well as standards issued by specific industries or individual companies, which are developed in the private sector but without the full consensus process. Moreover, codes of ethics or of personal conduct, issued by professional groups or trade associations, are often referred to as being standards.

On the other hand, a *specification* (as described above) is a document issued by a particular company as a technical definition that establishes their explicit requirements and acceptance criteria for components, packaging materials, labels, in-process goods, and finished products.

THE INTERIM-FINAL RULE

An interim-final rule (IFR) is a way under the Administrative Procedure Act that federal agencies can adopt and make regulations immediately effective, bypassing the usual need for time-consuming public comment when that for sound reasons seems to be neither necessary nor practical. The day that the final GMPs were published, FDA also published in the same issue of the *Federal Register* an IFR saying in effect that it might not always be necessary to do 100% identity testing of dietary ingredients received and establishing procedures by which a supplement manufacturer could apply for an exemption by way of submitting a "citizen petition" stating the scientific rationale and data to support that an alternative testing plan would be sufficient to ensure the correct identity. Since the publication of this IFR, it has become a Final Rule.

However, few (if any) firms responded to this concept, probably because they could see little or no economic advantages. Most of the major dietary supplement trade associations responded to FDA, saying in effect that they strongly opposed the IFR because they felt it would seriously weaken the overall system of needed process controls and would tend to undermine an important aspect of product quality since identity testing is critical to ensuring the proper composition of the finished products. The safety and liability risks outweighed any potential cost savings.

SUGGESTED READINGS

Bauer E. Pharmaceutical Packaging Handbook. Informa Healthcare, New York; 2009.
Bugay DE. Pharmaceutical Excipients. Informa Healthcare, New York; 1998.
Food Advisory Committee, Dietary Supplement Working Group, Ingredient Identity Testing, Records and Retention. FDA, 1999.

Hartburn K. Quality Control of Packaging Materials in the Pharmaceutical Industry. Marcel Dekker, New York; 1991.
Jenke D. Extractables and Leachable Substances from Plastic Materials. PDA J Pharm Sci Technol, 56, 2002:332–371.
Leonard E. Packaging Specifications, Purchasing, and Quality Control. 4th ed. Marcel Dekker, New York; 1996.
Paine FA, Lockhart H. Packaging Pharmaceutical and Healthcare Products. Springer, New York; 1995.
Piringer OG. Plastic Packaging: Interactions with Food and Pharmaceuticals. 2nd ed. Wiley-VCH, Hoboken, NJ; 2008.
Schatzman D. Product testing and the GMPs. Nutritional Outlook, July–August 2008: 257–259.
Sharp J. Evaluation of raw material supplies and suppliers. In: Good Manufacturing Practice: Philosophy and Applications. Interpharm Press, New York; 1991.
Smolinske SC. CRC Handbook of Food, Drug and Cosmetic Excipients. CRC, Boca Raton, FL; 1992.
Vesper J. Specifications. In: Documentation Systems: Clear and Simple. Interpharm Press, New York; 1998:193–198.

11 Sampling

The proper taking, handling, and interpretation of samples are important facets of GMP compliance. Critical decisions are often based on data obtained from samples. However, there are various approaches required, depending on what needs to be sampled and what use is intended for the information derived.

REPRESENTATIVE SAMPLES

The FDA's definition of a *representative* sample is "a sample that consists of an adequate number of units that are drawn based on rational criteria, such as random sampling, and that are intended to ensure that the sample accurately portrays the material being sampled." This definition does not necessarily imply that representative samples are always kept long term in that they might simply be used to test.

Under §111.80(a), it is mandated to collect representative samples of each unique lot within each unique shipment of components, packaging, and labels received from a supplier, as well as product received from a supplier for packaging and/or labeling as a dietary supplement for distribution, to determine whether they meet the specifications established in accordance with §111.70. This is in line with §111.73, which requires determination that established specifications are met, whereas §111.75(a) sets forth the *criteria* to be used to see to it that these specifications are met.

Similarly, §111.80(b) requires collecting samples of in-process materials of each manufactured batch at points, steps, or stages in the process as specified in the master manufacturing record, where control is necessary.

It is also required by §111.80(c) to collect representative samples of a subset of finished batches of each dietary supplement manufactured, identified through a sound statistical sampling plan (or if desired, *every* finished batch) before releasing for distribution, to verify that the finished products meet their established specifications. This gives the firm the flexibility of testing either *every* batch made or a properly selected *subset* of the batches. This verifies that the processes used in the manufacturing are under control.

Product received from a supplier for packaging or labeling as a dietary supplement must also have representative samples taken of each unique lot in each unique shipment to determine if the product received meets its specifications, as required by §111.80(d).

Under §111.80(e), it is required to collect representative samples of each lot of packaged and labeled dietary supplements to ensure they meet their specifications as established in §111.70(g), and (a) if applicable. This verifies that the packaging and labeling used were as called for in the master manufacturing record.

In summary, it is required to collect representative samples of components, packaging, labels, in-process materials, finished batches, product received from an outside source for packaging or labeling, and packaged and labeled dietary supplements. However, after the representative samples have served their intended purpose, they need not be retained.

RESERVE SAMPLES

On the other hand, by definition, a *reserve* sample is a sample that *is* to be held for a designated period of time. Such samples therefore differ from representative samples, which are not necessarily kept but are instead used primarily for examination and/or testing and thereafter may be disposed of.

Under §111.83(a), it is required to collect and hold reserve samples of each lot of packaged and labeled dietary supplements. Further, §111.83(b) states that such samples must be in the same container-closure system in which the packaged and labeled dietary supplement is distributed (or if the packaging and labeling is to be done elsewhere, the samples must be in a container-closure system that provides essentially the same characteristics to protect against contamination and deterioration as the one that will be used for distribution). The samples must be identified with the batch, lot, or control number and must be retained for one year past the shelf life date if such dating is used, or two years from the date of distribution of the last batch associated with the sample. These samples are for use in appropriate investigations (e.g., in the event of complaints received) and therefore must consist of *at least* twice the quantity necessary for all tests or examinations to determine whether or not the dietary supplement meets the product specifications. Although not mandated, it is prudent to retain *more than* double the amount needed to conduct the testing.

In §111.465, it is stated that reserve samples must be stored in a manner that protects against contamination and deterioration, and under conditions consistent with the product labels (or, if no storage conditions are recommended on the label, under "ordinary" storage conditions).

The regulations do not specify *who* must collect and hold the reserve samples, although §111.105(g) states that quality control personnel retain the *oversight* of the collection and holding of the required reserve samples, but the actual conduct of these steps does not necessarily need to be performed by QC personnel.

Although the regulations are silent on retaining samples of raw materials (components) packaging materials, and in-process samples, it is prudent to do so.

BASIC CONCEPTS OF SAMPLING

The need to take samples using a rational and sound basis is mentioned in several places in Part 111. The goal of sampling is to avoid the necessity of inspecting or testing 100% of the items in a large population or grouping of goods received or manufactured, to ascertain that the appropriate specifications have been met. This requires decisions on how many samples are needed to be truly representative and how such samples should be selected. The object of such decisions is to determine how best to use only a *few* items to examine or test to be reasonably certain that these few properly represent the entire group.

Sampling does imply risk, since some nonconforming items may exist in the portion of the lot that was not inspected.

SAMPLING PLANS

A plan must be devised or selected to ensure that an adequate number of units is taken, as accurately as feasible, to portray the totality of the material being sampled, yet to save time and effort (which have economic importance to the firm), the number of samples should be as small as feasible. If the sampling is being done to determine whether that material meets the required specifications, acceptance or rejection limits need to be established as part of the sampling plan.

Each sampling plan should be based on the specific objectives of the sampling as well as the possible risks and consequences from decision errors that might occur from the interpretation of data derived from samples that do not accurately reflect the characteristics of the items sampled.

Sampling plans provide a basis for determining the number and kind of samples required. This should be based in part on how critical the item to be sampled is to the quality of the finished product as well as the material variability and the quality history of the supplier. These are subjective considerations, but they need to be defendable and should be documented.

Random Sampling

It is critical that each sample is taken *randomly*, that is, each "member of the population" of units being sampled has an equal chance of being selected. Systematic bias must be eliminated. This is not always easy to achieve, but must be the goal. For example, when a large number of containers of goods are received and stacked in the warehouse, it may be physically difficult for the person assigned to sampling to reach a representative number of the containers in the stack, but even so, appropriate steps need to be taken to assure randomness. Failure to select truly representative samples from the overall population of items can seriously compromise the sampling plan's protection.

A poorly chosen sampling process, one that has a systematic error, is termed a *biased* sample. Nonrandom sampling is often subject to bias, particularly when entrusted to the subjective judgment of the person taking the sample. Such bias is due to a flaw in the sample selection process.

An excellent method to help ensure randomness (although not widely used due to the inconvenience involved) is to number each incoming container, and then select those containers from which samples are to be taken by consulting an appropriate table of random numbers.

ISO 24153, from the International Organization for Standardization, addresses procedures for random sampling and randomization.

A confounding dilemma can exist when there is the possibility that existing defects occur in *clusters*, for example, in the instance of cut labels where a printer inadvertently lets a "wrong" sheet of labels get mixed with the "good" sheets prior to die-cutting, stacking, and packing them for shipment, so that the distribution of the defective labels in the cases received is clustered and may not be detected in the sampling process. "Random" does *not* imply "haphazard" sampling. It requires a deliberately thought-out and well-executed approach to obtain random samples. To the extent that samples are not truly random, the results of the sampling may be meaningless.

Sample Size

Each sampling plan establishes the sample *size* (i.e., number of samples to be taken), together with the "accept" or "reject" criteria. The methods and techniques used to *take* the samples are *not* part of the sampling plan, but instead need to be detailed in sampling *procedures* (SOPs).

SAMPLING

Selecting the appropriate sample size is one of the most difficult parts of sampling. The optimal number of samples depends on many factors, chief among which is the degree of confidence required in the outcome. Fortunately, statisticians have studied this topic for many years, and have determined ways of selecting the number of samples for a variety of circumstances, as discussed below.

Acceptance Sampling

Acceptance sampling plans, sometimes termed lot acceptance sampling plans (LASPs), are used to determine whether or not a given lot of goods (e.g., packaging items) should be accepted or rejected. This approach was developed by the U.S. military during World War II, originally in connection with decisions needed for accepting or rejecting various lots of ammunition received from manufacturers, and was therefore originally known as "military standards." This method is based on inspection of random samples taken from each lot, and assessing those samples against specifications with prescribed accept or reject criteria.

It is important to note that acceptance sampling is *not* a technique for process control to prevent the occurrence of nonconformances, but it instead is a statistical method of deciding whether to accept or reject items that have *already been* manufactured.

Acceptance sampling is applicable where fairly large lots or batches are received, and is a statistically valid means of sampling if coupled with decision-making procedure as to whether the lot or batch is of acceptable quality.

The two forms of acceptance sampling are sampling by *attributes* and sampling by *variables*. Sampling by attributes involves "go–no go" or "yes-no" information determined either by visually examining the sample or by using simple gauges, that is, the item either does or does not meet the specifications. This plan is ANSI/ASQ Z1.4, also called ISO 2859 by the International Standards Organization, and was derived with only minor changes from the former MIL-STD-105E. This is usually the plan of choice for use with packaging materials and with solid dosage forms, when "conforming" and "nonconforming" can be defined. Sampling by attributes is the most commonly used form of acceptance sampling.

The other widely used type of acceptance sampling plan, sampling by variables, is useful when the data characteristics are *measured*, as opposed to being simply classified as good or "bad." Measurements may involve length, width, diameter, temperature, etc. For example, data on fill volumes or tablet weights, being expressed in terms of measurements, would be applicable to sampling by variables. Variable data can be ranked and averaged, such as average age, average temperature, or the average coating thickness. Such plans are covered by ANSI/ASQ Z1.9, also known as ISO 3951, the former MIL-STD-414.

If it is feasible to take measurements as opposed to simple pass/fail judgment, Z1.9 is usually preferable in that it allows decision-making with fewer samples. In using Z1.9, calculations are made of the averages and variability of the measurements, in accordance with specified procedures.

Actually, both Z1.4 and Z1.9 are *collections* of many individual plans. These include single-sampling, double-sampling, and multiple-sampling plans. In the single variety, one sample of the items is selected at random, and the disposition of the lot is decided upon on the basis of the resulting information obtained. Single sampling is the most common and easiest plan to use. On the other hand, with a double-sampling plan, after the first sample is taken and examined but no clear decision is possible, a second sample is taken and examined, and the combined results of both samples become the basis for the final decision. Double sampling can reduce the required sample size, which in turn reduces the costs involved, and therefore it is the plan of choice for many firms. Multiple-sampling plans are an extension of the double sampling concept, but where more than two different samples are needed to reach a conclusion.

In addition, the plans offer several three levels of inspection, *normal*, *tightened* where a history of poor quality exists, and *reduced* where the quality history has been very good. The plans also provide mechanisms (called switching rules) for adjustments to the sampling plans based on history, which can reduce the amount of inspection required. These plans establish the required sample sizes under a variety of conditions, listed in tables identified by letters. Although all of this seems a bit daunting, the detailed methods of using either Z1.4 or Z1.9 are explained in detail in the published standards and in a considerable body of literature both in print form and online. Moreover, helpful acceptance sampling software is available.

Handling Rejected Lots

If a lot is rejected, the question arises what to do with it. One possible answer may be to do 100% inspection of the lot to cull out the nonconforming items. Clearly, this is costly to whoever made the defective goods, so the costs involved in taking the necessary steps to *avoid the need* to rework or to scrap defective items is considerably less in the long run. Moreover, rejections can result in considerable disruption for the firm that acquired the defective lot, which may raise the question of whether it might be prudent to change suppliers. This points to the importance of process control and continual process improvement on the part of the supplier. At the same time, the fact that even the best of systems may occasionally fail also points to the usefulness of acceptance sampling despite the fact that it cannot totally ensure that nonconforming material may always be detected, it does greatly reduce that likelihood.

Acceptance Quality Limit

In addition to establishing the number of samples to be taken, each acceptance sampling plan provides a set of rules for decision criteria in determining whether the lot is to be accepted or rejected. Such decisions are based on the number of defects in the sample, where the term "defect" means any nonconformance from established specifications. Most manufacturing schemes are usually incapable of providing outcomes without *any* defects.

By definition, the acceptance quality limit (AQL) is the maximum number of defects per hundred units (DHU), which is acceptable as a process average. Lots, or specified portions of production, having a quality level equal to the AQL will be accepted about 95% of the time. In other words, AQL is the *percentage* of defects considered to be allowable, and this is usually established through negotiation between the vendor and the firm purchasing the items. AQL formerly meant acceptable quality level, but now *acceptance quality limit* is preferred.

In establishing an AQL for any given item, it is necessary to first define what constitutes a defect, and then for each type of defect, how many of such defects should cause a lot to be rejected. If there are several types of defects possible, an AQL needs to be assigned to each. These are commonly classified by the *significance* of the defects as critical, major, or minor. If *no* defects are allowable, 100% of the items in each lot would need to be inspected, but this is *not* acceptance sampling and is typically too time consuming and costly to be used. The AQLs tend to vary from industry to industry and company to company (even those regulated by FDA), but typical values are 0.25% for critical, 0.40% for major, and 0.65% for minor. Although not specifically mandated in Part 111, it is prudent to not only document AQLs being used but also to be prepared to defend the basis for having selected those values as well as to be prepared to demonstrate that responsible individuals within the firm understand the concepts of acceptance sampling plans and AQLs and that appropriate plans and AQLs are in use.

Another term related to acceptance sampling is lot tolerance percent defective (LTPD), which expresses the poorest quality in an individual lot that should be accepted. AQL refers to the long-term *average* quality level over a series of lots, whereas LTPD refers to a specific given lot.

For every sampling plan of this type, a graph can be drawn, with the percent defective on one axis and the probability of acceptance on the other. This is called an *operating characteristic* (OC) curve, which can be helpful in selecting the optimum plan for each situation.

In acceptance sampling, two types of errors may occur. One of these is considering that a lot is bad (and therefore should be rejected) when in fact it is good (and therefore should be accepted). This kind of false alarm is called a type I (or α) error, sometimes referred to as the "producer's risk." The other is type II (or β) when a bad lot is accepted as good. This kind of false-positive is sometimes called the "consumer's risk."

All sampling plans provide for making decisions as to whether to accept or reject a given lot. But such decisions are based on the examination or testing of a *sample* of the lot (not on looking at the entire lot), so there is always a chance of making the *wrong* decision.

In a practical sense, the firm acquiring items from a supplier usually rather quickly learns what to expect in the way of routine quality, based on the supplier's manufacturing methods and quality control system. The performance history of each vendor thus becomes a major factor in deciding on the best sampling plan to use in each instance, as discussed in chapter 29.

Sampling Using the Square Root of (N) plus One

The tables in standards Z1.4 and Z1.9 present one way of determining the number of samples required for a given sampling plan. Another useful and simple way is called the square root of N (where "N" is the population or the number of items in the lot or batch to be sampled), to which 1 is added. For example, of 25 drums of a component are received, the square root of 25 is 5, so six drums would be sampled.

This method was devised in the 1920s by the organization now known as the AOAC International, formerly the Association of Official Analytical Chemists. It has been used with good results these many years, although the mathematical basis for it is unclear. Over the years, many statisticians have denounced this method as not being scientifically sound, which has generated much debate. However, it happens that this simple approach gives sample sizes that closely match those specified in Z1.4. More importantly, the square root of (N) plus one is recognized as being a valid approach to determining sample sizes, not only by FDA but also by equivalent regulatory agencies in many other countries.

This does not constitute a sampling plan per se in that it establishes the necessary sample *sizes*, but does not include accept/reject decision-making criteria.

Personnel Aspects of Sampling

The individuals assigned to doing sampling should receive initial and ongoing training in the reasons for, and in the methods and techniques of, sampling. They should recognize the importance of the samples being truly representative of the lots or batches involved, and they should be familiar with the appropriate sampling SOPs.

For sampling of incoming components and packaging materials, they should also know the steps they must use to avoid the possibility of cross-contamination. For this reason, containers being sampled from may need to be cleaned. Care must be taken in this step to avoid dust and dirt spreading through the air. Blowing with compressed air is not satisfactory in this regard, making vacuum cleaning or wiping with a cloth appropriate when necessary.

After withdrawing samples from a container, that container should be resealed in a manner that prevents contamination. Markings or labels should be applied to indicate from which containers samples were taken.

The samples also should be clearly labeled as to what they represent, such as the name of the material, the lot number, the date the sample was taken, and the name of the person who collected the sample. It is advisable to maintain a log of such information, including to whom and when the sample was delivered.

Persons doing sampling should be aware of the importance of, and the need to report on, any unusual or unexpected circumstances encountered with any sampling performed, including information on anything visibly unusual about the condition of the containers from which samples were taken.

Sampling Dietary Ingredients and Excipients

Sampling incoming components typically is done for two purposes, the first being to ensure identity of components that are dietary ingredients as required by §111.75(a), and in addition (although not mandated), it is prudent to also check the identity of components that are *not* dietary ingredients. The second reason for such sampling is to enable meeting the requirements of §111.80(a) to determine that each lot of components meets its specifications, unless depending on a certificate of analysis as discussed in chapter 29.

As previously stated, §111.80(a) requires sampling from each unique lot within each unique shipment of components, packaging materials, and labels. To determine how many containers the samples should be drawn from is left to the discretion of the firm, using either Z1.4, Z1.9, the "square root of (N) plus one" method, or some other rational sampling plan is allowable to get representative samples.

For purposes of identity testing of components, it is prudent (but not mandated) to test a sample from *every* container, unless knowledge of the supplier's quality system and past history indicates that this clearly is not necessary. The reason for the wisdom of skepticism in this regard is the hazard of the possibility of mislabeled containers, particularly if the component is obtained from a distributor or broker as opposed to the original manufacturer, wherein repacking the possibility of mislabeling is real.

The methods of sampling of components for compliance with specifications (as opposed to identity testing) obviously differ according to the physical form of the component. The number of containers to be sampled is usually determined by the same criteria as for other types of acceptance sampling, as described above.

For dry powders and granular materials, the purpose of sampling is usually to determine not only whether the chemical specifications are met but also applicable physical properties such as particle size, crystal form, polymorphs, and others. Often, both regarding incoming components and during in-process testing, knowing the particle size distribution within the material is of importance, yet this is difficult to determine since powders and granules rapidly segregate with the smaller particles tending to go to the bottom of the container, while the larger ones go to the top (the so-called Brazil nut effect). Dry products typically arrive in drums or bags. Samples may be obtained by means of a plastic or stainless steel scoop, or a spatula, simply taking the sample from the top layer within the container, but this presumes that the material in the container is homogeneous throughout (which is frequently not the case, due to the segregation). Therefore, the use of a powder thief, for free-flowing powders and granules, is generally preferred. A sampling thief consists of two concentric tubes, one inside the other. The tubes have holes. One end of the outer tube is pointed to facilitate its insertion into the powder or through a bag. Rotating the inner tube opens and closes the holes, enabling capturing samples of the powder so that when the device is withdrawn, the samples representing various locations within the container can be transferred to a sample container. Some types of thieves are also appropriate for use with sticky or cohesive powders. Many different designs of powder thieves exist, some for sampling from drums and barrels and other from bags or sacks. Most are reusable (made of stainless steel or appropriate plastics), whereas others are intended for single use and are disposable. When sampling devices such as thieves are used, they must be scrupulously clean to avoid contamination. However, although better than simply sampling from the top layer, even the use of thieves or other sampling devices can result in misleading samples due to the physical action of the device on the powdered or granular material being sampled.

Another method of powder sampling is called coning and quartering. This consists of pouring a cone of the powder taken from the container, flattening it with a spatula, and then dividing it into four quarters. Two opposite quarters are discarded. The two remaining quarters are recombined, and made into another conical pile, and the process repeated. This process is continued until a sample of the desired size results. The final sample is reasonably representative of particle sizes in the material.

Still another way of sampling powder or granules, where particle size is of importance, is called rotary riffling. This method puts the powder or granules in motion in a stream, through spinning it in a device specifically designed for this purpose. Samples are then taken from this moving stream. Rotary riffling yields highly accurate results on particle size determination.

Liquid components can also be sampled with thieves designed for this application, or if the liquid is presumed to be homogeneous (which is usual), by simply pouring from the container into a sample jar. There are sampling devices for liquids such as pipettes fitted with suction bulbs as well as a variety of devices that are essentially small cups on a long handle that can be inserted into a drum through a top bung. Various types of stainless steel or plastic dip tubes exist for liquid sampling from drums and small containers. For sampling liquids from large tanks and storage vessels, a small container (e.g., a glass or plastic bottle) can be fitted into a basket made of an inert material, which can be lowered into the vessel by means of a lightweight chain or cord also made of an inert nonreactive material.

If dry products are frequently sampled, including various herbal and botanical materials, some firms find it advisable to perform sampling and weighing in a booth dedicated to such use. Such a booth should preferably have a good dust collection system, to the likelihood of avoiding cross-contamination. Samples should never be returned to the bulk. ASTM E 300, titled *Standard Practice for Sampling Industrial Chemicals*, contains useful suggestions.

Sampling Packaging Material and Labels

The approaches to sampling primary packaging (the immediate container into which the product is filled), closures, individual cartons, shelf cartons, shipping cases, labels, and inserts differ somewhat. However, determination of the number of samples to be taken of each unique lot is usually based either on Z1.4 or on the square root of (N) plus one scheme. For example, if a shipment of 10,000 labels is received, from Z1.4, it would be required to randomly select 325 labels for examination. If the 10,000 labels were in 10 cases of 1000 each, using the square root of (N) plus one method to get the 325 labels would mean 81 labels from each of four boxes (the square root of 10 being about 3, plus 1, makes 4).

The inspection of primary containers, closures, shelf cartons, and shipping cases is usually based on dimensions and aesthetic considerations. Many firms merely accept the supplier's certification of such items, plus visual examination.

The examination of printed material, such as individual cartons, labels, and inserts, requires careful comparison against standard copies. Printing errors on such items can cause significant problems, so careful checking is needed. Moreover, for marketing reasons, graphics and colors are also quite important, albeit more subjective and difficult to quantify.

Although the usual approaches to acceptance sampling are applicable to packaging material, setting the appropriate AQLs can be complex due to difficulty in defining defects. For example, if limits are put on scratches, the decisions are influenced by other subjective judgment considerations. Whenever it is feasible, however, it is prudent to define such defects in detail. A helpful way of doing so, for visual defects, is by using photographs to establish what constitutes acceptable and unacceptable spots, blotches, or scratches. Similarly, where colors are of importance, it is typically difficult to set meaningful AQLs, although approaches to this may involve standard color swatches or the use of standardized color systems such as the Pantone® Matching System (PMS) Color.

In-process Sampling

The requirement for in-process sampling is stated in §111.80(b), which calls for taking representative samples of each manufactured batch at points, steps, or stages in the manufacturing process, as specified in the master manufacturing record, *where control is necessary* to determine whether the in-process materials meet the specifications established in §111.70(c), and as applicable, §111.70(a).

As an example of in-process sampling, consider the need to ensure that the finished product actually contains the amount of the ingredients claimed in the labeling. Even if the correct total quantity is put into a batch, the amount present in each individual dose is of significant importance to the consumer. This is termed *content uniformity* and is not necessarily easy to achieve, particularly in the instance of solid dosage forms such as tablets and capsules. These are made by starting with a blending of several dry ingredients in powder form. The ingredients typically have different densities and a variety of particle sizes and are readily subject to segregation. Even in the mixing equipment, such as a ribbon blender or any of the several types of tumble-type mixers (e.g., V-blenders, double-cones, or drum mixers), uniform dispersion of the ingredients does not necessarily occur. Further segregation also typically happens when the mixture is transferred from the mixer to the tablet press or capsule filler, or to intermediate storage containers. Therefore, blend uniformity is the starting point for achieving content uniformity in the final product.

Blend uniformity is best addressed initially in the product development stages. This involves conducting blend analysis on batches by extensively sampling the mix in the blender and/or intermediate bulk containers (IBCs). This enables selection of the optimum blending time and speeds, and also identifies dead spots in blenders and locations of segregation in IBCs. Appropriate blend sampling techniques and procedures need to be developed for each product, depending on the physical and chemical properties of the blend components.

It is suggested to use individual samples about triple the size of a unit dose. The samples are typically taken at predefined time intervals from specifically targeted locations in the compression (for tablets) or filling (for capsules) operations.

After the powder blending process is well understood, it is advisable to establish appropriate in-process sampling locations and frequency criteria for use during routine manufacturing to achieve and confirm consistent content uniformity.

Similar plans should be considered for in-process sampling and testing, as appropriate, for other dosage forms, to ensure the finished products meet their established specifications.

Sampling Finished Bulk Products

After the manufacturing operations have been completed for bulk batches, representative samples need to be taken *either* from *each* batch *or* from batches selected by a statistical sampling plan (such as the square root of (N) plus one method). Appropriate examinations or tests need to be conducted on such samples. This is set forth in §111.80(c).

This diminished testing requirement for finished batches is based upon the concept that faithfully following written procedures and the other facets of the GMP regulations results in having the overall manufacturing process under proper control. Since quality cannot be "tested into" a product, following the GMPs ensures that established specifications will be consistently met since quality has been built-in throughout the process. The limited end-testing for selected specifications (as opposed to testing every batch for every specification) adequately confirms a high degree of integrity in the overall manufacturing process. This ensures that the identity, purity, strength, and composition goals have been met and that the products are free from contamination, which could lead to adulteration.

Sampling Finished Packaged and Labeled Products
The sampling of finished products is usually by what is called *systematic sampling*, where a sample is taken either at a time interval (e.g., once each hour) or each "*n*th" unit (e.g., each 100th unit coming off the line). Systematic sampling is considered to be a form of random sampling. This scheme is easy to use and gives assurance that the population is evenly sampled.

In addition to its frequent application to packaging lines, systematic sampling can also be useful in taking samples from a tablet press or a capsule filling operation, where samples may be taken, for example, every 30 minutes.

SUGGESTED READINGS
Allen T. Powder Sampling and Powder Size Determination. San Diego, CA: Elsevier, 2003.
Berman J, Planchard JA. Blend uniformity and unit dose sampling. Drug Dev Ind Pharm 1995; 21(11):1257–1283.
Brush GG. How to Choose the Proper Sample Size. Milwaukee, WI: American Society for Quality Press, 1988.
Chowan ZT. Sampling of particulate systems. Pharm Technol 1994; 81:48–56.
Cochran WG. Sampling Techniques. 3rd ed. New York: John Wiley, 1977.
Harwood CF, Ripley T. Errors associated with the thief probe for bulk powder sampling. J Powder Bulk Solids Technol 1977; 45(1):20–29.
Muzzio FJ. Sampling and characterization of pharmaceutical powders and granular blends. Int J Pharm 2003; 250(1):51–64.
Neubauer DV, Schilling EG. Acceptance Sampling in Quality Control. 2nd ed. Boca Raton, FL: Chapman & Hall, 2009.
Saranadasa H. The square root of (N) plus one sampling rule. Pharm Technol 2003; 27(5):50–62.
Schilling EG, Neubauer DV. Acceptance Sampling in Quality Control. 2nd ed. Boca Raton, FL: Chapman & Hall, 2009.
Schwerner WS. Product sampling—why bother. Pharm Eng 1990; 10(6):31–33.
Stevens KS. Handbook of Applied Acceptance Sampling: Plans, Principles and Procedures. Milwaukee, WI: ASQ Quality Press, 2001.
Thompson ST. Sampling. 2nd ed. New York: John Wiley, 2002.
Torbeck L. Statistical solutions: square root of (N) plus one sampling plan. Pharm Technol 2009; 33(10):128–129.

12 Deviations and corrective actions

DEVIATIONS
The usual meaning of the word "deviation" is a departure from established plans, goals, or routes. In the GMPs, while not specifically defined, the word means a failure to meet an established specification or to properly follow a prescribed procedure. Deviations can be either accidental or deliberate. An example of a deliberate deviation would be a necessary temporary and planned departure from existing SOPs to make emergency repairs.

Deviations typically are discovered through product complaints, returns, internal or external audits, employee feedback, management reviews, or FDA inspections.

The term "out of specification" (OOS) is synonymous with deviation, but is usually used in connection with the results of laboratory testing or examination.

Some firms classify deviations as critical, major, or minor, depending on their impact and importance. Such classifications, although necessarily subjective, are helpful in prioritizing corrective actions to be taken.

UNANTICIPATED OCCURRENCES
As used in the regulations, an "unanticipated occurrence" implies an unexpected happening that might lead to contamination or adulteration, for example leakage from a pipe onto a component. Operator errors and equipment failures are usually considered to be deviations triggered by unanticipated occurrences.

The two terms, deviations and unanticipated occurrences, are used in tandem at various points in the GMPs to describe conditions that may arise that could result in a failure to meet specifications. Such negative events must be avoided if at all possible. For example, this is why it is usual to have two persons involved with certain tasks such as weighing and adding components to a batch, one to do the work and the other to check and verify that it was done properly and to ensure that a manufacturing deviation or unanticipated occurrence is not overlooked.

Deviations and unanticipated occurrences should be reported and recorded as they occur, and trend data plotted. Trending can be an extremely useful tool.

The information on deviations and unanticipated occurrences should be shared within the firm at appropriate intervals. It is advisable to have a SOP on reporting, documenting, and reviewing deviations.

CORRECTIVE ACTIONS
While not specifically defined in the regulations, a "corrective action" is a reactive tool and remedial step taken following discovery of a deviation or an unanticipated occurrence. This is the process of fixing an existing problem or nonconformity. It is required by §111.75(i) to establish corrective action *plans* for use when an established specification is not met. Similarly, §111.210(h)(5) requires that the master manufacturing record include written instructions for corrective action plans for use when a specification is not met. Also, §111.140(b)(3)(iv) requires

identification of the action(s) taken to correct and prevent recurrence of a deviation or unanticipated occurrence.

Corrective actions need to be addressed in a timely manner, and should be assigned a date by which they are to be completed. The commitment date for corrective action should be related to the potential risk involved. Progress with agreed-upon actions should be closely monitored. After completion, the corrective action should be evaluated for effectiveness.

The FDA acknowledges that it may not be feasible to foresee *all* possible circumstances where corrective actions may be needed, but clearly the most probable situations need to be covered.

Moreover, §111.105(a) requires quality control personnel to approve or reject all processes, specifications, written procedures, controls, tests, examinations, and *deviations* from or modifications to them, while §111.140(b)(3)(i), (ii), (iii), and (iv) require records of all deviations and unanticipated occurrences, descriptions of the investigations into the causes of such deviations, and evaluations of whether the deviations or unanticipated occurrences could lead to quality failures or a failure to package and label a dietary supplement as specified in the master manufacturing record, and identification of action(s) taken to correct and prevent recurrence of a deviation or unanticipated occurrence. Similarly, under §111.113(a)(3), quality control personnel must conduct a material review and make a disposition decision if there *is* any unanticipated occurrence that may cause adulteration.

INVESTIGATIONS

Deviations from specifications and/or established procedures (SOPs) should of course be avoided as far as possible. However, critical deviations, when they occur, should be carefully investigated and the conclusions should be documented. It is essential to understand the problem, and in what ways the specification was not met or the procedure not properly followed. The nature of the failure, and the level of risk involved needs to be known. This usually involves data gathering and evaluation, after which the required corrective action can be selected and applied.

The investigation should also consider other batches of the same product and other products that might be impacted by the specific deviation. Such consideration is important whether or not the batch or batches involved have been distributed. It is advisable to have a standard format for investigation reports, as well as a SOP on conducting investigations and preparing the reports, including the types of deviations or failures that may be encountered, how the investigation should be conducted, and how the report should be prepared.

Such reports typically include a description of the event that triggered the investigation, a brief summary of the finding and recommendations for corrective action if such is needed, suspected causes of the problem, the rationale for why the suspected cause is likely or why it has been discounted, the root cause of the problem if such was found, the disposition made of the batch or batches involved, the impact on other batches or processes, plus any supporting data or minutes of meetings involving the investigation. If nonconforming components or packaging materials were involved, or any manufacturing processes needing further study or action, these should also be detailed in the report.

If no investigation is made of a deviation or an unanticipated occurrence, a record should be made as to why an investigation was considered to be unnecessary, with the name of the individual making such a determination.

The investigation report should explain what prompted the investigation, the findings, and any corrective actions needed. The report must be made part of the batch production records, according to §111.140(b)(3).

ROOT CAUSE ANALYSIS

The point of timely and competent investigation of deviations, nonconformances, and unanticipated occurrences is to determine their cause, and once that is known, to eliminate such causes to prevent their recurrence. Unfortunately, many investigations merely identify the symptoms or anecdotal evidence of problems, *not* the root causes, with the end result that no assignable cause can be established and therefore failures cannot be explained nor corrected. A useful technique for identifying the underlying reasons for *why* an event occurred is called *root cause analysis* (RCA). This helps enable investigators to determine the basic underlying reasons for the mistake, thereby allowing preventive measures to be taken to avoid repetition of the error. The RCA is a structured systematic approach, using a variety of multifunctional "tools" to arrive at the true root cause to avoid the same from recurring.

HUMAN ERRORS

Despite training and motivation, people are often the unintended cause of deviations. It is essentially inevitable that some such errors will occur, even with good supervision. It is prudent to have systems in place to detect, prevent, and correct human errors.

Human errors tend to be more difficult to trace than equipment failures. However, as mentioned above, it is easy to blame human errors even if they are not truly the root cause. Often the problems stem from the prescribed procedures and SOPs and their interpretation, or inadequate information being disseminated. Communication problems can also exist, as can insufficient experience with the tasks at hand. Performance can be compromised by environmental factors such as noise, lighting, and temperature. Stress and fatigue are also factors.

CORRELATION VS. CAUSATION

Not infrequently, when two events occur at or about the same time, it seems logical to assume that one may or must have caused the other. Our minds tend to connect the belief that something or someone must have caused an unwanted event to happen, and thus there must be one basic reason for its occurrence. For example, making tablets while a thunderstorm was raging outside does not necessarily account for the fact that the tablets were coming off the machine with imperfect surfaces: this might seem reasonable at the time but this could be a fallacy in reasoning. A basic rule in critical thinking is that *correlation* (two events happening at the same time) does not mean *causation*. To establish a causal connection between two phenomena, we need to be able to do a series of experiments that enable changing only one factor at a time. Thus, care must be taken in determining the actual causes of deviations and nonconformities.

CORRECTIVE VS. PREVENTIVE ACTION

The dietary supplement GMPs speak only of corrective action, while the drug and the medical device GMPs speak of corrective *and* preventive action (CAPA).

The concept of preventive action is to eliminate and prevent the potential causes of deviations to help ensure they do not happen. There are significant differences between corrective and preventive action. Corrective action focuses on fixing problems *after* they occur, while preventive action tries to *prevent* such problems from happening in the first place.

While not mandated by the GMPs, it is prudent to adopt a CAPA plan. This approach takes proactive steps to prevent deviations and unanticipated occurrences, as well as fulfilling the requirement for corrective action when problems do occur. Preventive action does not focus on past events. The CAPA, since it prevents errors, improves productivity and profits and eliminates poor practices and is, therefore, a good business approach.

SUGGESTED READINGS

Bredehoeft G, O'Hara J. A risk-based approach to deviation management. BioPharm International, April 2009:48–58.

Muchemu D. How to Design a World-Class Corrective Action Preventive Action System for FDA-Regulated Industries. Bloomington, IN: Author House Publishers, 2006.

Newslow DL. An ounce of prevention, an ounce of correction: a well-defined corrective and preventive action plan is essential. Food Quality, September–October 2002:55–57.

Oakes D. Root Cause Analysis: The Core of Problem Solving and Corrective Action. Milwaukee, WI: ASQ Quality Press, 2009.

Rooney JJ, Vanden Heuvel LN. Root cause analysis for beginners. Qual Prog 2004; 37(7):45–53.

Smith B. Closing the loop on nonconformance. Medical Device and Diagnostic Industry, January 1995:222–223.

Snyder JE. Corrective and preventive action planning to achieve sustainable GMP compliance. J GXP Compliance 2002; 6(3):29.

13 Incoming components, packaging materials, and labels

An important first phase of the manufacturing process is the receipt of dietary ingredients, excipients, containers and closures, labels and other labeling, as well as cartons and shipping cases. The receiving procedures need to be done in an organized and orderly manner, both for good business reasons and for GMP compliance. The receiving, quarantine, examination, and release of incoming materials must be conducted in a consistent manner to help ensure product quality.

INCOMING COMPONENTS

The regulatory requirements for receipt of components (dietary ingredients and other ingredients) are detailed in §111.155. Subsection (a) calls for visual examination of the immediate containers to check for the appropriate content label and any damage to the containers or broken seals that may have resulted in contamination or deterioration of the components. Damage to incoming containers of components, or any other problem that might adversely impact the quality of the material, should be recorded and promptly reported to quality control personnel.

§111.155(b) requires examination of the supplier's invoice, guaranty, or certification of the shipment to ensure that what has been received matches what was ordered.

§111.155 (c) requires that the components be *quarantined* before the items are used, until representative samples of each unique lot in each unique shipment (as discussed in chapter 11) are taken, and until quality control personnel have reviewed and approved the results of any tests or examinations and also have released the components for use. At that point, but not before these steps have been completed, the components may be released from quarantine. Further discussion of quarantine methods is in chapter 22.

Moreover, §111.155(d) requires that each unique lot within each unique shipment received must be given a *unique identity code* or number that allows tracing that lot to a specific supplier, the date received, the name of the component, and its status (i.e., whether quarantined, approved, or rejected). This unique identifier follows the component throughout the manufacturing process and is noted on the batch records. Later, from the batch production records (BPRs; as discussed in chapter 15), these identifiers allow tracing the history of each component used.

One possible (but not mandated) way of handling this detail is to provide the person(s) involved with receiving incoming shipments with prenumbered multipart receiving forms, to record the details of each unique lot of each item received. The number on such forms (typically called the "receiving report" or "RR" number or "laboratory number") can then serve as the required identifier. Copies of the receiving report should go to quality control personnel, both to alert them of the receipt and to establish with them the "RR number" (or other identifier) assigned, which will be used in various ways, including being part of BPRs. Other copies of the receiving report might also be routed to whoever does the purchasing and to whoever handles payments to the suppliers. The RR number (or something equivalent) should be clearly marked on *each* incoming container or package, either using a small sticker or label or a marking pen. Quality control personnel may add their own sticker or label, indicating the status of that container of material, such as "Sampled," "Approved," "Rejected," "Quarantined," etc. The GMPs do not specifically call for such a system, but something along these lines is frequently used. Preprinted and color-coded labels of this sort are readily available from a number of vendors. However, if such labels are used to indicate the status of each container, it is advisable that they be kept locked up and available only to authorized quality control personnel. However, to repeat for emphasis, the regulations give the firm flexibility of *how* to assign and use "unique identifiers," and the example cited is merely one suggestion as to how this might be done.

Under §111.75(a), it is required that before *using* a component, it is necessary to conduct at least one appropriate test or examination to identify a component that is a dietary ingredient, and also to confirm the identity of other components and also determine whether other applicable specifications established under §111.70(b) are met. The identity testing is usually performed shortly after the items have been received, while the testing to ensure meeting established specifications is usually either done in the laboratory prior to release or is obtained through a certificate of analysis from a qualified supplier as discussed in chapter 29. However, identity testing is critically important, as discussed in Comments 145 and 174 in the preamble to the regulations.

Methods of identity testing are discussed in chapter 19, and of course these vary depending on the type of component.

The regulations are not explicit as to whether *every* incoming container of components must be sampled and tested, but it is usually prudent to do so because of the possibility of a wrong label having been applied to one or more containers, particularly if they have been repackaged by a distributor. The inadvertent use of the wrong ingredient can result in devastating public health and product liability issues. However, the number of containers of each lot to be sampled is left to the discretion of the firm, based on knowledge of the component variability and the past quality history of the supplier.

It is advisable to actually confirm the quantity of each component received (as opposed to merely accepting the supplier's statement) to help ensure accurate inventory records as well as to aid in reconciliation of the disposition of the inventory.

As discussed in chapter 22 and according to §111.155(e), all components must be held under conditions that both protect against contamination and deterioration, and to help avoid mix-ups.

It is required by §111.153 to establish and follow written procedures for the receipt, sampling, storage, identification, quarantine, and release of components.

INCOMING PACKAGING MATERIALS AND LABELS

The details regarding receipt and handling incoming shipments of packaging materials and labels is set forth in §111.160.

The regulations regarding incoming packaging materials and labels closely follow the same requirements as discussed above for components. Subparagraph (a) requires visual examination of the immediate container(s) in an incoming shipment for appropriate content label, container damage, or broken seals, to determine whether the container condition may have resulted in contamination or deterioration of the packaging and labels.

Subparagraph (b) requires visual examination of the supplier's invoice, guarantee, or certification for the shipment to ensure that the packaging or labels are consistent with the purchase order.

Subparagraph (c) requires quarantining packaging and labels before they are used, until representative samples of each unique shipment and of each unique lot within each shipment are collected and quality control personnel review and approve the results of any tests or examinations conducted.

It is inadvisable to use gang-printed and cut labels, that is, labels derived from printing sheets on which more than one type of label is printed per sheet, requiring that labels for individual products be separated from the labels for other products. If the labels are of similar size, shape, and color, the possibility of mix-up is real, which can lead to costly recalls. It is prudent to avoid this possibility, although this is not mandated in the GMPs.

Similarly, it is advisable to store labels and other labeling material separately, in separate closed containers and with suitable identification, to minimize the potential for mix-ups to occur.

It is required to at least conduct a visual identification of the immediate containers and closures. Many firms obtain certification from the suppliers of containers and closures, making it unnecessary to repeat such testing, providing that the reliability of the supplier has been established. It is not required to test packaging proactively, relying instead on documentation from the supplier, as stated in §111.75(f)(1). However, container closure systems must provide adequate protection against foreseeable external factors in storage that might lead to deterioration or contamination of the dietary supplement product.

Quality control personnel must review and approve any tests or examinations conducted to ensure the items meet the appropriate written specifications established under §111.70(d). Printed items require careful examination and comparison to a standard, to ensure that no inadvertent printing errors have occurred.

If the items do meet the specifications, quality control personnel can then approve the items for use, and release them from quarantine. However, any such item that fails to meet specifications must be rejected.

Subparagraph (d) requires identification of each unique lot in each unique shipment received in a way that allows tracing the lot to the supplier, the date received, the name of the packaging and label, and its status (quarantined, approved, or rejected), that also allows tracing the dietary supplements in which these packaging items and labels were used. This unique identifier must be used whenever the disposition of these items is recorded. In the same manner as for components, this scheme provides a way to quickly and accurately trace the actual packaging and labeling items used in the manufacture of any given batch of product.

Subparagraph (e) requires that packaging and labels be held (stored) under conditions that protect them from contamination and deterioration, and in such a way as to avoid mix-ups.

While the regulations do not specifically say so, labels and other labeling material for each different dietary supplement product should be stored separately with suitable identification. Access to the storage area should be limited to authorized personnel. Moreover, rejected, obsolete, and outdated labels, other labeling material, should be destroyed to prevent their inadvertent use.

As with components, it is required by §111.153 to establish and follow written procedures for the proper handling of packaging materials, labels, and other labeling items.

LABEL AND LABELING DEFINED

The legal definition of the term "label" is in Section 201(k) of the FD&C Act, as "a display of written, printed, or graphic matter upon the immediate container of any article..." while the term "labeling" is defined in Section 201(m) as "all labels and other written, printed or graphic matter (1) upon any article or any of its containers or wrappers, or (2) accompanying such article." The term "accompanying" is interpreted to mean more than physical association with the product and extends to posters, tags, pamphlets, circulars, booklets, brochures, direction sheets, etc.

PRODUCT RECEIVED FOR PACKAGING OR LABELING

If the product is shipped to a firm to be labeled and/or packaged for distribution (as opposed to being returned to the supplier), the procedure required by §111.165 is quite similar to the methods described above for components, packaging materials, and labels. The incoming containers must be visually examined for appropriate labeling as to content, as well as for possible container damage that might have resulted in contamination or deterioration of the product received, and the supplier's invoice (or guarantee or certification) must also be visually examined to ensure that the product is what was expected. Then, the product must be quarantined until representative samples of each unique lot within each unique shipment have been obtained and quality control personnel have reviewed and approved the documentation to determine that the product meets the specifications for it that were established under §111.70(f). At that point, quality control personnel approve the product for packaging and/or labeling, and release it from quarantine. Again, a unique identifier must be assigned to trace the history of the lot, and care must be taken to hold the received product in such a way as to protect it from contamination, deterioration, or mix-ups.

HANDLING REJECTED ITEMS

Any component, packaging material, label or labeling, or any product received for labeling and/or packaging, that is rejected because the appropriate specifications were not met, must be clearly identified and held under quarantine for appropriate disposition, according to §111.170. This is further discussed in chapter 22. The intent of the quarantine is to prevent the use of unsuitable items in manufacturing or processing operations.

Rejected items are typically either destroyed or returned to the supplier for rework or replacement, although in some instances such items may be made suitable by culling (100% inspection by sorting and selectively removing the defective items), if this process is approved by quality control personnel. However, unless the defects are fairly obvious, the statistical probability of getting rid of *all* of the defective items through sorting is usually not encouraging.

INCOMING COMPONENTS, PACKAGING MATERIALS, AND LABELS

Whenever a rejection of an incoming item occurs, it is prudent that the supplier be promptly notified.

RECORD KEEPING

In addition to establishing and following SOPs for these operations, according to §111.153 mentioned above, §111.180 also requires keeping receiving records and documentation, such as the date that components, packaging materials, labels, or products for labeling and/or packaging were received, the name of the item, the name of the manufacturer or supplier, the manufacturer's lot number if appropriate, the quantity received, the name of the carrier, the "unique identifier" assigned by the firm, and the initials of the individual who performed the required receiving operations.

Receiving records should also contain copies of suppliers' invoices, and any certificates of analysis or suppliers' guarantees.

Moreover, documentation is required on the results of any tests or examinations conducted (including visual examinations), and any material review and disposition decisions reached.

THE ROLE OF QUALITY CONTROL PERSONNEL IN RECEIVING

Under §111.120, quality control operations have the responsibility of determining whether components, packaging, and labels and other labeling conform to the specifications established under §111.70(b) and (d), as well as conducting tests or examinations prior to approving and releasing from quarantine all such items before they are used. These tests or examinations must be conducted in a consistent manner, regardless of who conducts them or when they are conducted.

Moreover, quality control personnel must conduct a material review and make disposition decisions, if the specifications are *not* met.

STOCK ROTATION

While not specifically addressed in the regulations, items should be used on a "first-in-first-out" (FIFO) basis to the extent feasible. In other words, it is advisable to use components, containers, and closures so that the oldest approved stock is used first.

Moreover, careful inventory records should be maintained, either manually or by computer, to enable accurate reconciliation.

14 Master manufacturing record

It is required by §111.205 to establish written master manufacturing records (MMRs), a vital part of the production and process control system. These are documents that spell out in detail how each unique formula and each batch size of each dietary supplement must be made, that is, step-by-step instructions for the entire production process. MMRs can be considered as the equivalent of recipes, which spell out the ingredients, quantities of each needed, and then detailed ways of putting them together to make the finished product. A point where the analogy between an MMR and a recipe is not necessarily fully accurate in that, generally, in making a food product according to a recipe it does not matter whether the ingredients and procedure are followed with great precision, as some leeway is usually tolerable, but in the case of dietary supplement manufacturing, even minor variations are unacceptable.

The fact that a separate document is required for each formula and for each batch size means that most firms will have not just *one* but a number of MMRs since it is usual to have more than one product, and frequently to also have multiple batch sizes. The concept of having a separate MMR for each batch size is to lessen the likelihood of mistakes that can occur when a formula is "multiplied up" or "divided down."

The purpose of a MMR (sometimes called a "master formula") is to help ensure uniformity from batch to batch, that is, the identity, strength, and composition of each finished batch must be essentially the same, not only *between* batches but also throughout *each* batch, for example, at the beginning, middle, and end of a production run. From a practical point of view, "uniformity" does not imply that any two batches must be absolutely *identical* in all respects, since this is impossible due to the inherent variations that always occur, as described in chapter 9.

It is not only required to *prepare* MMRs, but §111.205(a) also requires that they be *implemented* and *followed*, whereas §111.205(c) requires *keeping* each MMR in accordance with §111.610. Moreover, §111.123(a)(1) requires that each MMR (and modifications to any MMR) be reviewed and approved by quality control personnel.

Similarly, §111.123(a)(6) requires that under the MMR, quality control operations must determine whether all in-process specifications established in §111.70(c) are met.

Under §111.205(b)(1), each MMR must identify specifications for each point, step, or stage of the manufacturing process where control is necessary. Subparagraph (b)(2) further requires establishment of controls and procedures to ensure that each of these specifications is met. This includes procedures for sampling, testing, and examinations, as well as other specific actions where control is necessary.

Although MMRs must be in writing and must be retained, there is no mandate to have a SOP for preparing MMRs. However, many firms do have such SOPs, and/or use checklists or templates in writing MMRs, to help ensure

MASTER MANUFACTURING RECORD

that all pertinent points are adequately covered. Moreover, it is common practice (although not mandated) to prepare drafts of MMRs and circulate them to the appropriate individuals within the firm to get additions or corrections. The final "official" version is printed as hardcopy for the necessary approval signatures. These approvals should consist of full hand-written signatures by at least two appropriately authorized individuals, although the regulations are silent on this detail. This document then becomes the actual MMR, which is later reproduced to become the batch production record (BPR), as discussed in chapter 15. The language used in MMRs (and consequently also in BPRs) should be specific, not vague. For example, it would be unwise to use such phraseology as "in a suitable container" or "dissolve ingredients with mixing" or "fill as required" and other such nonspecific wording.

By using the appropriate electronic security techniques, it is possible to maintain the MMRs on a computer (aside from the one manually signed "official" version) since paperless document management systems are frequently used. This requires adhering to 21 CFR Part 11, as discussed in chapter 25.

MMRs must be carefully checked, as any errors in them would be perpetuated and transmitted to the BPRs produced from them.

Under §111.123(a)(1), quality control personnel must review all MMRs and all modifications to them.

DETAILS OF WHAT MUST BE INCLUDED IN EACH MMR

According to §111.210, the following information must be included in each MMR:

- The name of the product to be manufactured and the batch size
- For each dietary ingredient used, its name, strength, concentration, weight, or measure in the batch (the regulations do not prescribe the units that must be used, giving firms flexibility in this)
- A complete list of components to be used
- The weight or measure of each component to be used
- The identity and weight or measure of each dietary ingredient that will be declared on the Supplement Facts label, and the identity of each ingredient that will be declared on the label's ingredient list
- A statement of any intentional overage amount of a dietary ingredient
- A statement of the theoretical yield at each point, step, or stage of the manufacturing process where control is needed, and the expected yield when manufacturing is finished, including the maximum and minimum percentages of the theoretical yield beyond which a deviation investigation is necessary and a material review is conducted and a disposition decision is made
- A description of the packaging and a representative label (or a cross-reference to the physical location of the actual or representative label)
- Written instructions, including the following:
  - Specifications for each point, step, or stage where control is necessary to ensure product quality, and that the dietary supplement is packaged and labeled as specified in the MMR
  - Procedures for sampling, along with a cross-reference to the procedures for tests or examinations

- Specific actions are necessary to perform and verify points, steps, or stages where control is necessary, and that the product is packaged and labeled as specified in the MMR. These actions must include verifying the weight or measure of any component plus verification that any component was, in fact, added. For manual operations, this would include one person weighing or measuring a component and another person verifying the weight or measure and one person adding the component and a second person verifying the addition
- Special notations and precautions to be followed
- Corrective action plans when a specification is not met

The use of intentional overages is covered by §111.210(e), since in some instances firms do deliberately use more of certain ingredients to ensure that the product will meet its specifications for the amounts of those ingredients at the end of its expected shelf life. It is not necessary to state in the MMR the reasons for adding such intentional excess amounts.

It is required by §111.210(f) to include in the MMR a statement of the *theoretical yield* expected at each appropriate step or stage, including at the end of the manufacturing process. With this must be the maximum and minimum percentages of the theoretical yield beyond which a deviation investigation is necessary and a material review and disposition decision must be made. The wording of this section gives the firm the flexibility of deciding specifically when such yield determinations are needed. However, it is the firm's responsibility to properly select the appropriate places in their processes for such determinations. These "theoretical yields" are the quantities that should be produced at any step in the absence of any loss or error in the production. They are, therefore, useful control mechanisms to ensure that the process is actually performing as intended.

§111.210(g) requires the MMR to include a description of the packaging, together with either a sample of the label to be used, or information on where an actual label or a representation of it can be found. If a representation is used instead of an actual label, it could be a picture or photocopy of the intended label, or a detailed and accurate description of what will be on the actual label. In the event that the product is being made by an outside contract manufacturer that does not have access to the product label, the contractor could cite the name and address of the firm that does have the actual label to comply with this requirement.

§111.210(h)(1) requires written instructions for having specifications for each point, step, or stage in the manufacturing process where control is necessary, while (h)(2) further requires instructions for *sampling plans* and procedures to enable collecting appropriate samples for tests or examinations, together with cross-references to the methods and procedures to be used in such testing. This does not imply that the actual testing methodology must be spelled out in the MMR, but instead permits including a cross-reference to such procedures to determine whether the appropriate specifications have been met.

§111.210(h)(3) requires written instructions for specific actions to perform and verify each point, step, or stage in the manufacturing process where control is necessary to ensure product quality, and that the product is packaged and labeled as specified. This includes verifying the weight or measure of each component used, as well as verifying the addition of each component (which, for manual operations, involves two persons, one weighing or measuring each component and the other verifying the weight or measure, and then one person adding each component to the batch with the second person verifying the addition as described in chapter 16). In some instances, instead of these steps being strictly manual, some firms use bar codes on containers to identify the components and their weights before and after weighing, and then contents of that container added to the batch, with appropriate scanning technology to verify the identity and weight of the components added. In other words, some of the steps may be partially under control of automated equipment, which may *not* require the usual "two-person" approach. The important point is that there must be verification that these steps are properly conducted, whether accomplished manually or at least partly by computer control.

§111.210(h)(4) calls for the MMR to provide for any special notations and precautions that need to be followed.

§111.210(h)(5) requires that the MMR include written instructions for corrective action plans for use when a specification is not met. The concept is to be prepared for corrective action to be promptly taken to correct, and prevent recurrence of, deviations that may occur. It is true that it may not be feasible to establish corrective action plans for *every* possible event that may occur, but at least most scenarios of this sort can be foreseen, and for those that can be predicted, it is expedient and useful to have corrective action plans ready.

BILLS OF MATERIAL

For business reasons, for example for inventory control, production planning, and purchasing purposes, it is common for firms to prepare and maintain bills of materials (BOMs), which are in some ways similar to MMRs. These are *not* required by the GMPs and are not mentioned in Part 111.

A BOM is a listing of all of the dietary ingredients, excipients, and packaging items, together with the quantities of each item required to manufacture a given quantity of goods. Each item on a BOM is usually given a unique part number to facilitate identification. The part numbers are typically also tied to the specifications that have been established for each item.

UNIVERSALITY MMRS

The MMR concept (by various names) is included in most GMPs worldwide, and is not limited to dietary supplements. For example, in drug manufacturing (both for prescription drugs and for over-the-counter drugs) in the United States, MMRs are required by §211.186, while similar requirements are also required in the European Union, Canada, Australia, and many other countries.

SUGGESTED READINGS

EudraLex, Rules Governing Medicinal Products in the European Union, Vol 4—Medicinal Products for Human and Veterinary Use: Good Manufacturing Practice Guidelines, Part 1, Chapter 4, Sections 4.14 and 4.15, Manufacturing Formula and Processing Instructions, European Commission, Brussels, 2008.

FDA. Current good manufacturing practice in manufacturing, processing, packing, or holding of drugs, 21 CFR 211.186, Master Production and Control Records, 1978.

Health Canada/Health Products and Food Branch Inspectorate, Good Manufacturing Practice Guidelines, Manufacturing Master Formula, Ottawa, Ontario, 2009, p. 29.

Pharmaceutical Inspection Convention (PIC/S), PE 009-9 Guide to Good Manufacturing Practice for Medicinal Products, Chapter 4, Manufacturing Formula and Processing Instructions, Geneva, 2009.

World Health Organization, Technical Report Series No. 908, Annex 4, Good Manufacturing Practices for Pharmaceutical Products, Section 15.23, Geneva, 2003.

15 Batch production record

It is required by §111.255(a) to prepare a batch production record (BPR) each and every time a batch of a dietary supplement is made.

The definition of the term "batch" is a specific quantity of a dietary supplement that is uniform; is intended to meet specifications for identity, purity, strength, and composition; and is produced during a specified time period according to a single manufacturing record during the same cycle of manufacture. Similarly, a "lot" is defined as a batch, or a specific identified portion of a batch. So, in effect from a regulatory point of view, a batch and a lot are the same. However, in general usage, the term batch usually refers to formulated bulk product in one or a few large containers, whereas a lot usually refers to packages of the finished product in the final containers. It is prudent for each firm to adopt standard nomenclature for these terms so that everyone in the organization clearly understands what constitutes a batch and a lot.

The BPR is often referred to as simply being the "batch record." Interestingly, these are not called "lot records."

A BPR must be prepared and followed whether the firm actually manufactures a batch, or whether the firm merely packages or labels a product received from an outside supplier (for final distribution, as opposed to being returned to the supplier). However, firms that just do packaging and/or labeling need to document only those parts of the process with which they are involved. For example, a labeler, under §111.260(e), would not need to include the identity and the weight or measure of each component used since that would be the responsibility of the firm that made the batch.

The batch record accompanies a dietary supplement product as it is being made, precisely directing all of the steps and stages that must be followed.

RELATIONSHIP BETWEEN MMR AND BPR

The BPR must accurately follow the appropriate MMR (as described in chapter 14), according to §111.255(c). When paper forms are used for this, they may be photocopies of the appropriate MMR, properly signed and issued. Similarly, if handled electronically, they stem from the approved computer-based MMR as an Electronic Batch Record (EBR) but using the proper controls required by 21 CFR Part 11, as discussed in chapter 25. Electronic and paper records both have advantages and disadvantages, but computer-based documentation is now frequently used, with some firms using a mixture or hybrid system having features of both. The GMPs are silent on this topic, aside from stating specific requirements in general for computer-based documentation.

The BPR must include *complete* information relating to the production and control of each batch, according to §111.255(b), and must be made and retained and available for review for at least one year beyond the shelf life date of the batch (if such dating is used), or for at least two years beyond the date of the distribution of the batch if there is no shelf life date used. This requirement is stated in §111.255(d) relating to §111.605.

INFORMATION THAT MUST BE INCLUDED IN EACH BPR

Each lot of packaged and labeled dietary supplement from a finished batch must have a batch, lot, or control number, according to §111.415(f). The batch record is *also* required to include a batch, lot, or control number, as stated in §111.260(a). When more than one party is involved in the operations, the requirements for each in this regard are spelled out in the subsections of §111.260(a), to ensure the ability to trace all the stages of manufacturing and control through distribution, as mandated by §111.410(d).

The BPR must also include the identity of the equipment and processing lines used in producing the batch, as stated in §111.260(b). The emphasis on such identification (including the condition of the equipment) is due to the possibility of improper functioning or introducing contaminants if any item is damaged or not properly cleaned. Similarly, the BPR, therefore, must show the date and time of the maintenance, cleaning, and sanitizing of the equipment and processing lines used, according to §111.260(c). However, this information can be cross-referenced to other records (e.g., individual equipment logs) where these data are retained. In other words, there is flexibility as to how such records must be handled.

Each component, unit of packaging material, and label must have a unique identifier assigned to it when received, as required by §111.155(d) and §111.160(d). The identifiers for the items used in production must be shown in the BPR, as called for in §111.260(d).

Further, the BPR must include the identity and weight or measure of each component used, according to §111.260(e).

Also, the BPR is required to include a statement of the *actual* yield as well as a statement of the percentage of the *theoretical* yield at the appropriate stages of processing as stated in §111.260(f). The "actual yield" is defined as the quantity actually produced at any appropriate step, whereas the "theoretical yield" is the quantity that *would* be produced at that step based on the quantity of materials used in the absence of any loss or error in the actual production. Limits should be established on the percentage of the theoretical yield actually attained since significant differences between theoretical and actual yields may signal that processing errors, mix-ups, or contamination have occurred. In the event of significant discrepancies, there should be a procedure in place to prevent approval and distribution of the batch in question until the questions are satisfactorily resolved.

§111.260(g) calls for inclusion in the BPR the actual results of any monitoring operations conducted during in-process steps.

The BPR must include the results of any testing or examinations performed during production of the batch, as stated in §111.260(h). However, if such results are retained elsewhere (typically in laboratory records), the firm has the flexibility of being allowed to simply provide a cross-reference in the BPR to those records.

§111.260(i) requires that the BPR include information indicating that the finished product meets its specifications as established in §111.70(e) and (g).

The requirements for documentation to be made and included in the BPR *at the time of manufacturing* are spelled out in §111.260(j). These include the date on which each step of the MMR was performed and the initials of each person performing each step. This includes weighing or measuring each component used in the batch and the person verifying that action, and similarly adding the component to the batch and the person verifying that. The person or persons who perform the steps should physically initial the batch record to acknowledge that the requirement was performed. Although not specifically stated in the regulations, the individuals weighing or measuring and adding components to the batch must ensure that those components have been released by quality control personnel. In fully computerized situations, electronic signatures are acceptable, as long as they comply with 21 CFR Part 11 as discussed in chapter 25.

Similarly, §111.260(k) requires documentation (again *at the time of performance* of the packaging and labeling), including the unique identifiers of the materials used, the quantity used, reconciliation between the number of labels issued and used (when such reconciliation is required), and either an actual label used or a cross-reference to where the actual or representative label can be found. The results of any tests or examinations conducted on packaged and labeled products (including repackaged or relabeled products) must also be recorded.

Label reconciliation is not required when 100% electronic or electromechanical inspection (machine vision) is performed during or after completion of the finishing operations.

Although it is true that some of the manufacturing steps frequently are carried out semiautomatically through the use of computers and appropriate software, even then many procedures may still involve human input and accordingly need manual entry into the BPR.

While the basic BPR is a true copy of the MMR, the final batch record, complete with supporting paperwork, laboratory reports, etc., typically consists of many pages and is therefore a thick document.

MATERIAL REVIEW AND DISPOSITION DECISIONS
Whenever material review and disposition decisions are made, they must be documented in the BPR, as required by §111.260(m).

REPROCESSING
If any reprocessing was required, under §111.260(n), the details of how this was accomplished must be detailed in the BPR. The procedure used for reprocessing may be unique, depending on the circumstances.

THE BATCH RECORD REVIEW
Quality control operations are required by §111.123 and §111.260(l) to *review* BPRs *after* each batch is made (MMRs are prepared *before* batches are made). This final review traces the complete cycle that was actually used in manufacturing a batch or lot of product to determine compliance with all of the established procedures. These records must "tell the whole story." This is an essential tool for assuring quality prior to releasing the product, as well as remaining as a clear record even long after the batch of product was made.

Checks must be made to see that *all* of the required records are complete, error-free, legible, and traceable to the appropriate raw data. It is of importance to ensure that the MMR has been accurately followed in its entirety in the BPR.

It is usual (but not mandated) to use a checklist to ensure completeness of the batch record review.

Unexplained discrepancies, or the failure of the batch or any of its components to meet established specifications, must be thoroughly investigated, and these investigations extended to other batches of the same product that may have been associated with the failure or discrepancy. This includes circumstances where the actual yield exceeds the established allowable maximum or minimum percentage of the theoretical yield.

It is advisable (but not required) to have a responsible person in the manufacturing group do a *preliminary* BPR review prior to turning the documents over to quality control personnel for their final review before the batch is released or distributed.

The quality control personnel who conduct the batch record review must be well trained in how to go about conducting the detailed review process. Moreover, the review by quality control personnel must *also* be documented, either on the batch record or on a separate document.

If a batch deviates from the MMR, including any deviation from specifications, or when an unexpected event occurs during a manufacturing operation that might lead to adulteration, or if there is an equipment problem or a calibration issue, quality control personnel must conduct a material review and make a disposition decision, as mandated by §111.113(a)(2) and (3), §111.123 (a)(4), as well as §111.87. Such decisions must be documented as required by §111.140(b)(3) and §111.260(m). The individual(s) designated to perform quality control operations who conduct a material review and make disposition decisions (as well as each qualified person who provides relevant information in this matter) must sign the documentation, as required by §111.140(b)(3)(vii).

The FDA considers the process of batch record reviews to be of extreme importance, and this is a topic that is usually looked at carefully during establishment inspections.

SUGGESTED READINGS

FDA. Current good manufacturing practice in manufacturing, processing, packing, or holding of drugs, 21 CFR 211.188, Batch Production and Control Records, and 211.192, Production Record Review, 1978.

Fish RC. Batch Record Review. Vol. 2, No. 17. Rockville, MD: FDA News & Information, AAC Consulting Group, Inc., 2001.

16 Manufacturing operations

Subpart K of Part 111 is titled *Requirements for Manufacturing Operations*. In this, §111.353 requires written procedures (SOPs) for manufacturing operations.

The logical and essential starting point for establishing such procedures is to clearly define the product(s) to be made. Once this is done, appropriate specifications can be established for the finished products, the components to be used, and the packaging and labeling. Knowing these items, the necessary manufacturing methods and equipment can be devised or selected. This is often referred to as the design phase, typically requiring input either from the firm's own Research & Development group, or outsourced as described in chapter 31.

Once the product specifications are established, it is required by §111.355 that manufacturing operations must be designed or selected to ensure that the product specifications are consistently met. This also follows the requirements of §111.70, §111.73, and §111.75.

§111.360 requires that manufacturing operations be conducted with adequate sanitation principles (as discussed in chap. 6).

§111.365 requires that precautions be taken to prevent contamination (as discussed in chap. 28). This includes the following:

- Using conditions and controls that protect against the potential for the growth of microorganisms and the potential for contamination
- Washing or cleaning components that contain soil or other contaminants
- Using water of appropriate quality
- Performing chemical, microbiological, or other testing (as necessary) to prevent the use of contaminated components
- Sterilizing, pasteurizing, freezing, refrigerating, controlling pH, controlling water activity, or using any other effective means to remove, destroy, or prevent the growth of microorganisms and prevent decomposition
- Holding components and dietary supplements that can support the rapid growth of microorganisms of public health significance in a manner that prevents the components and dietary supplements from adulteration
- Identifying and holding any components or dietary supplements, for which a material review and disposition decision is required, in a manner that protects components and dietary supplements that are *not* under a material review against contamination or mix-up with those that *are* under such review
- Performing mechanical manufacturing steps (such as cutting, sorting, inspecting, shredding, drying, blending, and sifting) by any effective means to protect the dietary supplements against contamination
- Using effective measures to protect against the inclusion of metal or foreign material in components or dietary supplements
- Segregating and identifying all containers for a specific batch of dietary supplements to identify their contents and, when necessary, the phase of manufacturing

- Identifying all processing lines and major equipment used during manufacturing to indicate their contents, including the name of the product and the specific batch or lot number, and when necessary, the phase of manufacturing

The requirement to establish SOPs for manufacturing operations ties with the requirement to establish a master manufacturing record (MMR) for each unique formulation and for each batch size under §111.205 and §111.210.

All of these sections of the GMPs require consideration of the facilities and equipment needed, together with the processes to be employed. The manufacturing operations should be robust and well understood and should provide consistent defect-free products. As mentioned in chapter 26, both a change control system and a program for continuous improvement are advisable (but not mandated).

The manufacturing operations established should provide for tracking the history of each batch of product. In addition to enabling investigations in the event problems occur, such histories may point to future changes that could result in improvements.

In particular, the initial commercial batches of any given product often yield useful information on the suitability and consistency of the manufacturing operations, including insight into the state of control.

The main goal in manufacturing operations should be to ensure finished product quality but with efficiency and at sustainable costs. This requires thorough process knowledge coupled with good decision-making skills.

DOSAGE FORMS

A "dosage form" is the physical type or way a product is presented to consumers for consumption. There are many such forms used for dietary supplement products, including tablets, various kinds of capsules, powders, liquids, and bars.

In the drug industry, dosage forms are often called "delivery systems," and this term may also be applied to dietary supplements.

By definition, in Section 201(ff) of the Federal Food, Drug, and Cosmetic Act (as mentioned in chap. 2), a *dietary supplement* is a product (other than tobacco), *taken by mouth* (i.e., not a topical product), intended to supplement the diet, which bears or contains one or more of the following "dietary ingredients":

- A vitamin
- A mineral
- An herb or other botanical
- An amino acid
- A dietary substance used to supplement the diet, by increasing the total dietary intake
- A concentrate, metabolite, constituent, extract, or combination of any of the above ingredients

Although legally considered as a special category of foods (not over-the-counter drugs), according to Section 411(c)(1)(B)(ii) of the Act, such products must *not* be represented as a *conventional* food and not intended as being the sole item of a meal.

MANUFACTURING OPERATIONS

The FDA has made it clear that *only* products intended for *ingestion* (meaning taken into the body through swallowing) can be marketed as dietary supplements. In addition to ruling out topical products applied to the skin, this definition also *excludes* products intended to enter the body through the mucosal tissues, meaning the inner linings of the cheeks, lips, and nose. This rules out not only transdermal products but also sublingual dosage forms, and many nasal products, as being considered to be dietary supplements. The use of chewing gum as a dosage form for dietary supplements has been controversial, but the FDA has sent some firms courtesy letters explaining that chewing gums are *not* dietary supplements. Firms interested in marketing products that might fall into these disallowed classifications would be wise to seek competent legal advice.

PARTICULATE SOLIDS

Some dietary supplement products are marketed as powders, while powders are typically also starting points in the manufacture of capsules and tablets. Therefore, particulate technology is important in many manufacturing operations. This is a complex topic that includes particle characterization, particle sizing, flow properties, separation issues, storage and handling of powders, agglomeration, and many other facets of dealing with particulate solids.

One important consideration is that many solids used in dietary supplement manufacturing can exist in different physical forms. This is referred to as "polymorphism," the result of certain materials being able to exist in more than one crystalline phase due to different conformations in the crystal lattice. Still other solid materials consist of disordered arrangements of molecules without *having* a crystal lattice and are therefore termed amorphous. In view of this, any given particulate raw material may exist with widely varying physical properties, such as melting point, density, solubility, and others, which can significantly impact not only the manufacturing operations but also such factors as stability, dissolution, and bioavailability in the finished products. Since most component specifications tend to be limited to *chemical* identity and purity, the particle size and shape (and polymorph form, if applicable) are often essentially ignored. This can lead to significant problems in manufacturing. This should be considered when establishing component specifications.

Powdered solids are actually a collection of discrete particles having a variety of sizes, shapes, and surface areas. Variations in particle shape mean that the particles typically have many points of contact with each other. This tends to result in the particles being held together by surface tension, which results in the powder being more or less cohesive. When the powder is forced to flow (in mixing, in feeding a tablet or capsule machine, or in transferring the powder from one container to another), these forces essentially always cause some segregation to take place. This can result in uneven distribution of the dietary ingredient and the excipients in the mixture and can result in problems in "flowability" in hoppers and feeders. To minimize such difficulties, a key step is to properly sample and characterize the bulk powder (Fig. 16.1).

As mentioned above, one of the most important characteristics of powders, in manufacturing operations, is the *flowability*, that is, the ability of a powder to flow through equipment reliably. Nonuniform flow can cause significant problems in storage in both bins and hoppers, in mixing and blending, and in feeding tablet and capsule machines, as described. Flow behavior is a function of

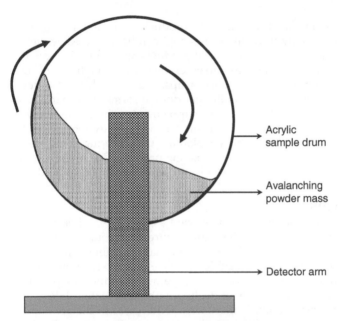

Figure 16.1 Several kinds of instruments are available for the characterization of powder flowability, including the avalanche type illustrated. The transparent rotating drum visually shows the powder behavior under dynamic conditions. *Source*: From Chan LW, Kou X, Heng PWS. Drug substance and excipient characterization. In Parikh DM, ed. Handbook of Pharmaceutical Technology, 3d ed., Informa Healthcare, New York, 2010.

both the characteristics of the powder and the design of the equipment being used.

Predictable flow may be impeded by the formation of an arch or a "rathole." Ratholes can occur when the central portion of a powder in a hopper flows freely, but the material at the hopper walls remains stagnant, leaving an empty hole through the powder. Arching (also called "bridging") in hoppers can occur when the forces acting on particles at the walls of the hopper equal the internal strength of the mass of the powder particles and is especially common with powders that tend to cake and pack. Both phenomena have been studied extensively since either can be disruptive to manufacturing. Both the flowability of the powder and the geometry of the hopper (including the size of its outlet) are important factors. Powder flowability is usually related to wall friction, shear strength, and bulk density of the powder.

In characterizing powdered solids, one factor is the *angle of repose*, which is determined simply by making a pile of the powder and measuring its slope from the horizontal. Other factors are moisture content, since moist powders tend to be more difficult to handle than are dry powders, and *cohesive strength*, measurement of which can be accomplished with instruments specifically designed for such use.

Particle size and shape distributions are often critical parameters, measured either by simple sieve analysis, or by microscopy, or through the application of more sophisticated laser-based instruments, or diffractometers, which study scattering patterns from a beam of light or radiation such as X rays.

There are laboratories that specialize in powder characterization. It is frequently useful to utilize such services during the design phase of a new product that is either a powder or uses a powder in a step in the manufacturing process. Thereafter, the vendor of the raw material should ensure furnishing equivalent material so that routine testing or examination of the distribution of particle shapes and sizes becomes unnecessary. This can be critical, since changes can cause difficulties and consistency in raw materials can significantly impact manufacturing operations as well as aspects of product quality.

SOLIDS MIXING

There are many types of mixing equipment used for mixing dry bulk solids. These are usually relatively simple devices, available in an extensive variety of designs and sizes from a large number of equipment manufacturers.

One type that is widely used is the horizontal ribbon blender. This style of mixer consists of a U-shaped horizontal trough fitted with an end-to-end rotating motor-driven shaft to which paddles (or more often metal helical ribbons) are attached. The ingredients are placed in the mixer to the appropriate fill level, and then the drive motor is started, which causes the material to be repeatedly pushed and turned, resulting in effective mixing (Fig. 16.2).

Tumble blenders, of various types, are also widely used. One common form has two cylinders mounted together in a "V" shape (sometimes called "twin-shell"), whereas another style is conical (or a so-called double-cone). Such mixers are rotated on an axis. These types come in many sizes, in stainless steel, and of sanitary design. These are the types of dry solids mixers most frequently used in both the pharmaceutical and the dietary supplement industries.

The order of addition of dry ingredients in a blender is usually not important. As long as the particle sizes are similar and the powders free-flowing, the blends tend to be satisfactory regardless of the order of addition.

From a practical point of view, tumble mixers typically require all ingredients to be placed in the blender before it is started, so it is a good procedure to initially distribute the ingredients, although this may not be necessary. It is prudent to add ingredients that are present in only small amounts near the center of the mixer to avoid them sticking to the walls of the equipment. Also, when some of the ingredients must be present in very small quantities, it may be useful to make a premixture of those ingredients with some of the other ingredients that are present in larger amounts. This technique is often called "master batching." This also helps minimize stratification by gravity, where some of the ingredients tend to settle toward the bottom of the equipment.

If powder mixing were perfect, any and all samples taken from a blend would contain the same proportions of each ingredient. Unfortunately, this rarely happens, and thus the usual goal is to at least achieve a *satisfactory* random mixture.

Solid particles that have significantly different sizes and/or densities tend to segregate. Even if well-mixed, "unmixing" (segregation) often occurs in handling, such as in transferring the mixture from one container to another, or in a hopper that feeds a filling machine, or in a tablet press or capsule-making machine. Segregation can even occur in containers of mixed powders left in a

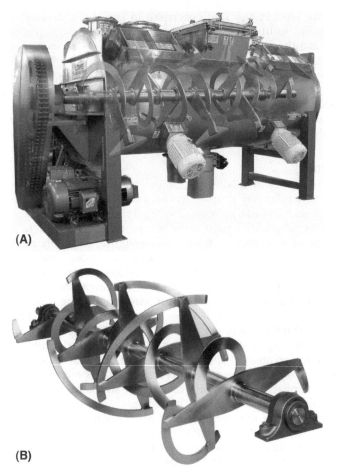

Figure 16.2 (**A**) Sanitary ribbon mixer and (**B**) internal ribbon. *Source*: Courtesy of Marion Mixers, Inc., Lowe Sanitary Equipment, 2011.

warehouse, simply from the minor vibration or "jostling" that typical occurs in the area. Small particles tend to "percolate" downward, passing through voids between the larger particles present. In some instances, adding a small amount of liquid to the mixture retards the tendency of segregation by rendering the particles slightly more cohesive.

Another form of segregation is the so-called Brazil nut effect, wherein large particles tend to rise to the top in a granular mixture of varying particle sizes, the name coming from the observation that when opening a can of mixed nuts, the large Brazil nuts are often on the top. This is more properly termed "granular convection."

However, segregation tends to occur at many stages in the production of final dosage forms involving particulate solids, which can cause significant problems in manufacturing operations.

AN OVERVIEW OF TABLET MAKING

Tablets (often called "pills," albeit technically a pill is spherical) are made by compressing mixtures of dietary ingredients and suitable excipients, fed into a cylindrical cavity called a die, through the action of piston-like punches. There are many variations of this theme.

Most tablet-making machines are of rotary design, employing multiple sets of punches and dies, which yield high rates of production (Fig. 16.3). Tablets can be made in a wide variety of shapes, styles, sizes, and colors to facilitate differentiation and use by consumers. Tablets shaped like capsules are sometimes referred to as "caplets."

Tablets can be designed to disintegrate and release the dietary ingredient(s) at specific locations in the digestive system.

Tablets can be made in two or more layers. For example, if two ingredients are chemically incompatible, a three-layer tablet can isolate them by using an insulating inert middle layer.

The so-called tablet tooling consists of the matched sets of punches and dies used in the tablet-making machines. Such "tools" need to be carefully handled and cared for, cleaned regularly, and properly stored. The integrity of the tooling is of significant importance in the proper production of good-quality tablets.

In addition to the dietary ingredients, tablets usually contain binding agents to hold the ingredients together (e.g., methyl cellulose), lubricants to reduce the friction in the punches and dies (such as magnesium stearate), as well as other appropriate excipients including diluents or fillers as bulking

Figure 16.3 A typical high-speed automatic tablet-making machine. Note the full enclosure to help prevent contamination while allowing clear viewing of the press in operation. Courtesy of Tabletpress.net.

agents (e.g., lactose), and disintegrants to help the tablet break up after ingestion (such as starch or cellulose). The actual formulations are carefully worked out by those with the appropriate expertise during the product development phase.

Uniformity of the content of the dietary ingredient(s) in tablets is obviously important, but often difficult to achieve. This is dependent in large part (but not entirely) on the content uniformity of the powdered or granular material being used. As mentioned above, segregation may occur at any of various stages. Analytical techniques such as near-infrared (NIR) spectroscopy (discussed in chap. 19), as well as frequent sampling of the tablets followed by chemical analysis, are steps often employed to help ensure content uniformity. A specification for content uniformity is not required, but it is advisable in the manufacture of high-quality dietary supplement solid dosage forms.

Tablets should be sufficiently strong to withstand fracture and physical damage as packaged, which typically is expressed in hardness and friability specifications. Hardness is measured by determining the load required to crush a tablet on end, for which hardness testers are commercially available. Friability is the tendency of a tablet to chip or crumble, as measured by tumbling tablets in a rotating drum, called a "friabilator," described in USP <1216> and in Annex 9 of ICH Q4B, titled "Tablet Friability."

The *disintegration* of tablets and capsules refers to how rapidly the dosage form breaks up and falls apart into particles after ingestion in that this is important in how the dietary ingredient(s) will be absorbed into the body. Similarly, the term *dissolution* refers to the rate and extent at which the dietary ingredient(s) go into solution. Testing techniques for these two factors are contained in USP <2040>. There are various types of testing apparatus for these characteristics. One kind consists of stainless steel wire mesh baskets on a rack, containing plastic tubes open at the top and bottom. The baskets dip into a beaker of a prescribed fluid held at a constant temperature. The baskets are raised and lowered into the fluid at a rate of about 30 cycles/min. For the tablet disintegration test, one tablet is placed in each tube, and the time to disintegrate and fall through the screen is noted. For the dissolution test (which can be applied to either tablets or capsules), the fluid is tested for the concentration of the dietary ingredients at specific intervals of time. Although such testing is useful for quality control, it is *not* mandated by the dietary supplement GMPs. However, products *labeled* as being "USP" must meet all of the requirements of the appropriate USP monograph, which may include dissolution testing.

Although the dietary supplement GMPs are silent on *bioavailability*, the term generally refers to the extent and rapidity of the dietary ingredients being absorbed in vivo in the gastrointestinal tract, to become functionally useful after the product is ingested. Many complex factors are involved in bioavailability, and standardized methods for measuring and expressing this property do not currently exist. "Bioequivalent" products are those that display *comparable* bioavailability.

To repeat for clarity, the dietary supplement GMPs do *not require* establishment of physical parameters such as disintegration and dissolution, but *if* such specifications *are* established, the firm should have data to support that the specifications are met. However, many (arguably most) firms do use these specifications, along with tablet appearance, allowable weight variation, thickness, hardness, and friability, to ensure product uniformity and integrity.

MANUFACTURING OPERATIONS

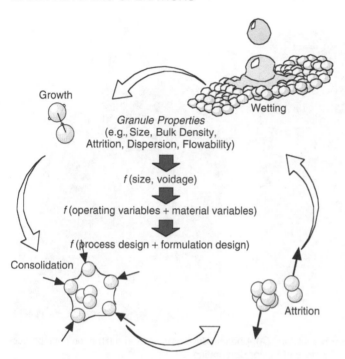

Figure 16.4 In wet granulation, particle agglomeration can be accomplished either by agitation or compression methods. As illustrated, this typically includes wetting, blending, drying, and milling (attrition) to obtain the optimum particle size, hardness, and porosity. *Source*: From Ennis BJ. Theory of granulation: an engineering perspective. In Parikh DM, ed. Handbook of Pharmaceutical Technology, 3d ed., Informa Healthcare, New York, 2010.

In tablet-making, although some types of powder mixtures are suitable for *direct* compression without requiring additional steps, usually an intermediate process termed "granulation" is employed. This can be either a dry or a wet step. The reason for preparing a granulation is to agglomerate fine powders into uniform larger and more dense particles (granules) that have improved flow characteristics, which can greatly facilitate the overall process.

The dry method of granulation may involve first making what are essentially very large tablets made from the dry powder, termed "slugs," which are then ground or milled into coarse particles for use in forming the final tablets. This requires both a tablet machine capable of making the slugs and an appropriate mill for reducing the slugs to granules. Another dry method of dry granulation involves putting the powdered mixture through closely spaced stainless steel rollers (called a "roller compactor") to partially compact the mixture, following which the resulting material is ground or milled to the optimum particle size for use in a tablet-making machine. Dry granulation is typically less costly than wet granulation (Fig. 16.4).

Wet granulation is the most frequently used process. In this, the blended dry ingredients are moistened with water or another binder liquid or a hot-melt, either by pouring or by spraying the fluid into the dry mixture. When a liquid is used, the amount needs to be carefully controlled since too much can cause the

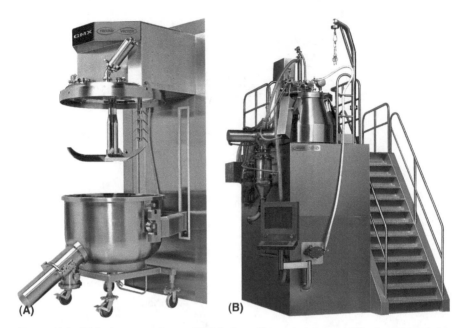

Figure 16.5 High-shear granulators. (**A**) A bottom drive granulator and (**B**) a top drive production-sized unit. *Source*: Courtesy of Vector Corporation, 2011.

resulting granules to be overly hard, whereas too little may result in them being too soft and subject to crumbling. After this step, the resulting wet mass is forced through a coarse screen. The material is then dried, usually in a conventional tray-type dryer or in a fluid-bed dryer. The dried material is thereafter milled to the appropriate granule size for feeding to the tablet-making machine(s).

Systems and equipment for granulation continue to evolve. High-shear mixers are now often used for preparing granulations, employing rotating impeller blades. Still another approach is the use of the so-called single-pot granulators that incorporate mixing, granulating, and drying in just a single vessel. Some designs of single-pot granulators use vacuum to speed the drying, while still others employ microwave drying (Fig. 16.5).

Although granulation and other steps have been studied extensively, tablet-making continues to be a rather empirical process based on experimentation to obtain the desired results. The entire process is based both on science *and* on the experience and skills of the persons doing the manufacturing.

TABLET COATING

Coatings are often applied to tablets (and some capsules) for any of a variety of reasons, among which are their protection from the adverse stability effects of light or humidity, or to mask unpleasant tastes, or to help ensure optimum bioavailability by controlling where the dietary ingredients are released within the digestive system. Coatings are also used simply to make these dosage forms distinctively attractive for brand recognition. Moreover, solid dosage forms may be imprinted with logos or letters and numbers to enhance identification.

The historically oldest but still widely used process is *sugar coating*. In this process, it is usual to first seal the tablet cores with a waterproofing substance such as "pharmaceutical grade" shellac in a suitable solvent, or with synthetic polymers. A syrup of sugar is applied to the sealed tablet cores, coat by coat, with drying between each addition. If the finished tablets are to be colored, it is usual to apply the color in the final coating stages, by adding it in a thin sucrose syrup. Finally, polishing the coated tablets is accomplished to achieve a glossy finish, typically through the application of waxes such as beeswax or carnauba. Successful sugar coating tends to depend on the skills and experience of the personnel conducting the operation.

Film coating is also widely used, resulting in a durable and less bulky coverage of the tablet cores than sugar coating, albeit this also tends to more easily show any surface defects in the cores. The film coating process is usually accomplished in a single step, often via a spraying technique, and is therefore less labor intensive than sugar coating. The film is typically a thin cellulose polymer layer to which plasticizing ingredients are added. Film coats tend to improve both the visual appearance and the "swallowability" of tablets. Types of film coatings may be used for immediate-release or delayed-release applications, for example, enteric coatings. Enteric coatings protect the dietary ingredients from the stomach acids and instead deliver them for absorption in the intestines, which in some instances enhances bioavailability.

A common method of applying coatings to tablet cores is through the use of conventional coating pans, which are usually spherical and mounted on an angle, although other configurations also exist. The tablet cores are tumbled in the pan, with the coating liquid being applied either by pouring or by spraying. Often the drying of the liquid is hastened by warm air being blown into the rotating pan. Generally, each coat is allowed to dry before the next coat is applied. Such pans are classified as solid-wall (nonperforated) or perforated types, with the perforated version having the advantage of permitting the introduction of higher volumes of drying air. Both types typically containing internal baffles facilitate movement of the tablets to help ensure evenly distributing the coating materials onto all of the surfaces.

Successful tablet coating is usually largely dependent on the skill of the operator conducting the process. This is looked on as being more of an art than a science.

Coating can also be done in fluid beds (described below) or by *compression coating* done by placing the tablet cores surrounded by coating powder into the dies of specially designed tablet machines, and then again compressing. This in effect produces a tablet within a tablet in a complex process.

Methods of continuous coating also exist, which are useful for very large production volumes.

CAPSULE MAKING

Capsules have long been used as an alternative to tablets as an oral solid dosage form, both for pharmaceutical products and for dietary supplements. One reason for this is that it tends to be faster and easier to produce a new product through encapsulation. This is in part due to the relative simplicity of capsule formulation, if the ingredients used are compatible with the material of which the empty capsule shells are made. In fact, this ease is used by some consumers to make their own dietary supplements on a "do it yourself" basis, using small and relatively inexpensive "home-style" capsule filling equipment.

Some consumers consider capsules to be easier to swallow than tablets.

Some herbal ingredients are difficult to compress into tablets, not infrequently causing capsules to be the optimum solid dosage form for botanical-based products.

A number of different types of equipment can be used in commercial-scale filling of hard two-piece capsules. Two of the most common types are termed "dosator" and "tamping." Some such equipment can fill 200,000 or more capsules per hour.

The majority of hard capsule shells are made of gelatin, although hydroxypropyl methylcellulose (HPMC) and other materials are also used. HPMC is considered to be vegetarian, as opposed to gelatin, which is bovine sourced. This can be a marketing factor with certain ethnic and religious groups as well as to others opposed to consuming animal-based products. Gelatin and HPMC tend to differ in their dissolution properties due to the way moisture penetrates and dissolves the capsule shells. Some varieties of HPMC exhibit pH-dependent solubility.

Two-piece hard-shelled capsules are typically filled with dry powdered or granulated ingredients, although liquids or semisolids also can be (and often are) satisfactorily encapsulated. Moreover, filled capsules can be coated after filling, for example, by enteric coating processes, using polymers that are resistant to gastric fluids but are soluble or permeable in intestinal fluid, which allows passage through the stomach to the small intestine, where the dietary ingredients are released.

Empty capsules are available in a wide range of colors, some even being "two-toned" with the two pieces being of different colors. The variety of colors and imprints can provide marketing advantages.

Empty hard-shell capsules are available in a wide variety of sizes, usually designated by numbers from 000 to 5. Printed tables exist for guidance to dietary supplement manufacturers as to the weight or measure of ingredients appropriate for each given shell size.

The capsule shells are usually cylindrical, with hemispherical ends, with the two pieces being of slightly different diameters to enable them to fit together. The cap portion is usually shorter than the base. Some types of hard-shell capsule bodies are made to essentially lock together to help avoid bursting or leakage, whereas in some instances, shrink-band sealing is employed both to minimize possible leakage and to aid in preventing deliberate tampering with the filled capsule contents.

Soft gelatin capsules are frequently used when the dietary ingredients are dissolved or dispersed in oils, in the form of liquids or suspensions. These capsules are one piece and are hermetically sealed. The flexibility of soft-shell capsules results from plasticizing ingredients used in the material from which the shells are made. The manufacturing process typically involves the use of two ribbons of gelatin passing between two rotating die cylinders, with the liquid to be encapsulated injected as the capsule halves are sealed. This process is unique, with few manufacturers equipped to handle it. However, soft gelatin capsules have been in use for more than a century. These tend to readily dissolve in the gastric fluids in the digestive tract, which may enhance bioavailability (Fig. 16.6).

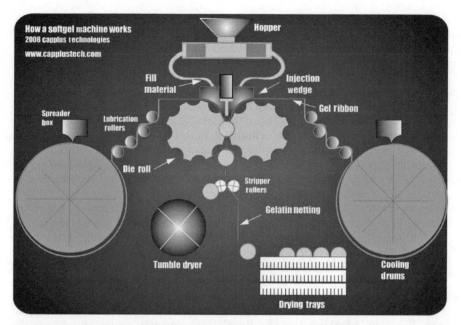

Figure 16.6 Manufacturing process of soft gelatin capsules. *Source*: Courtesy of CapPlus Technologies, AZ, U.S.A., 2011.

FLUID-BED PROCESSING

When a stream of an essentially inert gas (usually air) is introduced into the bottom of a bed of particulate solids, the gas moves upward through the empty spaces between the particles. Depending on the density of the particles, at some certain velocity of the flow of the gas stream, the downward gravitational pull on the particles will be overcome, causing the particles to essentially "float" in the gas and behave in a manner similar to that of a liquid. This is termed "fluidization," which presents useful process possibilities, such as an efficient way particles can be dried and/or heated or cooled. Coatings can be applied to fluidized particles or tablets, or the particles can be agglomerated into pellets or granules through a binder liquid being sprayed in. Therefore, fluid-bed processing is often the process of choice for certain steps in the manufacture of dietary supplement products.

The early equipment used for fluid-bed processing was flat in shape, with a screen or perforated plate having appropriately sized opening to hold the particulate solids, and a plenum below for introduction of the stream of gas. Over the years, however, the design of fluid-bed equipment has advanced greatly, with a wide variety of types, sizes, styles, and shapes now commercially available for batch or continuous use, some equipped with state-of-the-art computerized controls.

LIQUID DOSAGE FORMS

Many dietary supplement products are in liquid forms.

Although it is more of a matter of labeling regulations than GMPs, there has been considerable discussion over the importance of, and the means of, differentiating between liquid supplements and beverages, which are generally considered to be conventional foods. The distinction is of importance from a regulatory point of view.

The manufacturing processes for liquid products tend to be considerably simpler than those for solid dosage forms. The majority of the mixing of liquids can be accomplished with low-shear agitators, generally propellers on a motor-driven rotating shaft within a cylindrical tank (which can be fitted with an external jacket for heating or cooling, if appropriate). This type of equipment results in thorough blending of the liquid ingredients. It is also possible to disperse or dissolve solid ingredients in this manner. The introduction of particulate solids into liquids requires consideration of how these will initially be "wetted out." Dispersion of solid ingredients usually occurs only *after* the wet-out and after the breaking-up of any clumps that may have agglomerated.

In some instances, the use of *baffles* is helpful. These are straight flat sheets or plates of metal set vertically in a mixing vessel with one edge on or near the cylindrical wall. Usually four such baffles are used, although the number can vary. The object is to break up or stop the tendency for *all* of the fluid being mixed to spin in the tank due to the forces exerted by the impellers.

Particle size reduction in liquids, and emulsification of liquid-liquid ingredients by breaking up the droplets to prevent coalescence, typically requires *high-shear* mixing equipment, such as a four-bladed rotor turning at high speed within a stator enclosure.

A variety of stator designs exist, many of which are slotted, whereas others have round or square holes or fine screens.

In-line high-shear liquid mixing equipment also exists. These can be designed for either single pass or recirculation use. The selection of the optimum in-line equipment tends to be based on consideration of the viscosity of the liquids involved and the flow rate. Cavitation may occur in high-shear mixing, which can result in bothersome equipment abrasion problems.

The selection of the best high-shear equipment for any given application can be daunting, although the major vendors of such items can be very helpful. Some equipment suppliers have in-house laboratories equipped with small-scale models of their various types of mixers in order to run experiments for potential customers. They, and others, also use the relatively new but highly advanced field computational fluid dynamics (CFD) technology to evaluate the many engineering factors involved, to aid in the selection of the best type of equipment for complex mixing, blending, and dispersion situations.

WEIGHING AND DISPENSING

An important step in manufacturing operations is the weighing and dispensing of ingredients, sometimes referred to as "charge-in of components." Accurate dispensing of raw materials is a fundamental requirement for producing high-quality dietary supplement products. Errors must not occur such as dispensing the wrong ingredient (or one not properly released by quality control personnel) or dispensing the wrong weight or quantity. This is covered in §111.210(h)(3)(i), which requires specific actions including verifying the weight or measure of any component as well as verifying the actual addition of any component.

§111.210(h)(3)(ii) further requires one person weighing or measuring a component with a second person verifying this, plus one person adding a component with another person verifying that action. However, in Comment 261 of the preamble, the FDA agreed that a computer-generated weight record plus a bar code system on containers could be used to verify the material's contents and weight, plus the addition could be adequately controlled and verified through scanning technology. This gives firms the flexibility of either handling these details manually, or using a system partially under the control of automated equipment.

§111.210(h)(4) requires that the MMR include written instructions for special notations and precautions to be followed.

§111.260(j) requires documentation, at the time of performance, of the date on which each step of the MMR was performed, together with the initials of the person responsible for weighing or measuring each component used in a batch and the initials of the person verifying this. Similarly, the initials are required of the person responsible for adding the component to the batch and the initials of the person verifying this. The intent of such use of initials is for the persons involved to acknowledge that they performed the requirement.

Although not specifically addressed in the regulations, the area in which raw materials are weighed and placed into clean containers is usually near the warehouse where the components are stored, and typically includes space for staging the components and for weighing or measuring them. It is usual (but not required) to place all of the weighed components for a given batch on clean pallets. The weighing is usually done manually with the operators using appropriate scoops and drum-tipping devices. Small quantities are generally weighed on bench scales, whereas larger quantities are handled on floor scales.

As discussed in chapter 28, appropriate steps need to be taken at all stages in the manufacturing operations to avoid contamination and cross-contamination, particularly in the handling of powdered ingredients, some of which may become airborne or may be tracked on the shoes of personnel moving from one area to another. Care also needs to be taken whenever dusts are generated, not only because of the possibility of causing contamination but also due to the possible risk of explosions or flash fires.

SUGGESTED READINGS

Allen LV Jr., Popovich NG, Ansel HC. Ansel's Pharmaceutical Dosage Forms and Drug Delivery Systems. 9th ed. Baltimore, MA: Lippincott Williams & Wilkens, 2011.

Anderson J. Computational Fluid Dynamics. New York: McGraw-Hill, 1995.

Barnum R. Ebb and flow: understanding powder flow behavior. Pharmaceutical Processing, March 2009:18–21.

Berman J. Blend uniformity and unit dose sampling. Drug Dev Ind Pharm 1995; 21 (11):1257–1283.

Gotoh K, Finney JL. Representation of the size and shape of a single particle. Powder Technol 1975; 12(2):125–130.

Chowan ZT. Drug substance physical properties and their relationship to the performance of solid dosage forms. Pharm Technol 1994; 81:45–60.

Daveswaran R. Concepts and techniques of pharmaceutical powder mixing process: a current update. Res J Pharm Technol 2009; 2(2):245–249.

Food & Drug Administration, Guide to Inspections of Oral Solid Dosage Forms, 1993, Guide to Inspections of Dosage Form Manufacturers, 1993, and Guide to Inspections of Oral Solutions and Suspensions, 1994.

Franzke RS. Mixing powders into liquids. Powder and Bulk Engineering, January 2010:47–53.
Gad SC. Pharmaceutical Manufacturing Handbook: Production and Processes. Hoboken, NJ: John Wiley, 2008.
Gupta CK, Sathiyamoorthy D. Fluid Bed Technology in Materials Processing. Boca Raton, FL: CRC Press, 1998.
Harwood CF. Errors associated with the thief probe for bulk powder sampling. Powder Bulk Solids Technol 1977; 1(2):20–29.
Holman PR. Master batching: a basis for uniform mixtures. Powder and Bulk Engineering, January 1994:45–48.
Hoyle W. Powders and Solids. London: Royal Society of Chemistry, 2001.
Jeon I. Pros and cons of roll compaction. Pharm Technol Eur 2011; 23(3).
Jones BE. How gelatin and hypromellose capsules differ in product release during dissolution testing. Tablets & Capsules, January 2010:16–19.
King R. Fluid Mechanics of Mixing. Dordrecht: Springer, 2010.
Lee SW. The fundamentals of optimizing machine uptime and product yields in capsule filling operations. Tablets & Capsules, July 2010:4–6.
Liberman HA. Pharmaceutical Dosage Forms. 2nd ed. New York: Marcel-Dekker, 1996.
Lu Wei-Ming. Effects of baffle design on liquid mixing. Chem Eng Sci 1997; 52(21):3843–3851.
Meeus L. Direct compression vs. granulation. Pharm Technol Eur 2011; 23(3):14–18.
Mohan S. Unit dose sampling and blend uniformity testing. Pharm Technol 1997; 21:116.
Nink F. Advantages of encapsulating liquid products. Tablets & Capsules, September 2007:44–47.
Perry J. Addressing combustible dust hazards. Chem Eng Prog 2011; 107(5):36–41.
Podczeck F. Pharmaceutical Capsules. 2nd ed. London: Pharmaceutical Press, 2004.
Porter SC. The role of high-solids coating systems in reducing process costs. Tablets & Capsules, April 2010:10–15.
Rantanen J. Process analysis of fluidized bed granulation [Article 21]. AAPS PharmSciTech 2001; 2(4):21.
Rhodes MJ. Principles of Powder Technology. New York: John Wiley, 1990.
Roth G. Solid dosage manufacturing trends. Contract Pharma/Nutraceuticals World, March 2011:S4–S10.
Tousey MD. The granulation process 101: basic technologies for tablet making. Pharm Technol 2002:8–13.
Yetly EA. Multivitamin and multimineral dietary supplements: definitions, characterization, bioavailability, and drug interactions. Am J Clin Nutr 2007; 85(suppl):269S–276S.

17 Packaging and labeling operations

A review of product recalls and of citations in the FDA establishment inspection reports (EIRs) makes it clear that packaging and labeling operations are where most mix-ups and errors occur. These operations are usually the *last* steps in manufacturing dietary supplement products, so errors or mistakes in packaging and/or labeling can spoil all of the good work that has gone before. For these reasons, avoiding problems at this point in the process is a topic particularly worthy of careful attention.

The purposes of packaging are to contain and hold a given quantity of product and to protect the product from damage, contamination, or deterioration. Labeling is to identify the product, to attract the attention of prospective purchasers, to convey important information such as the purpose of the product, the use directions, the caution statements (if any), the quantity in the package, the name of the manufacturer or distributor, and other such pertinent information. The legal and regulatory details of *what* must appear on labeling, and *how* such information must be presented, is a separate topic *not* directly covered by the GMP regulations but is instead elsewhere in Title 21 of the *Code of Federal Regulations*.

LABELS VS. LABELING
Section 201(k) of the FD&C Act defines the term "label" as a "display of written, printed or graphic matter upon the immediate container of any article," while Section 201(m) defines "labeling" as consisting of "all labels and *other* written, printed or graphic matter upon any article or any of its containers or wrappers, or accompanying such article." Printed matter that promotes or is used to promote the use of a product generally is considered to be labeling within the meaning of Section 201(m). This may, in some cases, even include books, journal articles, etc., that either are affixed to or *accompany* a product, although the definition of the term "accompany" has been the subject of considerable litigation. The Dietary Supplement Health and Education Act (DSHEA) added Section 403B to the Act, exempting certain publications related to dietary supplement products from the definition of labeling under Section 201(m), providing they meet certain specific criteria.

REQUIREMENTS APPLICABLE TO PACKAGING AND LABELS
The basic GMP requirements applicable to packaging and labels are stated in §111.410; subparagraph (a) of which says that it is necessary to determine whether the packaging meets the established specifications to help ensure the quality of the finished product. This includes ensuring that the packaging per se will not contaminate the products nor cause them to deteriorate. This reiterates the requirement in §111.70(a) and (d) to establish satisfactory packaging specifications and in §111.75(f) to determine that those specifications have been met before using the items. This does *not* require the firm to test the packaging

proactively, but instead allows reliance on documentation from the supplier, such as a continuing product guarantee combined with a statement of the intended use of the item(s).

To repeat for emphasis, it is up to the *firm* to establish the suitable specifications and to substantiate that they are met. This of course depends in large part on the dosage form and the characteristics of the specific product to be packaged.

REQUIRED SOPS FOR PACKAGING AND LABELING OPERATIONS

Under §111.403, it is required to establish and follow written procedures for packaging and labeling operations. Furthermore, under §111.430, it is required to make and keep records of the written procedures; in other words, all such activities must be carefully documented.

COMMON TYPES OF PACKAGING USED FOR DIETARY SUPPLEMENTS

It is usual to refer to primary and secondary packaging items, where the *primary* package is what actually contains and holds the product (typically a jar, bottle, blister pack, etc.). In other words, in the primary packaging, the product and the package are in physical contact. The *secondary* packaging is typically a printed carton (if such is used) and the label, that is, all packaging other than the primary container.

In the instance of jars and bottles, the *closure* (which may also contain a liner) is considered to be part of the primary package since it is also in contact with the product.

Primary packaging must not be reactive, additive, or absorptive to the product in such a way as to alter the safety, identity, strength, quality, or purity of the dietary supplement product, including through the mechanisms of leaching (extraction of ingredients from the container material into the product) or migration (movement of a low-molecular-weight substance product from the container into the product). Among the factors to consider in selecting the optimum primary packaging material are considerations of possibly needed protection from light exposure, moisture, or oxygen.

Primary packaging materials include several grades of glass and many types of plastics, such as high- or low-density polyethylene, polypropylene, polyvinyl chloride (PVC), polyethylene terephthalate (PET), polystyrene, and other polymeric materials.

Secondary packaging typically encloses the primary package, for example, cartons printed with marketing and labeling information. Additional secondary packaging frequently includes making bundles of 6 or 12 individual packages, or placing such quantities into a larger carton (often called a "shelf carton") to facilitate handling by the retail trade. Larger quantities are typically overpacked in shipping cases or boxes made of corrugated board.

For products subject to degradation from trace amounts of moisture, very small packages of a desiccant (silica gel, a molecular sieve, or other such material) are sometimes added into the primary package along with the product.

Capsules and tablets are sometimes packaged in blister packs composed of individual pockets formed in a sheet of aluminum or plastic film, such as PET,

PVC, PVdC (polyvinylidene chloride), COC (cyclic olefin copolymer), or PCTFE (polychlorotrifluoroethylene), covered with another film of aluminum, plastic, or coated paper or paperboard as lidding. In using aluminum foil, the hardness or temper can be an important factor. The choice of materials is usually largely dependent on the moisture barrier properties needed as well as cost considerations. Moreover, the various material choices also impact the required machine settings, including temperature, dwell time, and pressure, which in turn affect production speeds and exposure of the product to heat. These "push-through" forms of packaging offer convenience for the consumer and good protection for the product.

ISSUANCE AND RECONCILIATION OF USAGE

§111.410(b) requires careful control of the issuance and use of packaging materials and labels, with *reconciliation* of the actual use. In other words, procedures must be established to reconcile the quantities of packaging materials and labels issued, used, and returned to stock, versus the number of packages of product actually produced, and to evaluate any discrepancies.

Although this portion of the regulation refers to *both* packaging materials *and* labels, concern is generally focused mostly on *label* reconciliation. From a practical point of view, this requires accurately counting the number of labels issued and, at the end of the run, comparing this count with the number of packages produced (including the number of labels returned to stock and any damaged labels that were destroyed). However, the regulation is silent on the allowable *deviation* in such reconciliation. Of course the ideal is *no* deviation, although there are usually a few labels "not accounted for." In establishing the SOP on reconciliation, reasonable limits should be established, and the firm should be prepared to defend the limits set, in the event this should ever be challenged. These limits, if set as a percentage, need to take into account the size of the lots produced since (for example) a limit of 0.1% for a lot of 1000 units is quite different from that for a lot of 100,000 units. Moreover, reasonable limits of reconciliation may also depend on the *type* of labels used since some types are easier to handle than others.

The most usual causes of unsatisfactory reconciliation are errors in the initial counts and final counts of labels or cartons (including rejects, returns, or samples not being properly considered), or errors in the count of final packaged product produced. However, there is also the possibility of mixed or rogue labels or cartons being involved.

§111.410(b) does state that label reconciliation is *not* required if 100% online inspection is performed by appropriate electronic or electromechanical equipment ("machine vision") during or after completion of finishing operations.

Such automated inspection can essentially eliminate labeling errors, at least when functioning properly. However, like all equipment, machine vision must be properly maintained and frequently checked to ensure correct performance, particularly since there is usually a tendency to assume that since such a system is in use, the probability of malfunctioning is zero. This creates obvious risks, which must be avoided.

In general, *roll* labels are much easier to reconcile than are *cut* labels, since automatic counting devices are usual on equipment using roll labels. However, both maintenance and calibration of such equipment need consideration.

One problem that has occurred on occasion with roll labels has been the human error of inadvertently splicing different label reels together.

Reconciliation is also required by the *drug* GMPs, although this has triggered much discussion over the years as to whether reconciliation is a truly effective means of eliminating (or at least minimizing) labeling errors. The point has been made that if some incorrect labels get mixed in prior to issuance to the packaging line, they will not be detected by reconciliation, and this unfortunately occurs all too often. Reconciliation is also done after the fact, often after the labeled packages are already in the sealed shipping cases. Moreover, some companies tend to overly rely on counting labels before and after a run, often at the expense of better alternatives. In the manufacture of both drugs and dietary supplements, *in addition to* reconciliation (which is mandated), it is advisable to use labels of unique size, shape, and color for each different product, making it easy for line operators to spot a "wrong" label if such should occur. This, plus machine vision mentioned above are excellent approaches to minimizing labeling errors.

If labels are precoded with lot or control numbers (as opposed to being marked with such numbers online), the excess labels bearing the lot or control numbers should be promptly destroyed at the end of the run.

Labels returned to stock at the end of the run should be maintained and stored in a manner that helps prevent mix-ups and provides proper identification.

Although not mentioned in the regulations, it is also prudent to reconcile the amount of bulk product (e.g., the number of tablets or capsules) issued and the number of primary containers filled. Many firms also reconcile the number of primary containers filled with the number of secondary packed units produced.

EXAMINATION PRIOR TO STARTING A RUN

As required by §111.410(c), *before* packaging and labeling operations, the packaging and labels for each batch of dietary supplement must be examined to ensure that they agree with what is called for in the master manufacturing record (MMR).

MANUFACTURING HISTORY

There is a requirement, in §111.410(d), to be able to determine the complete manufacturing history and control of the packaged and labeled dietary supplement through distribution. Although not specifically required, the usual way of accomplishing this is to put the batch, lot, or control number on the product label or on the primary container, usually referred to as "coding" the labels or packages. Coding can be accomplished either online or off-line. This gives the ability to trace the manufacturing history, for example, in the instance of a customer complaint or a report of an adverse event. This would enable taking any appropriate corrective action, including a recall if necessary, and also could extend any investigation to associated lots that might be implicated.

FILLING, ASSEMBLING, PACKAGING, AND LABELING OPERATIONS

It is required by §111.415 that filling, assembling, packaging, labeling, and other related operations be performed in ways that ensure the quality of the products and that their packaging and labeling are as specified in the MMR. The regulations say that this must be done by any effective means.

As discussed elsewhere in this book, the design and construction, as well as the maintenance and calibration of the equipment, and cleaning and sanitation are of paramount importance as are the physical and spatial separation of equipment and the protection against all forms of contamination (including airborne sources).

LINE CLEARANCE PROCEDURES

In most facilities, at least some of the packaging and labeling lines are used for more than one product or product strength. That makes it imperative that between runs, the lines must be carefully cleaned and all vestiges of items from previous production be removed to eliminate possible mix-ups or contamination. Total cleanout and inspection of the packaging and labeling areas are of great importance. Steps must be taken to ensure that the work area, filling machines, packaging lines, coding machines, and other equipment are totally free from any products (e.g., a few tablets or capsules) or packaging or labeling materials from previous runs that are not required for the current operation.

It is prudent to check any ledges, indentations, tops of equipment, and other such places that could harbor even one "foreign" label or tablet.

It is advisable to conduct line clearance according to specific and detailed written procedures and with the aid of a checklist to help ensure nothing has been left to chance. Following the line clearance, all equipment should be inspected both by production and by quality control personnel, and such inspection should be documented and made part of the batch production records. Where production is conducted on more than one shift, particular care should be taken at shift changes. If there is a significant time interval between runs, even if the line was cleared at the end of one run, it should be reinspected immediately before the start of the next.

Inadequate attention to line clearance details is a major source of errors, mix-ups, and resultant recalls.

BRITE STOCK

§111.415(e) requires identifying, by any effective means, filled dietary supplement containers that are set aside and held in *unlabeled* condition for future label operations. This refers to the production of what is sometimes called "brite stock" for made-to-order (MTO) production runs, where a batch of a product is made and packaged but not labeled, and later when orders are received for relatively small quantities of "private label" goods, portions of the batch are accordingly labeled for that purpose. This method of operating enables a firm to economically produce short runs of products for specific accounts.

The term brite stock originated in the food industry where some products are filled into cans prior to cooking and are then heated to a high temperature in a retort to cook and/or sterilize the contents. Labels are then applied after cooling, but meanwhile the unlabeled shiny cans are called brite stock. The unlabeled containers must be somehow identified to avoid applying the wrong labels.

Since unlabeled containers of dietary supplements are sometimes held for labeling later, §111.415(e) properly requires having a suitable means of identifying the contents of such containers to avoid mix-ups.

BATCH, LOT, OR CONTROL NUMBERS

It is required by §111.415(f) to assign a batch, lot, or control number to each lot of packaged and labeled dietary supplement (including goods sent to another firm for packaging and/or labeling). However, the regulations do *not require* that this number be placed on the product label or the immediate container, giving flexibility as to *how* the manufacturing history can be traced. The firm may be able to devise another acceptable way of tracing the history of any given lot.

Although not mandated, the concept of placing the lot or control number on the product label or package is widely used, convenient, and effective. This is useful, too, for the trade and for consumers in the event of a recall, or in the event of wanting to communicate with the manufacturer about a specific lot of product, for example with a complaint or suggestion.

EXAMINING SAMPLES OF THE FINISHED PRODUCT

It is required by §111.415(g) to examine a representative sample of each batch of the packaged and labeled dietary supplements to determine that they meet the specifications for packaging and labeling established in accordance with §111.70(g). Such examinations should be documented in the batch production record.

OBSOLETE OR INCORRECT LABELS

As stated in §111.415(h), labels and packaging for dietary supplements that are obsolete or incorrect must be suitably disposed of to ensure they are not used in any future packaging and labeling operations.

REPACKAGING AND RELABELING

If it becomes necessary to repackage or relabel dietary supplements, §111.420(a) requires that this be done only after quality control personnel have approved doing so. After such steps have been taken, §111.420(b) requires examination of a representative sample of each batch that has been repackaged or relabeled to determine if the products then meet the specifications that were established in accordance with §111.70(g). Moreover, quality control personnel must approve or reject each batch that has been repackaged or relabeled prior to its release for distribution, as required by §111.420(c).

REJECTED PACKAGED AND LABELED PRODUCTS

Under §111.425, packaged and labeled products that have been rejected for distribution must be clearly identified and held under quarantine for appropriate disposition.

TAMPER-EVIDENT PACKAGING

Following the deaths that occurred in the 1982 tampering incidents, the FDA instituted a requirement that OTC drug products (with certain exceptions) must use tamper-evident packaging that provides visible evidence if tampering has occurred. The term "tamper evident" was chosen instead of "tamper proof" after it was established that it is nearly impossible to ensure that any form of packaging is truly tamper *proof*. Moreover, the GMPs for OTC drugs also require that two-piece hard gelatin capsules also be sealed using an acceptable tamper-evident technology. The dietary supplement GMPs are silent on this topic, although such products obviously could be subject to tampering too. It is a management decision, not a regulatory matter, whether to consider tamper evidence for dietary supplements.

CHILD-RESISTANT PACKAGING

Under the Poison Prevention Packaging Act, certain consumer products must be packaged in child-resistant packaging (CRP). The regulations for this are administered by the Consumer Products Safety Commission (CPSC) and are stated in Title 16 of the *Code of Federal Regulations*.

Although several types of OTC drugs require CRP, the only dietary supplements that fall into this category are those that contain 250 mg of elemental iron per container from all sources, except those in which iron is present solely as a colorant. The details of this are stated in 16 CFR 1700.14(a)(13).

It is possible that product liability considerations (not regulations) might make CRP prudent for use with certain dietary supplements due to their toxicity profiles.

ONLINE CONTROLS DURING PACKAGING AND/OR LABELING

- It is advisable to have the name and batch number of the product being handled clearly displayed.
- Different products should not be packaged at the same time in close proximity, unless there is adequate physical segregation.
- Products, packaging materials, and labels delivered to the line should be checked to ensure they are in agreement with the MMR.
- Samples taken away from the packaging line should not be returned.
- Both the production personnel and the quality control personnel should be alert for and conduct periodic checks for the general appearance of the packages, and should ensure that fill weights or volumes, or the counts of tablets or capsules being packaged, are correct.
- Any unusual events should be promptly reported to supervisory personnel.
- Primary containers should be clean before filling.
- Closures must be properly applied, and if of the screw type, they should be tightened to the appropriate removal torque.
- Batch production records should at least include the name and batch, lot, or control number of the product, the date and times of the packaging operations, the name of the supervisor of the line, a record of the line clearance procedures, records of checks conducted, references to specific equipment used, if feasible samples of printed packaging materials and labels used, the quantities of printed packaging and labels issued, used, and returned, details of any coding applied, details of the bulk product used and its release by quality control personnel, a record of any unusual or unexpected occurrences during the run, and the quantity produced.
- Finished products should be quarantined pending completion of examination or testing and release by quality control personnel.

AVOIDING CROSS-CONTAMINATION DURING PACKAGING OPERATIONS

It is important to assess and control the risks of cross-contamination that might occur while packages are still open. Possible sources of such contamination include improperly cleaned containers and equipment, personnel, airborne sources with emphasis on HVAC, and other operations being conducted nearby.

SHELF LIFE AND EXPIRATION DATING

The use of expiration dating, shelf life dating, or "suggested use by" statements is *not* required by the GMPs at this time. The FDA made the decision to not incorporate this into Part 111 because in some instances there may not be adequate methods available to assess the strength of a dietary ingredient. However, the product must provide 100% of the labeled amount of each quantified ingredient for the entire time the product is on the shelf in the marketplace. For this reason, intentional overages may be used, although the amount of such overage should be limited to the amount needed to ensure that the product does not fall below its label claim due to deterioration with time. The MMR is required to carry a statement of any intentional overages in the amount of a dietary ingredient, according to §111.210(e), although the FDA does not require any statement of the *reasons* for adding the intentional excess amount.

Some firms feel that consumers and the trade prefer to have such dating even if not mandated. The FDA's position on this is that firms *may* use expiration dates or "best if used by dates" on the product labels if they wish to do so, but if they do, such dates should be supported by data. However, they have declined to offer guidance on the types of data that are acceptable for such support.

SUGGESTED READINGS

Casola AR. FDA's guidelines for pharmaceutical packaging. Pharmaceutical Engineering, January/February, 1989:15–19.
Consumer Healthcare Products Association. Guidelines for the Stability Testing of Non-Prescription Pharmaceutical Products. Washington, DC: Consumer Healthcare Products Association, 2004.
FDA. Center for Drug Evaluation and Research, Guidance for Industry: Submission of Documentation in Drug Applications for Container Closure Systems, June 1997.
Hartburn K. **Quality** Control of Packaging Materials in the Pharmaceutical Industry. New York: Marcel Dekker, 1991.
Jenke D. Extractable/leachable substances from plastic materials used for product containers. PDA J Pharm Sci Technol 2002; 56(6):332–371.
Jenkins WA, Osborn KR. Packaging Drugs and Pharmaceuticals. Lancaster, PA: Technomic Publishing Co., 1993.
Kania K. Package integrity testing. Pharmaceutical & Medical Packaging News, July 2004:34–36.
Leonard EA. Packaging: Specifications, Purchasing, and **Quality** Control. 4th ed. New York: Marcel Dekker, 1996.
Summers JL. Dietary Supplement Labeling Compliance Review. Hoboken, NJ: Wiley-Blackwell, 2004.
Swan E. Picking the right blister material. Pharmaceutical & Medical Packaging News, September 2002:26–31.
Swan E. Consistent, repeatable package testing. Pharmaceutical & Medical Packaging News, July 2003:32–36.

18 Quality control responsibilities

The GMPs define *quality* to mean that the dietary supplements consistently meet their established specifications for identity, purity, strength, and composition, including limits on contaminants, and have been manufactured, packaged, labeled, and held under conditions to prevent adulteration (as defined in the FD&C Act). *Quality control* (QC) is defined as a planned and systematic operation or procedure for ensuring the quality of the dietary supplements, whereas the term *quality control personnel* refers to any person, persons, or group within or outside of the firm who has/have been designated to be responsible for the firm's QC operations.

There are many possible ways of approaching the required tasks. One way is always laboratory testing as discussed in chapter 19, which is of course important. However, QC involves much more than just testing, since quality must be "built-in" and cannot be "tested-into" products.

As the regulations say, QC involves a complete and carefully planned *system* that encompasses the many facets that can impact the products that reach the marketplace. Among other topics, this typically includes oversight of seeing that sound and workable specifications are established, linked with optimum supplier selection, integrity, and management, to help ensure using only "good" components and packaging/labeling materials. Moreover, this also involves ensuring that all of the many steps are properly conducted in manufacturing, packaging, labeling, warehousing, and physical distribution.

Clearly, it is not up to QC personnel to actually perform or supervise any or all of these functions, which are typically done by a variety of other persons or groups within the firm, but it *is* QC's responsibility to help ensure that these steps *are* properly handled. This of course requires that the QC personnel understand what is expected of them. Moreover, they must receive necessary and accurate information in a timely manner. Also, it is critical that all employees understand, acknowledge, and respect the QC function. Top management must see to it that everyone accepts this. However, at the same time, QC personnel must avoid a "traffic cop–like" role.

Many items in the GMPs state that "You must ..." but by definition, in this usage, the term "You" refers to the *firm*, not necessarily to specific individuals or organizational groups within the firm.

In practice, most product defects that do occur are the result of simple human errors. This is precisely why having well-written SOPs and insisting they be properly followed are so critically important. It is a matter of consistently "doing it right the first time." Although QC personnel should insist on this, it is usually up to the supervisors to enforce proper conduct of the employees.

In the strictest sense, however, QC does not really *control* the quality, since this is in fact done by the manufacturing personnel. QC monitors the manufacturing procedures and sees to it that appropriate sampling and testing is conducted, and then either approves or rejects. It is important that the manufacturing group understand *their* vital role in ensuring quality rather than having a false belief that *quality* is actually someone else's responsibility.

QUALITY CONTROL VS. QUALITY ASSURANCE

In some large firms, particularly where both drug products and dietary supplements are made, there may be two separate departments, one called "quality control" and the other "quality assurance" (QA). These terms are not synonymous, but instead refer to different functions and responsibilities.

It is of course up to the top management of the firm to define the specific functions of each organizational unit. However, in general, where a QA function exists, that group typically has an *overall* responsibility to ensure that products are consistently made to the agreed-upon and appropriate levels of quality. This is of course closely interrelated with the QC function, both in turn being based on full compliance with the GMPs, which of course encompass many activities. However, QA tends to employ a broader and overall point of view of the management aspects of quality, whereas QC is usually more directly involved in the day-to-day and hour-to-hour aspects, including (but not limited to) inspection and testing. QA often focuses on improvements in an effort to lead to more consistent quality by avoiding failures. QA attempts to ensure the effective and efficient use of the available resources. Where both functions exist, in military terminology, QA focuses on *strategy*, and QC on *tactics*. The regulations do *not* mandate having *both* QA *and* QC, although they do detail the requirements for QC as discussed below. To repeat for clarity, there is *no* regulatory requirement to have a QA function, and in practice many (arguably most) dietary supplement firms do not have.

STATISTICAL QUALITY CONTROL

As discussed, variability is common and essentially unavoidable, although it can be minimized. Techniques have been developed for sampling and testing based on statistical probability theories to determine acceptability versus rejection. These methods are parts of what is often referred to as *statistical quality control* (SQC), and are, indeed, very useful and practical tools. These involve the inspection of random samples to establish whether an item's characteristics fall within a predetermined range, as discussed in chapter 11.

SQC can be helpful not only for the inspection of incoming goods and supplies, but also during *production processes*, where *control charts* are often used. A control chart is a graph used to monitor the extent to which the ongoing production meets or deviates from specifications. Such charts have a central line indicating the ideal value, above and below which are parallel lines indicating the upper and lower specification limits. Samples of the product are taken and measured or tested at predetermined time intervals, and the results are plotted on the chart in time sequence to graphically show the state of control of the process. This enables appropriate adjustments or corrections to be made if and when necessary. Such charts are often used to control such process variables as tablet weights or thickness, or fill volumes, for example (Fig. 18.1).

SQC was originally introduced in the 1930s by Walter A. Shewhart, an engineer and statistician, and has since been adopted by many manufacturing industries. There is voluminous literature available on this topic. The GMPs are

QUALITY CONTROL RESPONSIBILITIES

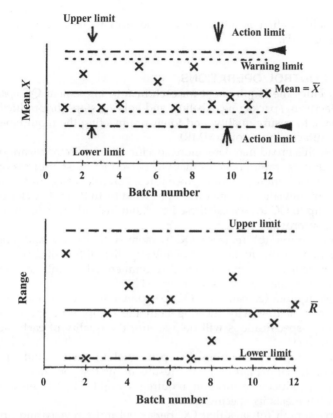

Figure 18.1 Quality control $\overline{X}$ and range charts. *Source*: From Bolton S, ed. Pharmaceutical Statistics: Practical and Clinical Applications. 5th ed. Informa Healthcare, New York, 2009.

silent regarding SQC, and therefore the formal use of it is *not* required, although for specific applications SQC may be useful to employ.

QUALITY CONTROL PERSONNEL

In some industries regulated by FDA, there are requirements for establishing a QC unit or department. However, instead, the dietary supplement GMPs, in §111.12(b), call for naming a person (or persons) identified as being responsible for QC operations. Such individuals must have *separate* responsibilities related to QC, which are distinct from the responsibilities those individuals have when *not* performing QC operations. That means that QC personnel need not exclusively *just* perform QC functions. Such individuals may have additional assigned tasks unrelated to QC, particularly in smaller firms. However, no employee should ever be in a position to review or approve their own work.

QC personnel must have appropriate education, training, or experience (or some combination of these) to perform the assigned functions, according to §111.12(c). The specific details of such qualification are left to the firm's discretion, although it is advisable for the firm to be able to articulate the established criteria.

QUALITY CONTROL OPERATIONS

§111.65 specifies the requirement for each firm to implement QC *operations* for the manufacturing, packaging, labeling, and holding steps for producing dietary supplements, to ensure quality, and to ensure meeting the requirements of the master manufacturing record (MMR).

Under this broad mandate, more specific QC requirements are set forth in §111.105. Subparagraph (a) states that QC personnel must approve or reject all processes, specifications, written procedures, controls, tests, and examinations as well as any deviations or modifications from or to them. For clarity, it is not necessarily up to QC to *establish* these items, but instead, their function is one of review and approval.

Subparagraph (b) requires QC personnel to review and approve the documentation for the qualification of any supplier, although such qualification (which is extremely advantageous and recommended) is only mandated when relying on certificates of analysis, under §111.75(a)(2)(ii).

Subparagraph (c) calls for QC personnel to review and approve the documentation stating why meeting both the component specifications and the in-process specifications will help ensure the quality of each dietary supplement product.

Subparagraph (d) requires QC personnel to review and approve the documentation that explains why the results of the tests or examinations called for in each product specification required by §111.75(c)(1) ensure that the finished batch meets its specifications.

Subparagraph (e) says that QC personnel must review and approve the basis for (and the documentation of the reason for) why any product specification mentioned in (d) above is exempted from verification, and explain what alternative steps will be used to ensure that periodic testing of the finished batches involved will instead suffice.

Subparagraph (f) requires QC personnel to see to it that *representative* samples are collected, and similarly subparagraph (g) mandates that QC personnel *ensure* that the required *reserve* samples are collected and held. However, QC personnel need *not* actually be the ones to collect the samples, but instead have the responsibility to see to it that this is properly done.

Subparagraph (h) makes it QC personnel's responsibility to determine whether all of the specifications called for in §111.70(a) are met.

Subparagraph (i) is a "catch-all" statement requiring QC personnel to perform the other operations required in Subpart F, which includes having SOPs covering *all* of QC's operations, including conducting material reviews and making disposition decisions (discussed below), and approving or rejecting any reprocessing as mentioned in §111.103. Having such SOPs is essential to ensure that all QC operations are conducted in a consistent and appropriate way.

LABORATORY OPERATIONS

As discussed in chapter 19, QC personnel are responsible under §111.110 for laboratory operations, including reviewing and approving the laboratory control processes, ensuring that all of the tests and examinations required by §111.75 are conducted, and reviewing the results of such testing.

MATERIAL REVIEW AND DISPOSITION DECISIONS

Under §111.87 and §111.113, QC personnel must conduct all required material reviews and make disposition decisions, although neither the term "material review" nor the term "disposition decision" is defined.

A material review is essentially the same as an investigation of a deviation or of a failure to meet specifications. This would involve a complete and careful review of the applicable production records and specifications in an effort to determine whether a given batch must be rejected or whether a treatment or reprocessing step could make it acceptable. In other words, following a thorough review of the facts and circumstances, a decision must be made on the "disposition" of the batch in question, either it must be rejected and destroyed, or acceptable steps can be taken to make it "releasable." In this usage, the term disposition refers to what must be done with the material in question.

As mentioned above, §111.103 requires an SOP for conducting such reviews and making the necessary decisions. Further, under §111.113(c), the person conducting the material review and disposition decision process must, at the time of performance, document the details.

Moreover, §111.140(b)(3)(vii) requires the signature(s) of the individual(s) who conduct each such review and decision-making process, as well as those of the persons who provide relevant information. The FDA does allow qualified individuals *other than* QC personnel to *contribute to* the material review, for example, an individual from the production department might prepare a report containing all of the required documentation and information, which could be given to the designated QC person as background information to assist in the process of the material review followed by the disposition decision. This is in accordance with §111.12(b), which allows individuals other than designated QC personnel to participate in QC operational matters. However, the final responsibility for material reviews and disposition decisions rests with QC personnel.

Further information on the circumstances that trigger material reviews and disposition decisions is contained in §111.113. This includes situations when established specifications are not met, deviations from what the MMR requires, certain unanticipated occurrences, calibration-related problems, and for returned goods.

EQUIPMENT, INSTRUMENTS, AND CONTROLS

QC operations regarding equipment and instruments are detailed in §111.117. This includes reviewing and approving all calibration activities and processes, including periodic reviews of all calibration records, inspections, and checks of automated, mechanical, or electronic equipment and ensuring that such equipment functions in accordance with its intended use. The term "periodic" is not specifically defined and is therefore left to QC's discretion.

QC is not necessarily mandated to be directly involved in these activities, but instead must see to it that these activities are in fact done, and thereafter review and approve the documentation involved.

REPROCESSING

By definition, "reprocessing" means using clean, uncontaminated components in manufacturing, which were previously disallowed for use, but have been successfully reconditioned and thereby made suitable for use.

Under §111.90(a), it is *not* permissible to reprocess a rejected dietary supplement, or make an in-process adjustment to a component, packaging, or label to make it suitable for use if an established specification is not met, *unless* QC personnel conduct a material review and approve the reprocessing, treatment, or in-process adjustment that is permitted by §111.77. Moreover, similarly, under §111.90(b) and §111.140(b)(3)(vi), in such corrective steps, QC's decision must be based on scientifically valid reasoning.

To repeat for clarity, in-process materials *can* be reprocessed if necessary, when a suitable method is available, but QC personnel must give appropriate oversight of such steps to ensure that the quality of the dietary supplement is not compromised. This may include reprocessing needed due to contamination with microorganisms, aflatoxins, heavy metals, pesticides, or other types of contaminants.

According to §111.90(c), any batch that is reprocessed and contains components that have been treated or that have been made suitable by in-process adjustments must be approved by QC and must comply with §111.123 (b) before releasing for distribution.

The decision to approve reprocessing must be made in a consistent manner, regardless of who conducts the operation or when it is conducted, and it is necessary to have a clear basis to decide that reprocessing will actually correct the problem.

RETURNED GOODS

When returned goods are received, according to §111.510, they must be identified and quarantined. They must then undergo a QC material review and disposition decision-making process as required by §111.130. Unless that decision approves salvage for either redistribution or reprocessing, under §111.515 the goods must be suitably disposed of such as by destruction. If, however, QC approves reprocessing in an effort to meet all of its specifications established under §111.70(e), the reprocessed goods may either be released for distribution or be rejected by QC, according to §111.525.

The point here is that it is in QC's province to make these decisions. This is also discussed in chapter 20.

If the reason for a product return may implicate other batches, under §111.530 it is required to conduct appropriate investigations of those batches to determine compliance with specifications.

PRODUCT COMPLAINTS

QC must review and approve decisions about whether to investigate a product complaint and must approve the findings and follow-up actions on any investigations performed, according to §111.135. This is discussed in chapter 21.

BATCH PRODUCTION RECORDS

As discussed in chapter 15, QC has the responsibility for reviewing batch production records (BPRs). This is further mentioned in §111.260(l) and §111.123(a)(2).

MASTER MANUFACTURING RECORDS
According to §111.123(a) and (b), it is required that QC review and approve all MMRs, as discussed in chapter 14.

PACKAGING AND LABELING OPERATIONS
QC must review the results of the visual examination and documentation for packaging and labeling operations to ensure that specifications established by §111.70(f) are met, for all products the firm receives for packaging and labeling (for distribution rather than being returned to the supplier), according to §111.127(a). This is also mentioned in chapter 17.

Similarly, under §111.127, products received from a supplier for packaging or labeling for distribution (rather than for return to that supplier) must be approved by QC and released from quarantine before they are used for packaging or labeling.

§111.127 also requires QC to review and approve all records for packaging and labeling operations and to determine that the finished packaged and labeled dietary supplement conforms to the specifications established in accordance with §111.70(g). QC has the task of approving or rejecting any repackaging of a packaged product or the relabeling of a packaged and labeled product, and in short, approving for release for distribution, or rejecting, any packaged and labeled dietary supplement.

REPACKAGING AND RELABELING
According to §111.420, any repackaging or relabeling operations may only be conducted with QC approval, and QC must examine representative samples from each batch of repackaged or relabeled products to ensure compliance with §111.70(g).

COMPONENTS, PACKAGING, AND LABELS
As stated in §111.120, QC is responsible for reviewing all receiving records for components, packaging, and labels and for determining whether they conform to their specifications established under §111.70(b) and (d) before their use.

Specifically for components, §111.155(c)(2) and (3) require QC personnel to review and approve the results of any tests or examinations conducted and approve the components for use. This includes the approval of any treatment or in-process adjustments of components to make them suitable for use. QC must release approved components from quarantine.

Similarly, QC must conduct any required material review and make a disposition decision, under §111.120(c), and under (d), QC must approve or reject any treatment or in-process adjustments of components, packaging, and labels to make them suitable for use.

Under §111.160(c)(2), QC personnel must review and approve any tests or examinations conducted on packaging and labels, and under subsection (c)(3), they must approve them for use and release the packaging and labels from quarantine. Similarly, under §111.120(e), QC must approve and release from quarantine all components, packaging, and labels before they are used. Obviously, QC personnel can also *reject* unsatisfactory components, packaging, or labels.

§111.77(a) and (b) also state that if the specifications established for components, packaging, and labels (as well as with dietary supplement products) under §111.70 are *not* met (unless QC personnel approve a treatment, in-process adjustment, or reprocessing that will ensure the quality of the finished dietary supplement), QC must make a rejection. No finished batch may be released for distribution *unless* it complies with §111.123(b), which covers component identification, batches that fail to meet the specifications of §111.70 (e), batches not manufactured, packaged, labeled, and held under conditions to prevent adulteration, and products received from an outside source for packaging or labeling for which sufficient assurance is lacking to adequately identify the product and to determine that it is consistent with the relevant purchase order.

CONCLUSION

From the above discussion, it is clear that the duties of QC personnel are extensive and diverse, albeit clearly defined. The number of individuals required is of course dependent on the size of the firm and the variety and types of products handled. However, it is evident that the people involved in QC must be rather special to competently handle their many and varied assigned tasks. This requires both broad and detailed knowledge plus management skills. Moreover, QC personnel must be of impeccable integrity and able to forcefully impose their will when the occasion demands, without needlessly disrupting operations.

SUGGESTED READINGS

Defo J, Juran JM. Juran's Quality Handbook. 6th ed. New York: McGraw-Hill, 2010.
Grant E, Leavenworth R. Statistical Quality Control. 7th ed. Columbus, OH: McGraw-Hill, 1996.
Gryna FM. Quality Planning and Analysis. New York: McGraw-Hill, 2001.
Haider S, Syed EA. Quality Training Manual. Boca Raton, FL: CRC Press and Taylor & Francis Group, 2011.
McCormick K. Quality. Woburn, MA: Butterworth-Heinemann, 2002.
Montgomery DC. Introduction to Statistical Quality Control. 6th ed. Hoboken, NJ: John Wiley, 2009.

19 Laboratory operations

As discussed in chapter 18, it is the responsibility of the designated quality control (QC) personnel to take the necessary and appropriate steps to ensure that the firm's dietary supplement products consistently and properly meet all of their established specifications and are neither actually nor legally adulterated. As outlined, this is a complex challenge consisting of many facets, one of which involves laboratory operations. This portion of the QC duties is stated in §111.110, albeit somewhat briefly.

In essence, QC personnel must ensure that all tests and examinations required under §111.75 are conducted, and then must review and approve of the results. Many of these tests and examinations require laboratory operations, and thus it is required to have access to one or more adequately staffed and equipped laboratories.

Some firms elect to have the necessary facilities and staff in-house to handle these activities, whereas, as mentioned in chapter 31, other firms choose to outsource all or part of their laboratory operations. The use of contract analytical laboratories (CALs) as an extension of the firm's other functions is acceptable, provided this is properly done and with adequate oversight. However handled, as stated above, *all* appropriate tests and examinations must be conducted and must be reviewed and approved by QC personnel. However, there is no requirement that the QC personnel be actually directly involved in carrying out the testing.

The term "Laboratory GMPs" is often used, but this is a misnomer. Although obviously some of the dietary supplement GMPs do pertain to laboratory operations, there are no separate and specific regulations identified as being Laboratory GMPs.

However, in FDA facility inspections, considerable attention is usually given to the laboratory, test methods, and documentation. As mentioned in chapter 33, the FDA frequently sends investigators well trained in analytical chemistry to conduct that portion of the inspection, and therefore laboratory personnel should understand how to conduct their portion of such inspections.

LABORATORY MANAGEMENT
The overall management of laboratory operations is of obvious importance. This includes the assignments of duties and responsibilities of personnel. This also includes oversight of the training and experience of analysts in the specific methods and instruments in use and in establishing work schedules to best coordinate with manufacturing needs. Moreover, laboratory managers should see to it that periodic audits are conducted as discussed in chapter 30, and any needed corrections attended to as promptly as feasible.

Adequacy of staffing for the assigned operations is an important consideration. If the laboratory has an insufficient number of personnel, component and product releases are likely to be delayed, excessive overtime work may result, and this often results in low morale.

Laboratory managers should oversee the keeping of pertinent test and examination records. Supervisory personnel should be watchful to ensure that complete and accurate documentation is maintained, including the raw data and any necessary calculations.

ANALYTICAL CHEMISTRY

Although some of the required tests and examinations are merely visual or only involve taking physical measurements, other tests require some form of chemical analysis. Conducting such testing requires the participation of one or more *analytical chemists*. These are persons knowledgeable of, and experienced in, the field of analytical chemistry. This is a specialized subdivision of the broad scientific field of chemistry. Becoming an analytical chemist first requires a general knowledge of chemistry followed by specific training in analytical methods, techniques, instrumentation, and interpretation of the data involved. Many accredited colleges and universities offer curricula in this field. There are also frequent symposia, seminars, and training courses to help keep analytical chemists informed, including some on the Internet. Moreover, there are specialized technical journals as well as a plethora of books in the field of analytical chemistry.

Analytical chemists need, in addition to the above, a propensity for detail, good computer skills, and problem-solving skills. Good oral and written communications skills are also valuable assets for their functions.

The Examinations Institute of the Division of Chemical Education of the American Chemical Society offers examinations in analytical chemistry, including study guides.

Assistants to qualified analytical chemists are typically referred to as "laboratory technicians." These are paraprofessional positions, helping the analytical chemists through such activities as logging-in and storing samples, maintaining and calibrating laboratory equipment, preparing needed solutions and reagents, keeping the laboratory and glassware clean, and other similar functions.

CERTIFICATION PROGRAMS

Some firms have implemented certification programs for their laboratory personnel, although this is not required by, nor mentioned in, the GMPs. This is a process in which laboratory management reviews and evaluates individuals as to their backgrounds, their knowledge and skills, and their demonstrated abilities to perform specific tests or duties. Firms using such a plan should develop objective criteria for certification. Some firms evaluate analytical chemists and laboratory technicians through written or oral examinations or quizzes, through demonstration of techniques, or by comparing their analytical test results with those of more experienced analysts. Certification, either "overall" or for specific laboratory procedures, can be used as qualification for promotion or higher salaries, and as such may be incentives for personnel to work toward becoming certified. Such certifications are usually strictly related to the firm involved, not implying any rights or privileges beyond the firm.

LABORATORY FACILITIES

§111.310 requires having *adequate* laboratory facilities to perform the testing. Testing may involve examinations or tests necessary to determine whether components and packaging materials meet their specifications, in-process specifications are met as established in the master manufacturing record (MMR), and/or whether finished product specifications are met.

"Adequate" is of course *not* defined in the regulations since this would necessarily be determined by many factors such as the number and type of items involved and by the complexity and the instrumental requirements of the required testing procedures. The object is to ensure that the laboratory facilities used are designed, constructed, maintained, and equipped suitably for carrying out the necessary tests and examinations. This is related to the subjects discussed in chapters 4 and 5. Since the topic is subjective and essentially impossible to explicitly define, it is prudent to keep in mind the fact during an inspection, the FDA *would* judge the *adequacy* of the facilities. Warning letters have been issued to firms for having inadequate laboratory facilities, including lacking necessary testing equipment.

The laboratories should have the appropriate physical environment and utilities. The space should be suitable for personnel conducting the tests, allowing for orderly placement and storage of equipment and supplies and avoiding mix-ups and cross-contamination. Good lighting is important. Temperature and humidity conditions should be considered. Safe storage of toxic materials should be provided. If hazardous volatile or flammable solvents must be used, fume hoods should be provided. Provision should be made for the proper handling of potentially toxic or dangerous waste, which may require special plumbing considerations. Emergency showers and/or eyewash facilities may be appropriate. Depending on the tests to be conducted, compressed air, gas, steam, and/or vacuum facilities may be needed. It may be desirable to provide separate nontesting areas for sample receipt and storage and for "office-type" activities, such as performing calculations and preparing records and reports. Instrumentation may require segregation from wet chemical analyses.

As stated above, the laboratory facilities may either be in-house or at a contract laboratory. However, if a contract laboratory is used, the firm is responsible for ensuring that *their* laboratory facilities are adequate and that tests and examinations are properly performed, as required by §111.310.

LABORATORY CONTROL PROCESSES

The concept of laboratory control processes is in effect "quality control of laboratory operations," meaning that it is required to have sound operating procedures and practical work methods to ensure that the QC laboratory (*not* the research and development laboratory) performs properly. Laboratory control *processes* are *established by* laboratory personnel, then reviewed and approved by QC personnel as stated in §111.315(b).

The laboratory control processes must be in the form of written procedures and must be carried out in a consistent manner, as stated in §111.303. This requires having SOPs covering *each* of the test and examination procedures conducted. Routine deviation from SOPs should not be allowed, and if a "modified method" is perceived as being better, the SOP should be formally revised. Supervisory personnel should ensure that the laboratory procedures are properly followed.

SPECIFICATIONS

As discussed in chapter 10, the firm initially determines precisely the characteristics of the products it wants to make, and then under §111.70 establishes the necessary and appropriate specifications to achieve this. Thereafter, §111.73 requires that those specifications be met, whereas §111.75 sets forth what must be done to ensure the specifications *are* met. §111.315 identifies the minimum steps the *laboratory personnel* must take to ensure that the necessary tests or examinations are completed, reviewed, and recorded in a timely fashion before products are released for distribution to the public. The steps outlined include the use of criteria for selecting the examination and testing methods and the use of appropriate criteria for selecting standard reference materials (SRMs) as needed.

§111.315(a) requires the establishment of laboratory control processes that include the use of criteria for establishing appropriate specifications, implying that laboratory personnel need to be involved in setting the specifications.

To repeat for clarity, the firm is responsible for establishing specifications for identity, purity, strength, and composition related to incoming components as well as limits on those types of contaminants that may adulterate or lead to adulteration of the finished product. Moreover, specifications must also be established for packaging materials, such as containers and closures, blister packs if used, cartons, labels, and other items of packaging. Similarly, specifications must be established for in-process steps, as well as for finished batches, and for finished packaged goods. Once established, such specifications are also incorporated into the MMR as discussed in chapter 14.

SAMPLING PLANS

§111.315(b) covers having sampling plans for obtaining truly representative samples, as discussed in chapter 11 and in §111.80. Such sampling plans are required for components, packaging, labels, in-process materials, finished batches, product received for packaging and/or labeling, and for packaged and labeled finished goods. Again, it should be noted that laboratory personnel should be involved in developing appropriate sampling plans, which should then be reviewed and approved by QC personnel.

SELECTION OF APPROPRIATE TESTING METHODS

§111.315(c) calls for establishing and following a laboratory control process for selecting appropriate *examination* and *testing* methods. This of course does not preclude the possibility of considering the recommendations of a contract laboratory or other sources in establishing such methods, but it is the *firm* that bears the final responsibility to actually *select* the appropriate tests.

The testing methods used must be *appropriate for their intended use*, as stated in §111.320(a), and should also be specific and selective. Also, according to §111.320(b), the methods must be *scientifically valid* to determine whether each specification is met, although these terms are not defined in the regulations. This has generated considerable discussion within the dietary supplement manufacturing industry to help ensure compliance with FDA's expectations. However, the FDA has made it clear that their interpretation of "scientifically valid" does *not* necessarily require formal analytical methods validation procedures such as is the rule in the pharmaceutical industry. Although most compendial and Association of Official Analytical Chemists (AOAC) methods *have* been formally validated, the FDA also accepts most test methods published in scientific journals and textbooks as being scientifically valid. The FDA has said that a scientifically valid method is one that is accurate, precise, and specific for its intended purpose, and consistently does what it is intended to do.

§111.75(h) also states that test and examination methods to determine whether specifications are met must be appropriate and scientifically valid but do provide some flexibility in their choice. The FDA has also said that firms have the flexibility to adopt methods that are the most suitable for what they are testing.

The important aspects of any analytical method are *accuracy* (how close the measurements are to the true value) and *precision* (the ability to get the same results when the testing is repeated).

If a firm decides to modify an "officially validated" method, the reason for the modification should be documented together with data to establish that the modified method produces results at least as accurate and reliable as the established method.

In practice, it is advisable to use methods that appear in the USP or other such compendia *if* such exist, or that have been validated by the AOAC, or that have been published in well-recognized, peer-reviewed journals. However, on occasion, in-house methods may be deemed "appropriate" if it can be established that they consistently produce reliable results.

In general terms, most test methods involve either "wet chemistry" or instrumental methods. The comments below are of course well known to laboratory personnel but are briefly explained here to aid management personnel in communicating with their technical colleagues.

Wet chemistry, sometimes called "bench chemistry" or "classical chemistry," techniques historically were originally all that were available prior to the development of more sophisticated instrumental methods briefly described below. Wet chemistry is typically used in teaching at the high school and beginning college levels, but useful analytical wet chemistry methods still have a role in many dietary supplement QC laboratories. These methods are called "wet" since they typically are done in the solution phase, using gravimetric (weight-based measurements) or titrimetric (volume-based measurement) techniques, with the analyst physically observing the results and making appropriate calculations using the data obtained. Although most wet chemistry analytical methods use only relatively simple apparatus, some do involve more elaborate automated equipment.

Most of the *instrumental* methods are classified as being either *spectroscopic* or *chromatographic* techniques.

In those that are spectroscopic, the prepared sample is exposed to a beam of electromagnetic radiation, such as a beam of light of a prescribed frequency or wavelength, and the sample will then absorb, scatter, or emit a modified form of the radiation. If the radiation employed is a light source, it may be in the ultraviolet (UV), visible, or infrared (IR) range. The resulting complex spectra can then be interpreted through the use of *chemometrics*, a practical statistical and mathematical computer-based scientific discipline developed in the 1970s to extract useful information from the graphical data generated by some instruments used in analytical chemistry.

IR illumination is frequently used. This is subclassified as near-IR (NIR), mid-IR (MIR), and far-IR (FIR).

IR *absorption* spectroscopy methods are particularly useful where the atoms or molecules in the sample absorb the IR light, resulting in transitions between energy levels, yielding absorption spectra. In these instruments, the light is usually first passed through a grating monochromator to select the optimum wavelengths, and this beam then passes through both the sample and a reference material. As discussed below, NIR spectroscopy is frequently used in raw material identification.

Still another form of spectroscopy is *X-ray fluorescence* (XRF), which is based on the behavior of atoms when they interact with X rays. When a sample is illuminated by a strong X-ray beam, some of the energy is scattered and some is absorbed. The sample may then emit a spectrum of wavelengths characteristic of the types of atoms in the sample. The advantages of this technique are that it is nondestructive, and relatively little sample preparation is needed. The method is fast, often giving information within seconds. This method is sometimes used for the determination of calcium and for detecting arsenic, mercury, and lead contamination in dietary supplements.

In addition to the many forms of spectroscopy, another useful analytical technology is *chromatography*. The term literally means "color writing."

Chromatographic methods of analysis have been in use since the early 20th century. There is an entire range of such techniques, usually involving a liquid or gas *mobile phase* and a *stationary phase*, which is usually either a solid or a liquid coated onto a solid. The mobile phase is caused to flow through the stationary phase, carrying all of the parts of the mixture with it, but resulting in the separation of the compound mixture into each of its constituent parts. Different parts travel at different rates, which is what enables such separation to occur. Once the separations have been accomplished, there are several ways each part can be identified, for example, visually, by conduction of electricity or heat, or by other methods.

Thin layer chromatography (TLC) is a technique performed by spreading a thin layer of an absorbent (such as silica gel moistened with water and a binder) on a plate or sheet of glass, plastic, or aluminum foil, which is then oven-dried. This becomes the stationary phase. The sample is dissolved in a suitable solvent, which becomes the mobile phase. A small amount of the mobile phase is applied to the dry film on the plate, near the bottom. The plate is placed in a shallow pool of a solvent, with only the bottom of the plate in the liquid. The mobile phase is slowly drawn up into the stationary phase by capillary action, but since the various constituent parts in the mobile phase ascend at different rates, separation is achieved. Various techniques are available to identify the individual parts once they have been separated. In some instances, the various compounds are colored, making visualization easy. Or, UV light may be used. TLC is fairly quick, simple, and inexpensive, and therefore often a useful analytical tool. Moreover, an understanding of TLC helps introduce other forms of chromatography, which operate on similar principles.

There is also an enhanced form of TLC, automating various steps, and employing increased resolution, called high-performance thin layer chromatography (HPTLC), considered by some to be an excellent method for botanical identity testing.

In paper chromatography, the stationary phase is a uniform absorbent paper, whereas the mobile phase is a suitable solvent. The paper is suspended in a container with a shallow layer of a solvent, and the container is covered so that the atmosphere surrounding the paper is saturated with the solvent vapor. The solvent slowly travels upward, causing separation to occur as described above for TLC.

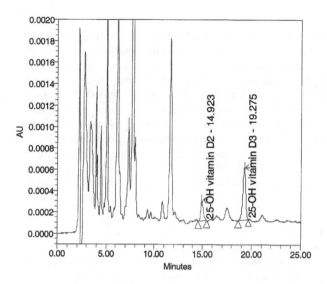

Figure 19.1 This is a typical HPLC chromatogram illustrating the complexity of interpreting the results. Experienced analysts can frequently do so visually, but in other instances must rely on chemometric methods.

The same basic concepts as described above for TLC and paper chromatography can be used on a larger scale in what is termed *column chromatography*. This is done by packing the materials into a vertical column. The most frequent application of this is termed HPLC. The mobile phase is forced through the packed column under high pressure, and the stationary phase is composed of small particles. The sample is injected via an automated system. The time from which the sample is injected to the time a compound travels through the column to the detector is termed the *retention time*, which varies depending on the pressure used, the type of solvent, and the details of the composition and particle size of the stationary phase. The *detector* may be based on absorption of a beam of UV light passing through the liquid coming out of the column, although there are several other approaches sometimes used. In any event, the output from the detector is recorded as a graph with a series of peaks, each representing one of the constituents of the mixture passing through. There are many equipment and procedural details in HPLC, known to experienced analytical chemists, but which are beyond this overview. However, it is important to keep in mind that HPLC is an important analytical tool quite frequently used in dietary supplement QC laboratories. (Fig. 19.1).

Similarly, there is another type of chromatography used in analytical chemistry called *gas chromatography* (GC), where the stationary phase is either a solid or a high-boiling liquid and the mobile phase an inert gas such as nitrogen or helium. A small amount of the sample is injected into the instrument and is volatilized in a hot chamber, and then it is swept by the carrier gas through a heated column containing a stationary high-boiling liquid. As this travels through the column, the various components of the sample travel at different rates, separating into pure components that exit the instrument passing a detector, which sends an electronic message to a recorder.

SYSTEM SUITABILITY

In some instrumental methods of analysis, particularly chromatographic techniques such as HPLC, it is necessary to establish that the equipment is performing properly and can generate results of acceptable accuracy, precision and repeatability, and with the required specificity. A testing regime may be conducted prior to starting an analytical procedure, and repeated if necessary throughout the run, to ensure the system's performance is acceptable. This is called *system suitability* testing, as described both in the USP and in an ICH (International Conference on Harmonization) guideline, and elsewhere. The entire procedure is considered to be a "system" in that it includes the samples, the equipment, the electronics, and details of the analytical method in use. The dietary supplement GMPs are silent on this, but analytical chemists are familiar with the concept.

IN-PROCESS TESTING

The GMP defines an "in-process material" as any material that is fabricated, compounded, blended, extracted, sifted, sterilized, derived by chemical reaction, or processed in any other way for use in a dietary supplement.

As previously mentioned, §111.70(c) requires establishment of in-process specifications for any point, step, or stage *where control is necessary*, but otherwise it does not identify specific in-process tests that must be performed. Such specifications may include a list of tests, references to analytical procedures, and appropriate acceptance criteria. It is the responsibility of the firm to ensure that appropriate in-process specifications do exist. Accordingly, it becomes one of the requirements of the laboratory operations to conduct the suitable in-process *testing* to confirm that those specifications are met.

An example of such in-process testing could be for making tablets, to periodically conduct tests for weight variation, hardness, and friability (albeit the regulations do not mandate those specific tests). In that instance, laboratory personnel would confirm that those in-process tests were done and that the results were within the specifications. The in-process specifications may differ from the release specifications, and some of the tests done may differ from those done at release. For example, a firm might decide to do disintegration testing of tablets as an in-process test, but dissolution testing as a release test.

As mentioned above, §111.315(b)(2) calls for the use of sampling plans to obtain *representative* samples of in-process materials, and, as previously mentioned, it is very important that the samples are truly representative.

FINISHED PRODUCT TESTING

Originally, during the writing of Part 111, the FDA considered requiring 100% testing of finished batches. However, it was decided that in view of the requirements established for meeting component and in-process specifications, it would be satisfactory for firms to test either *every* finished batch *or* only a subset of such batches identified by a sound statistical sampling plan. Similarly, the regulations allow testing for only *selected* specifications as opposed to every specification. This flexibility can greatly reduce the testing burden.

Under §111.75(c) and (d), it is permissible to select and test for meeting only some specifications in a *subset* of batches to demonstrate that the process control system does in fact produce satisfactory products. However, it is required to document an explanation for *why* meeting only the selected specifications ensures that a batch does meet *all* of its product specifications. QC personnel must review and approve this documentation. This is selective end-product testing justified by good statistical sampling.

Moreover, it is similarly permissible to exempt certain of the finished product specifications from requiring being tested if it can be established that there is no scientifically valid method of testing for that (or those) specification(s) at the finished batch stage. Again, this must be documented, which must be reviewed and approved by QC personnel.

REFERENCE MATERIALS

§111.315(d) covers establishing and following a laboratory control process on the selection of SRMs to be used in performing tests and examinations, when such are needed. SRMs are used for accurate calibration purposes.

The preamble to the GMPs comments on FDA's recognition of two general types of reference materials, compendial reference standards that do not require characterization (detailed description), and noncompendial standards that should be of the highest purity reasonably available but need to be thoroughly characterized. The FDA prefers the use of compendial reference standards when possible, but the use of appropriately characterized in-house materials prepared from representative lots can be satisfactory if no compendial reference standard exists. The FDA believes that reference materials need to be appropriate to the assay procedure for which they are used.

The National Institute of Standards and Technology (NIST) is collaborating with the Office of Dietary Supplements (ODS) of the National Institutes of Health (NIH) to help ensure adequate reference materials and reference standards are available since they are critical for GMP compliance.

It is advisable to keep complete records of any testing and standardization of laboratory reference standards, reagents, and standard solutions used.

As further discussed below, for botanical or herbal materials the "reference materials" used in *organoleptic* examinations (i.e., based on the human senses of sight, feel, smell, or taste) may be materials authenticated as being the correct plant species and part(s). Similarly, for chemical and microscopic tests, the reference materials should be well characterized.

CERTIFICATES OF ANALYSIS

§111.75(a)(1)(ii) states that a firm may rely on a certificate of analysis to determine whether specifications are met, provided certain steps have been taken to verify the reliability of the supplier's certificates of analysis. However, it is necessary to conduct at least one appropriate test or examination *to verify the identity* of a component that is a dietary ingredient, and also to confirm the identity of other components.

OUT-OF-SPECIFICATION RESULTS

Although Part 111 is silent on out-of-specification (OOS) laboratory test results, the FDA recognizes these will in fact occasionally be generated. It is their expectation that all such instances will be carefully and thoroughly investigated using scientifically valid principles in a timely manner. Such investigations and assessments should be well documented. During FDA inspections, the handling of OOS situations frequently comes into focus.

OOS test results typically are caused by an error on the part of the analyst, by an error in manufacturing, or due to a nonrepresentative sample. It is prudent for the firm to have an SOP, preferably including a checklist of steps to be taken, for conducting investigations when OOS results occur.

The usual starting point is for the analyst and the supervisor to verify that all calculations were correct and that the test method was properly followed. Having the analyst repeat the test, preferably with the *same* sample, and/or having another analyst run the test, may pinpoint the cause of the failure.

It is *not* acceptable to continue several repeat tests, hoping to eventually get and use favorable results. Similarly, it is generally considered unwise to average the results of multiple tests since this tends to hide the variability among individual test results.

If it is determined that the incident was due to an error on the part of the analyst, the root cause may be that the wording of the testing procedure was inadequate or that the analyst was not properly trained. Consideration should also be made of the possibility of problems with the equipment or its calibration, or with solutions, reagents, or reference standards used. However, it is unrealistic to expect that the cause of analyst errors can *always* be satisfactorily determined. For this reason, many firms have the policy that the analyst should not discard test samples, nor standard materials used in the analysis, until test results have been reviewed and approved.

In general, firms should not rely on resampling to release a product that has failed testing unless the failure investigation clearly discloses evidence that the original sample was not representative or that the sample was improperly handled.

In the pharmaceutical manufacturing industry, OOS test results have posed significant problems in the past. This included an important lawsuit involving Barr Laboratories in 1993, in which the Court and FDA established principles regarding resampling and retesting, plus the role of scientific judgment as the basis of releasing batches for distribution. This eventually resulted in FDA releasing a document titled *Guidance for Industry: Investigating Out-of-Specification (OOS) Test Results for Pharmaceutical Production*. Although this, of course, does not directly apply to dietary supplements, it does reflect FDA's thinking on the topic.

TRENDING

Although not mentioned in the regulations, it is advisable to plot the results of laboratory analyses over time, for components, in-process batches, finished batches, and any stability assays. Control charts are useful formats for this, enabling looking for trends. Monitoring trends can help with anticipating and preventing problems.

STABILITY TESTING

The regulations do *not require* expiration dates or other forms of shelf life statements such as "best if used by" dating to appear on the labels of dietary supplement products. In writing the GMPs, the FDA did consider this issue but decided not to address it since there may not always be explicit analytical methods available to determine the shelf life of some dietary ingredients (especially botanicals). However, for marketing reasons, some firms prefer to use such label statements. Moreover, lacking an expiration date means that the product label content claims must remain correct for the entire time the product is on retail shelves.

If the firm decides to use some form of expiration dating, the FDA has said that such dating should be supported by appropriate data, although the FDA has declined to offer guidance on the *type* of data they would consider to be acceptable. The preamble to Part 111 says (in part), "... if you use an expiration date on a product, you should have data to support that date. You should have a written testing program designed to assess the stability characteristics of the dietary supplement, and you should use the results of the stability testing to determine the appropriate storage conditions and expiration dates."

A working group of representatives from the trade associations involved, dietary supplement manufacturers, and contract analytical laboratories is developing a stability guideline intended to be acceptable to FDA.

DOCUMENTATION

Laboratory operation SOPs are required by §111.303 and §111.325(b)(1). These should be complete, accurate, and up-to-date. They should include all general laboratory procedures and methods, including equipment calibration and system suitability. The SOPs should be approved by appropriate management personnel, and there should be a system for periodic review.

Any laboratory policies should be documented, including those related to "What to do if...."

Documentation that the proper methodology is followed is required by §111.325(b)(2), which also requires that the person who conducts the tests and examinations document, at the time of performance, that the proper methodology was used, and must include the results obtained.

§111.260(h) requires that batch production records include the results of any testing or examination performed during the batch production, or a cross-reference to such results, which provides flexibility as to how this should be handled. If an outside laboratory performed all or part of the testing, copies of their test results should be on file. §111.260(i) also requires that the batch production record includes documentation that the finished dietary supplement meets specifications.

During FDA inspections, the investigators often review laboratory analysts' notebooks for accuracy and authenticity and to verify that raw data are retained to support conclusions found in laboratory results. They typically also review laboratory logs for sequence analysis versus the sequence of manufacturing dates.

Although the regulations are silent on the details of analysts' notebooks, it is generally best to keep raw laboratory data in hardbound books, preferably with prenumbered pages. Electronic notebooks may be used. Correction fluid should not be used in paper-based notebooks, nor should results be changed without explanation. Blank spaces should be lined out. Any deviations from the usual procedure should be explained. Major instruments used should be identified and their calibration status noted.

All chromatograms should be kept and clearly labeled. Instrument printouts should also be retained. References should be made to all standards, solutions, and chemicals used. Calculations should be shown in full, with the appropriate units of measure stated.

Many firms have adopted the use of Laboratory Information Management Systems (LIMS) software. These tend to increase security in terms of compliance by requiring proper sign-offs plus LIMS facilitates audit trails. The use of LIMS may reduce the number of employees required in the laboratory, although it is questionable if this alone would be a significant financial saving. Moreover, LIMS requires a high level of staff training. LIMS does provide advantages in sample handling, including login, chain of custody, and tracking. Data entry from instrumentation is greatly facilitated, and equipment maintenance and calibration programs enhanced. As mentioned in chapter 22, traceability is easier when using LIMS. There are many types of LIMS available, so if considering making this move, careful evaluation is advised in making the decision of which system to purchase.

CARE OF LABORATORY EQUIPMENT

As discussed in chapters 5, 7, and 8, it is important to keep *all* equipment and instruments (including those in the laboratory) in good repair and properly calibrated at regular intervals. Some laboratory equipment requires specialized procedures for maintenance and calibration, which may make it desirable to have some of these tasks performed via a service contract with the vendor of such equipment.

Analytical balances, even those with internal "autocalibration" systems, should be checked frequently through the use of standard weights. "Autocalibrators" should be periodically verified. The frequency of check weighing should be dependent on the frequency of use of the balance. For example, an analytical balance that is used only once a week should be checked before each use, whereas one that is used six times a day needs to be checked only once a day. If a check is performed and the balance is found to be out of tolerance, that raises the question of all weighing made on that balance from the time is was found to be out of calibration back to the last time it was known to be functioning properly.

Other examples include pH meters, which should be verified with a standard solution at least daily. Spectrophotometers, and other forms of "sophisticated" electronic instruments, require frequent verification and calibration. Although not specifically addressed in the regulations, these details should be covered in laboratory SOPs, and logs should be kept indicating when these steps were taken and by whom.

RAW MATERIAL IDENTIFICATION

§111.70(b)(1) requires establishment of an *identity* specification for each component used in manufacturing dietary supplements. §111.75(a) states the necessity of conducting at least one appropriate test or examination to verify the identity of any component that is a *dietary supplement*. This is of critical importance since mix-ups or inadvertent use of the wrong material can have disastrous consequences. However, selecting the most appropriate identification tests, having sufficient specificity, can be challenging, particularly for botanical or herbal components, but also in some instances even for *single molecular entities* (defined as any distinct atom or molecule identifiable as a separately distinguishable entity).

In identity testing dietary ingredients, all samples *must* be lot representative, and the sampling must be performed using a scientifically valid sampling plan, as discussed in chapter 11. However, to ensure the samples *are* lot representative may require a large number of individual samples. For this reason, the choice of identity testing methods can be important from a time and cost point of view. Rapid, easy-to-use methods, but with the required specificity, are usually the optimum types.

Many firms elect to conduct identity tests for most or all of their components, not *just* for dietary ingredients. However, for components that are *not* dietary ingredients, it is permissible to rely on a certificate of analysis, provided certain steps have been taken to qualify the supplier's certificates of analysis as covered in §111.75(a) and as discussed in chapter 29. Thus, to repeat for clarity, under §111.75(a)(2), the firm has the option of either conducting tests or examinations to confirm the identity of components that are *not* dietary ingredients, or depending on a certificate of analysis from the supplier.

Obviously, in addition to identity, the firm must also ensure that *all* established specifications for incoming components have been met.

TLC or HPTLC, as well as HPLC, are methods often used for identity testing. However, these methods tend to take significant amounts of time. For vitamins and minerals, USP or AOAC identity testing methods exist. Ingredient suppliers and instrument vendors can often be helpful in suggesting appropriate methods, or a search of the literature may be useful.

In many instances, various types of NIR spectroscopy can be useful, since these are fast and easy to use. The identification is based on a comparison between the spectral data from the sample and spectral data of multiple samples in a reference library, using automated chemometrics. Handheld instruments are commercially available and convenient to use. However, the *proper* use of NIR is very important, and several serious problems have resulted from *inappropriate* application of this technology. Laboratory personnel not fully familiar with the details of NIR testing would be well advised to seek expert help in establishing the methods. The instrument vendors can often be quite helpful.

According to §111.75(h)(2), the test methods to ensure whether established specifications are met (which would include identity specifications) must include at least one of the following:

Gross organoleptic analysis
Macroscopic analysis
Microscopic analysis
Chemical analysis
Other scientifically valid methods

Botanicals and herbs often require the use of more than one test to ensure identity, often referred to as "orthogonal" testing. Since the advent of the dietary supplement GMPs, a considerable amount of work has been done on the establishment of reliable botanical identification methods (BIMs). In particular, AOAC, in cooperation with the NIH, has made significant progress on this.

Methods include melting or boiling points, "fingerprinting" using TLC and HPTLC or GC, NIR, UV, and several other methods, now also including genetic testing such as polymerase chain reaction (PCR) testing. New analytical technology and modifications of existing technology for identity testing are continually being developed.

Obviously, identity testing of botanicals is complicated by the fact that such ingredients are available in a variety of physical forms, including whole or parts of plants, dried and "ground-up" material, and extracted botanicals. Moreover, even for a specific item, there are significant variations based on the geographic source of the material, when it was harvested, and the weather during its growing period. Storage conditions can also impact identity verification. The training and experience of the personnel involved in collecting the plants can be a factor, since similar-looking plants might be inadvertently collected.

Both organoleptic (using the senses of sight, small, or taste) and microscopic techniques are useful in botanical identification, although both require good reference standards for comparison, and both require skills on the part of the analyst. It is also important to develop a trusting relationship with suppliers to help ensure identity of the correct plant species.

In practice, there tends to be variability lot to lot and within lots of botanicals, which tends to complicate identity testing.

The test methods should be sufficiently specific to distinguish the correct plant species and plant part(s) from potential adulterants.

AOAC

AOAC International has frequently been mentioned above. This is a nonprofit scientific association headquartered in Gaithersburg, Maryland, devoted to excellence in analytical chemistry. It was founded in 1884 as the Association of Official Agricultural Chemists under the auspices of the U.S. Department of Agriculture (USDA). One of the founding fathers was Dr Harvey Wiley who was involved with the early days of the FDA, as discussed in chapter 32. In 1965, the name of the organization was changed to the Association of Official Analytical Chemists, and later to AOAC International.

AOAC is a worldwide provider and developer of validated analytical methods and laboratory quality assurance programs. It also offers a laboratory accreditation service and a laboratory proficiency testing program. It has a number of useful publications, and offers training courses in laboratory management. The organization has close ties to the FDA and 120 years of experience in validating and approving analytical methods. Although not the organization's actual name, some refer to it as the Association of Analytical Communities. There is a monthly periodical titled *Journal of the AOAC*.

SUGGESTED READINGS

Ahuja S, Dong M. Handbook of Pharmaceutical Analysis by HPLC. San Diego, CA: Elsevier, 2005.
Burns DA. Handbook of Near-Infrared Analysis. 3rd ed. Boca Raton, FL: CRC Press, 2007.
Consumer Healthcare Products Association. Guidelines for the Stability Testing of Nonprescription (OTC) Pharmaceutical Products. Washington, DC, 2004.
FDA. Guidance for Industry: Testing Glycerin for Diethylene Glycol. Rockville, MD, 2007.

Fitz JC. Laboratory instrument qualification: solving the puzzle. Pharmaceutical Engineering, March–April 2006:24–34.

Gentry AE. Methodical evaluation of OOS results—a problem solving approach. Am Pharm Rev 2001; 4(4):78–81.

Jaksch F, Roman M. Handbook of Analytical Methods for Dietary Supplements. Washington, DC: APhA, 2005.

Kemsley J. Improving metal detection in drugs. Chemical & Engineering News, December 8, 2008; 8(49):32–34.

Lanese J. Out of Specification Investigations Still Perplex Labs, Pharmaceutical Regulation and Quality, February–March 2011; Anatomy of an OOS Result, April–May 2011: 30–32; and OOS: The Last Resort, June–July 2011; 13(3):29–31.

Lunn G. HPLC Methods for Pharmaceutical Analysis. Vols 2–4. New York: Wiley-Interscience, 2000.

Mullin R. LIMS in the cloud. Chemical & Engineering News, May 24, 2010:12–16.

Nilsen CL. The use and misuse of analytical balances. Pharmaceutical Formulation & Quality, February–March 2000:51–52.

Reich E, Schibli A, eds. High-Performance Thin-Layer Chromatography for the Analysis of Medicinal Plants. New York: Thieme Medical, 2006.

Rodriguez-Diaz R. Common practices for analytical methods transfer. Pharmaceutical Outsourcing, July–August 2010:52–57.

Singer DS. Laboratory Quality Handbook of Best Practices and Relevant Regulations. Milwaukee, WI: Quality Press, American Society for Quality, 2001.

Stoeppler M. Reference Materials for Chemical Analysis. New York: Wiley-VCH, 2001.

Sutton SVW. Laboratory Design: Establishing the Facility and Management Structure. Bethesda, MD: PDA Books, 2010.

Thurston C. LIMS in food safety traceability. Food Quality, April–May 2010:33–35.

Waksmundzka-Hanjos M. High Performance Liquid Chromatography in Phytochemical Analysis. Boca Raton, FL: CRC Press, 2011.

Zschunke A, ed. Reference Materials in Analytical Chemistry: A Guide for Selection and Use. New York: Springer-Verlag, 2000.

20 Returned goods

Finished products may be returned to the manufacturer for a variety of reasons, including perceived quality defects, and also possibly because of slow retail sales, overstocks, ordering the wrong quantity, a retailer going out of business, or any of a number of other reasons. Also, in the event of a recall, goods are returned to the manufacturer as discussed in chapter 23. Therefore, §111.503 mandates the establishment and following of a written procedure (SOP) for fulfilling the requirements of Subpart N of the GMPs titled *Returned Dietary Supplements*.

In addition, it is usual (but not required) for each manufacturer to have a Return Goods Policy to inform their customers of procedures to be used in making returns. Such policies typically define which items are returnable, where reimbursement will be paid, and which items may be returned but will not be reimbursed. It is also usual to define claims for products invoiced but not received, as well as for claims for obvious damage encountered in transit. Most firms also provide a form that must be filled out by the customer, and must accompany all merchandise being returned. The forms typically include the customer's name, address, and account number, the invoice number and date, the product identification (including lot number, if applicable), the date of the return, the quantity being returned, and the reason for the return. Some but not all firms require prior authorization for returning products. It is also usual to define who is to pay for transportation charges involved (including insurance, if applicable).

RECEIVING AND HANDLING RETURNS

Under §111.510, when a returned dietary supplement is received, it must be properly identified and quarantined until quality control personnel conduct a material review and make a disposition decision.

In some instances, the appearance of the returned goods may imply that they have been stored or shipped under unfavorable conditions that may impact their quality or safety. The condition of the product, its container, carton, or labeling may cast such doubt. As discussed in chapter 22, improper storage that might negatively impact quality include factors such as temperature and humidity, but could also be associated with the age of the product or exposure to smoke or fumes.

In other situations, the goods may appear to be fully intact and satisfactory to be returned to stock. This decision can be aided by determining *why* the customer decided to return the product(s).

In either event, under provisions of §111.113(a)(5) and §111.130, it is the responsibility of quality control operations to conduct a material review and make a disposition decision. This may, or may not, involve conducting tests or examinations, and may or may not require certain steps of reprocessing, at the discretion of quality control personnel. A full range of options exist, yielding flexibility, but it is clear that quality control personnel must choose between approving, or rejecting, any reprocessing that may be considered.

Once the disposition decision has been made by quality control personnel, the returned goods can be released from quarantine. Under §111.515 and §111.520, the goods may be salvaged for redistribution, or may be reprocessed. Otherwise, the product(s) must be suitably disposed of, including destruction if appropriate. Under §111.525, any dietary supplements that are reprocessed must meet *all* of the specifications established under §111.70(e), and quality control personnel must either approve or reject any returned dietary supplement that has been reprocessed, according to §111.130(d).

It is interesting to note that §111.515 does *not require* testing since the combination of written procedures and oversight by quality control personnel is considered by the FDA to be adequate to determine the appropriate disposition of returned goods without the necessity in every case to prove that the returned products meet specifications. Flexibility is given to quality control personnel as to *how* they make their decision.

POSSIBLE IMPACT ON OTHER BATCHES

If the reason for a dietary supplement being returned implicates *other* batches, it is required to conduct an investigation of the manufacturing processes and each of those other batches to determine compliance with specifications, as mandated by §111.530.

REQUIRED RECORDS OF RETURNED GOODS

Under §111.535, it is required to have SOPs for fulfilling the requirements of Subpart N, as well as records of all material reviews and disposition decisions reached by quality control operations on returned dietary supplements, and the results of tests or examinations conducted to determine compliance with product specifications established under §111.70(e). Similarly, it is required to make and keep records of the documentation of the reevaluation by quality control personnel of any dietary supplement that is reprocessed, including the determination by quality control personnel that the product meets all applicable specifications.

It is advisable to record the name and description of each returned product, the lot number, the reason for the return, the quantity returned, the date of the disposition decision, and the ultimate disposition of the returned product.

SUGGESTED READINGS

American National Standards Institute, NSF/ANSI Standard 173, Dietary Supplements, Section 8.7.4, Returned Products. 2010.
Food and Drug Administration. Current Good Manufacturing Practice, 21 CFR 211.204 Returned Drug Products.
World Health Organization, Good Distribution Practices for Pharmaceutical Products, Chapter 18, Returned Products, Geneva, 2009.

2.1 Product complaint handling

IMPORTANCE OF COMPLAINTS

Complaints are of significant importance to manufacturers of dietary supplement products, for regulatory compliance reasons, and also for consumer feedback on the information they yield that can be used as part of the program of continual improvement. It is good business to be in touch with consumers, and when appropriate, to make use of the information they provide to tailor the products in ways to make them more fully acceptable. Proper handling of complaints can be part of a firm's customer relationship management system to help sales growth. It is important to satisfy consumers, both to retain current users and keep them coming back, and to win new customers.

Care should be taken to make it easy for consumers to contact the firm, both with complaints, and with comments and suggestions that might be useful. In essence, it is good business to essentially invite consumers to contact the firm, making it easy for them to do so by the wording on packaging citing toll-free phone numbers and Internet addresses. Customers contacting the firm by telephone should be able to reach a "real live person," not merely a recording.

The individuals within the firm who receive messages from consumers by telephone, e-mail, or in writing should be well trained, courteous, pleasant, and fully responsive. The proper and efficient handling of incoming comments, inquiries, and complaints is of significant importance in attracting and keeping satisfied customers.

In addition to the necessary SOPs on complaint handling, it is prudent to also have a company policy stressing the importance of the proper handling of consumer inquiries and complaints, stating how incoming messages along these lines should be routed, so that all employees are fully aware of both the importance of and the correct procedures for the handling of such contacts.

COMPLAINTS DEFINED

The Part 111 definition of a product complaint (often referred to as *consumer* complaints) is any communication that contains any allegation, written, electronic, or oral, expressing concern, for any reason, with the quality of a dietary supplement that could be related to current good manufacturing practice. Examples of product complaints are foul odor, off taste, illness or injury, disintegration time, color variation, tablet size or variation, underfilled container, foreign matter in the container, improper packaging, mislabeling, products that are superpotent or subpotent, products that contain the wrong ingredient, or contain a drug or other contaminant (e.g., bacteria, pesticide, mycotoxin, glass, or lead). Clearly all of these issues would be important alerts of quality problems requiring assessment and correction. Certainly, such complaints point to failure of the firm's quality system to detect and ensure that all prudent steps have been taken to avoid such defects from occurring. Therefore, the FDA puts considerable emphasis on this during establishment inspections, reviewing the complaint files in detail, and studying whether proper handling and action on such complaints is taken.

Product complaint issues are in addition to the requirements for reporting *adverse events*, which are covered by other regulations as described in chapter 27.

Some messages from consumers may be related to the aesthetics of the product or its package, the price, late deliveries, incorrect billing, the retail outlets carrying the brand, or any of many other topics not related to product quality. Such topics, that are unrelated to GMP compliance issues, are *not* considered to be product complaints. This is an important distinction, and must be carefully considered, and it is advisable to make it clearly defined within the firm as to what constitutes a complaint.

SOURCES OF COMPLAINTS

Apart from direct contact from consumers, complaints are sometimes registered with retailers or with health care professionals, and are then passed along to the manufacturer, either directly or through the firm's sales force or agents. Moreover, the FDA has mechanisms whereby individuals, medical personnel, pharmacists, hospitals and nursing homes, trade associations, and other interested parties may make complaints, through the MedWatch program or through the Consumer Complaint Reporting System at FDA offices throughout the United States and Puerto Rico. Such reports received directly by the FDA may trigger for-cause establishment inspections.

Very often, consumers will not bother to take the time and effort to contact the manufacturer about their dissatisfaction with a product. Instead, they will simply not buy that product again. Moreover, many consumers will air their dislikes on blogs and social networking Internet sites as well as with family members and friends. This can quickly have an extremely negative impact on sales, so it is actually advantageous to the firm to hear *directly* from their consumers, both to alert them to take corrective action if needed, and for the opportunity to transform the consumer from being unhappy to being a loyal and supportive customer through being treated properly in the complaint-handling process. Therefore, some manufacturers actually look on complaints as being a welcome "gift," alerting them to take corrective action in place of losing sales.

Both manufacturers and distributors are subject to the requirements related to the review and investigation of a product complaint that they receive, as noted in FDA's response to Comment 28 in the preamble to the final regulations.

ACTIONS IN HANDLING COMPLAINTS

The regulations, at §111.560(a), require that a *qualified person* must review *all* complaints that involve a possible failure to meet any specifications or other requirements of the GMPs, or any other requirements that, if not met, may result in a risk of illness or injury, and then quality control personnel must review and approve decisions about whether to investigate a product complaint, and also approve the findings and follow-up action of any investigation performed, as stated in §111.560(b). The requirement for quality control operations to review and approve decisions about whether to investigate a product complaint, and reviewing and approving the findings and follow-up action of any investigation performed, is also contained in §111.135 to emphasize that the investigation of a product complaint has the potential to uncover problems with the production and process control system and therefore quality control personnel must exercise appropriate oversight of investigations of product complaints.

Although quality control personnel are required to perform the *oversight* function for the review and evaluation of product complaints, they are expressly *not* required to do the actual investigations.

The review by a qualified person and quality control personnel about whether to investigate a product complaint (including the findings and follow-up action on investigations performed) must extend to *all* relevant batches and records.

The term "qualified person" is not defined in the regulations, but it implies someone with the appropriate education, training, and experience to perform the necessary tasks properly. This person is responsible for deciding whether a complaint might involve failure to fully meet specifications or other GMP requirements that, if not met, may result in a risk of illness or injury, *before* involvement of quality control personnel. However, it is then the task of quality control personnel to review and approve (or disapprove) the decisions made by the qualified person. In other words, quality control personnel do not necessarily conduct the investigation.

This procedure, including the findings, and any follow-up action triggered by any investigation performed must also extend to all relevant batches and records.

An appropriate SOP is required, outlining specifically how these requirements are to be handled as stated in §111.553 and §111.570(b)(1).

RECORDS RELATED TO COMPLAINTS

There must be a written record or log kept on every complaint that is related to GMP compliance, which must have the name and description of the dietary supplement, its batch, lot, or control number if available, the date the complaint was received, the name, address, or phone number of the complainant (i.e., the person making the complaint), if available, the nature of the complaint (including how the product was used, if known), the reply to the complainant (if any), and if an investigation is performed, the findings of the investigation together with any follow-up action taken. This is outlined in §111.570(b)(2).

The maintenance of records of complaints and investigations should be assigned to specifically designated persons or departments, as it is important that such records be up-to-date and readily accessible.

KEEPING MANAGEMENT INFORMED

Since complaints reflect the state of control being exercised, it is important that key management personnel be kept fully informed about them at frequent periodic intervals. It is prudent for management to take an active interest in assuring that complaint handling procedures are properly handled. Commitment to effective complaint handling is important both for business reasons and for full GMP compliance.

As mentioned in chapter 12, plotting *trends* related to complaints and deviations can be a useful tool. Computer software programs are commercially available to facilitate both the record-keeping and the trending associated with complaint handling.

SUGGESTED READINGS

Barlow J, Moller C. A Complaint Is a Gift: Recovering Customer Loyalty When Things Go Wrong. San Francisco, CA: Berrett-Koehler Publishers, 2008.

Braga GK. Complaint handling in pharmaceutical companies. Qual Assur J 2007; 11(1):16–21.

Goodman J. Manage complaints to enhance loyalty. Qual Prog 2006; 39(2):28–34.

Stauss B, Seidel W. Complaint Management. Mason, OH: Thompson South-Western Publishing, 2005.

22 Holding and distributing

HOLDING

In GMP parlance, the term "holding" means storage or warehousing, whether short term or long term.

§111.455(a) requires components and dietary supplements to be held under *appropriate* conditions of temperature, humidity, and light to ensure that their identity, purity, strength, and composition are not adversely affected. However, the "appropriate conditions" are not defined, leaving this to the discretion of the firm.

The *temperature* at which items should be held is generally in agreement with any storage conditions mentioned on the products' labels, if any such mention is made. A few products must be refrigerated, although most of the others should be stored at "controlled room temperature" (CRT) generally defined as 68°F to 75°F (20°C–25°C), with brief excursions or deviations from these limits being acceptable. The USP definition of CRT is more complex, being based on what is called the "mean kinetic temperature," although there is no requirement to follow the USP. It is prudent to monitor warehouse temperatures either manually or with automatic recording devices.

It is usual to monitor temperatures in several locations within the warehouse since variations may occur. The use of more than one recording instruments also provides redundancy in the event one fails.

Similarly, regarding *humidity*, it is advisable to establish an acceptable range, and then periodically check the actual relative humidity (RH) in the warehouse at predetermined intervals. RH is a comparison between the actual amount of moisture in the air versus the maximum possible at any given temperature since warm air can "hold" more water vapor than can colder air. RH can be manually determined by comparing the temperatures read on a "dry-bulb" thermometer and a "wet-bulb" thermometer, and then using a psychometric table or chart, which can be found in various textbooks and on the Internet. A dry-bulb thermometer is usually simply a suitable glass-stem thermometer, whereas the wet-bulb type is the same except that the bulb is surrounded by a wick of cotton (or equivalent) that is saturated with water. It is necessary to take such readings with a suitable amount of airflow passing the two thermometers, and this is typically achieved by mounting the two thermometers on a holder with a handle that allows manually whirling them through the air. This type of instrument is called a *sling psychrometer*. However, fortunately, many types of automatic hygrometers are also commercially available that greatly facilitate measuring and recording RH.

Provision should be made to periodically review the records of warehouse RH and take corrective action if necessary. For most items, the packaging provides adequate moisture protection, although as mentioned in chapter 28, undesirable mold growth can occur when botanicals are stored under high humidity conditions, which may lead to contamination from mycotoxins or aflatoxins.

11.455(a) also requires consideration of *light* impacting dietary supplements and/or components. Although a few items may be photosensitive and subject to degradation if overexposed to light, in most instances the packaging provides adequate protection from light in storage. However, since this is a GMP issue, it is worthy of due consideration.

Similarly, §111.455(b) requires that packaging materials and labels must be held under appropriate conditions so that they will not be adversely affected. The same factors as discussed above must be given consideration regarding warehousing and storage of such items.

§111.455(c) states that dietary supplements, packaging, and labels must be held under conditions that do not lead to *mix-up, contamination,* or *deterioration.* This implies the need for adequate space for proper storage, including the use of racks or cabinets when appropriate. It further implies the need for a manual or computer-based system for keeping track of the location of the items being stored. Certain packaging items, as well as labels, should be kept in secure areas, such as locked cages or cabinets, to which only specifically authorized personnel have access.

Rejected components, packaging materials, labels, in-process lots, and finished goods must be held under quarantine, preferably in a locked room or caged area with a locked gate, or on a top rack accessible only by a forklift truck, while awaiting a disposition decision. However, with proper electronic status identification, spatial segregation and physical barriers may not be required for temporary storage of rejected items.

According to §111.460, *in-process* materials must be held under conditions that protect against mix-up, contamination, and deterioration, including appropriate conditions of temperature, humidity, and light.

As mentioned in chapter 11, according to §111.465 (and §111.83), *reserve samples* of dietary supplements must be held in a manner that protects them against contamination and deterioration, under conditions consistent with product labels, or if no storage conditions are recommended on the label, under ordinary storage conditions. Such samples should be stored in the *same* container-closure system in which the packaged and labeled dietary supplement is distributed. The regulations do *not* require holding samples of *components* since components can be traced back to their manufacturer or supplier in the event an in-depth investigation should ever be triggered involving specific components.

Finished goods held in the warehouse should be physically counted at periodic intervals, and the counts should be compared to what the records show as being the quantity of each item that should be on hand. Any significant deviations should be investigated, and such investigations documented.

RECEIVING

One of the warehouse functions is the receiving of incoming goods, as discussed in chapter 13. Receiving records are of significant importance in that they provide information on the traceability history of components, containers/closures, labels, and other packaging materials. Such records should include as a minimum the specific identification of the item and the vendor, the date of receipt, the name of the carrier, the firm's purchase order number (if such was used), the manufacture's lot or batch number, the quantity received (verified, if appropriate), and the control number assigned by the firm.

Quality control personnel should be promptly informed of receipts that need to be sampled and disposition decided upon. Certain items must be quarantined until released, and provision should be made to apply stickers or other means of identifying the status, for example whether sampled, released, or rejected.

SOPs covering the details of handling incoming goods should exist under §111.453.

WAREHOUSE SANITATION

The warehouse is often an area at the back of the facility, with docks to facilitate loading and unloading goods. This area is usually not often visited by most of the employees, and in fact it is advisable to restrict access to storage areas to those working there and to other authorized personnel. However, the warehouse does need to be kept clean and sanitary. Such areas should be frequently inspected for cleanliness and good housekeeping.

Many of the dietary supplement finished products and components are inherently attractive to insects and rodents. This requires careful pest control. Ultraviolet lights are useful for identification of traces of rodent urine. Utilizing the services of a certified pest control service may prove useful in enhancing relative freedom from pests in warehouse areas, albeit such services should be clearly instructed regarding GMP issues, and should be closely monitored to avoid the possibility of unintended contamination.

Care should be taken to avoid entry of birds into warehouse areas since dock doors need to be opened frequently. Eliminating nesting and roosting nearby is one useful step that should be taken care of.

Loading and unloading docks should be kept free from accumulation of spillage and debris. Docks should preferably be equipped with dock seals so that trucks backing into the docks engage the seals to prevent pest entry. It is preferable to provide roofs or overhead canopies over dock doors.

Insect electrocutor light traps (sometimes called "bug zappers") placed near doors may help reduce the number of flying insects entering the warehouse.

Spillages should be cleaned as soon as possible to minimize the possibility of cross-contamination.

PALLETIZING

In warehousing and transporting goods, it is common to use *pallets* (sometimes called "skids") that are flat structures specifically designed for such use. Openings in the pallets facilitate moving them (and if desired, lifting them to racks) through the use of forklift trucks or pallet jacks.

Wooden pallets can harbor insects, spores, bacteria, and other forms of contamination, and can become moldy if exposed to moisture. They tend to be difficult to keep clean, and are subject to splintering and occasionally may have exposed nails that can damage shipping cases or harm employees. A recall was once triggered by wooden pallets having been treated with Tribromophenol (TBP) that degraded to Tribromoanisole (TBA), which has a disagreeable musty mildew-type odor detectable even at parts per trillion. These issues have triggered an intensive ongoing debate as to the relative advantages and disadvantages of using wooden versus plastic pallets.

The issues include costs, weights, environmental impact, fire protection matters, and other such considerations. Regarding costs, plastic pallets can either be owned or rented. Plastic pallets can also be tagged for radio-frequency identification (RFID) tracking.

There are no universal standards for pallet dimensions, and these vary from country to country, complicating international shipping and receiving. The International Standards Organization (ISO) lists six different sets of pallet dimensions, and the National Wood Pallet and Container Association (NWPCA) has published their own uniform standards for wood pallets. However, within the dietary supplement industry, the most commonly used pallets are those conforming to the Grocery Manufacturers Association (GMA) standards, which include 48 × 40 in. overall dimensions.

Any stacking done should give due care to safety issues. Pallets of goods should be put into prescribed orderly places, spaced to allow proper ventilation. It is advisable to store goods on clean pallets, at least 18 in. from walls to facilitate cleaning and inspection. Storage off the floor on pallets also allows for better cleaning.

DISTRIBUTION

The term *distribution* refers to the steps taken to get the finished products from the manufacturer to the ultimate consumer. The steps involved differ depending on circumstances, but often the products go first to a wholesaler who in turn fills orders from and ships to retailers, while the ultimate users purchase the product from a retailer. Large retail chains often elect to bypass the wholesales and order directly from the manufacturer.

Regardless of the steps involved, however, distribution is an important activity, in part because the manufacturing firm's warehouse is the last point at which the goods are still under the firm's control. Moreover, under §111.470, it is required to distribute dietary supplements under conditions that will protect them from contamination and deterioration.

Wholesalers who hold dietary supplements are *also* subject to compliance with §111.470, as well as with other applicable portions of the GMPs such as those related to their physical plant and grounds, personnel, equipment, returned goods, customer complaints, etc. The regulations *per se* do not list all such requirements since individual operations vary.

It is advisable to distribute the oldest approved goods first, although deviations from this are acceptable when appropriate and temporary.

Transportation of finished products should be done under conditions that continue to protect the products from physical, chemical, or microbial contamination, as well as from deterioration of either the products or the shipping containers. Product identification should be maintained, and precautions taken to avoid spillage, breakage, theft, or tampering.

Although the term is not used in the GMPs, the guidelines for proper distribution procedures are often referred to as *good distribution practices* (GDP).

TRACEABILITY

Traceability is the ability to retrieve the history of the manufacturing and distribution of products. The data recorded in a traceability system enables efficient investigation if an incident occurs that requires doing so. This also enhances the ability to conduct a recall if necessary.

In the event of discovering after the fact that a specific lot of a component was defective or contaminated, it is important to be able to quickly determine which batches of finished product were involved and to whom (and when) these lots were shipped. Computerizing batch records clearly is superior to the difficult and tedious task of manually searching through records.

The use of RFID technology, mentioned above, can be quite useful in tracking and tracing.

§111.475(b) requires making and keeping procedures (SOPs) and records for holding and distributing operations, and specifically records of product distribution for tracing.

SUGGESTED READINGS

Bailey LC. Mean kinetic temperature: a concept for storage of pharmaceuticals. Pharm Forum 1993; 19(5):6163–6166.
Brittingham M. The case for product protection at the dock. Food Quality, December–January 2010:32–34.
Cutlier TR. A traceability reality check. Food Quality, February–March 2009:46–47.
Deakins J. The truth about lot tracking. Contract Pharma, March 2009:80–82.
Food and Drug Administration. Current Good Manufacturing Practice for Finished Pharmaceuticals, 21 CFR 211.150, Subpart H, Holding and Distribution.
Forcinio H. Particle-free pallets. Pharmaceutical Technology, September 2004:32–38.
Gay M. Track, trace technology drives business improvements. Food Quality, August–September 2010:43–45.
Siddiqi Z. Past and present: over time, pest control has undergone a sea change. Food Quality, August–September 2010:28–32.
United State Pharmacopeia, <1079> Good Storage and Shipping Practices, 2009.
World Health Organization. Guide to Good Storage Practices, WHO Technical Report Series, No. 908, Annex 9, Geneva, 2003.
World Health Organization. Good Distribution Practices for Pharmaceutical Products, Geneva, 2006.

23 Handling recalls

DEFINITION OF A RECALL
Occasionally, consumer products (including dietary supplements) that are in some way defective or violate the laws administered by the FDA do get onto the market. The best and most expedient way of getting such items *off* the market again and protect the consuming public is a recall. Although the GMPs are silent on this topic (other than for a mention in the preamble) and therefore nothing is specifically mandated, it is prudent for each firm to be prepared for this event, preferably with a written procedure in place and appropriate training accomplished so that if a recall becomes necessary, it will go smoothly.

Preparedness is far better than acting extemporaneously, particularly under pressure in a crisis. It is important to reduce delays to the extent possible, which is another reason to have a recall policy in place.

Most recalls are undertaken voluntarily by manufacturers or distributors, but in urgent situations they may be requested by the FDA. However, if a firm refuses to cooperate, the FDA has various regulatory procedures that can be used, including having the U.S. Marshals physically seize the goods and remove them from retail shelves and from stocks in warehouses and distribution centers.

Recalls are necessarily costly, embarrassing, and damaging to the reputation of the firm involved, in that both the retail trade and consumers are typically aware of the situation both from the media and from the Internet. There are more reasons why compliance with the GMPs is essential, although even that does not always insulate firms from the need for a recall.

Some firms offset at least part of the enormous costs involved through recall insurance, which is now available through a few carriers.

Recalls usually result in FDA inspections to ensure the underlying causes have been corrected. Moreover, recalls may trigger product liability litigation.

MARKET WITHDRAWALS
In addition to recalls, a product that is found to have a minor defect, not subject to FDA regulatory action, may be removed or corrected, and such actions are termed *market withdrawals* or *stock recoveries*, neither of which needs FDA involvement. A market withdrawal implies a marketed product already in distribution channels, whereas a stock recovery involves products not yet in the marketplace and that have not yet left the direct control of the company. Only situations that would trigger the FDA legal action are considered to be recalls. Quiet market withdrawals are of course preferable to recalls, when feasible.

REASONS FOR RECALLS
The most usual cause of such problems is human error. This, too, can be minimized through good training and supervision, having and following

well-written specifications and procedures, providing good working conditions, and motivation through good leadership skills. **Quality** control personnel also play a vital role in reducing the need for recalls, both through ensuring that procedures are adequate and by careful review of production records prior to releasing goods for shipment. When errors and out-of-specification results are detected, they must be thoroughly investigated and corrective action taken.

All personnel must be constantly alert for their own and others' careless errors, and point out to supervisory management where improvements are needed. The cooperation of everyone in the workforce is needed to eliminate possible causes of errors. This requires having a company-wide culture that emphasizes product quality and constant faithful compliance with both the letter and the spirit of the GMP regulations, as discussed in chapter 26.

Periodic internal audits are also helpful in reducing the source of potential errors. Operators tend to become very familiar with their tasks and not infrequently will take shortcuts to save time or effort, believing their techniques are equal to or better than those in the existing written procedures. Therefore, it is important to see to it that SOPs are diligently followed, or if better ways of conducting any given task are found, to change the appropriate SOPs and train the workers involved in the revised versions. This is one facet of the desired continual improvement process, but it involves having a well-planned and executed change control procedure in place. However, personnel at all levels should be encouraged to make suggestions for improvements if they perceive better ways of conducting specific tasks and functions, but any such improvements must be carefully evaluated and documented prior to implementation.

Preventing contamination that can lead to adulteration is another key to helping prevent the need for recalls.

RECALL CLASSIFICATION

The FDA classifies recalls by numerical designations, depending on the relative degree of health hazard presented by the product being recalled. Class I is for a situation in which there is a reasonable probability that the use of, or exposure to, the violative product could cause serious adverse health consequences or death. Class II is a situation where the use of, or exposure to, the product may cause temporary or medically reversible adverse health consequences or where the probability of serious health consequences is remote. Class III is the least serious, where the product is not likely to cause adverse health consequences.

As soon as the FDA becomes aware of the need of a recall, a board of experts is convened to evaluate the health hazard involved, and the classification number assigned.

INFORMING FDA

When the firm concludes that a recall is necessary, it is prudent to initiate a voluntary recall and to immediately inform the applicable FDA district office of the circumstances.

Starting the recall process promptly can help reduce the costs by catching goods still in the distribution chain, before they have been purchased by the consuming public. This also helps reduce the potential product liability exposure.

RECALL STRATEGY

The FDA defines a recall strategy in §7.42 as a program tailored to suit the individual circumstances, but including the results of a health hazard evaluation, the ease in identifying the defective product (which is why individual packages and shipping cases need to clearly and legibly show the batch, lot, or control number), the degree to which the product's deficiency is obvious to the consumer or user, the amount of product remaining in the marketplace (which is one of the reasons why §111.475(b)(2) requires maintaining records on distribution), the *depth* of the recall (i.e., the level in the distribution chain to which the recall is to extend), whether and how public warnings are needed, and if needed, how they will be handled, how consignees of the goods will be contacted, how effectiveness checks will be conducted, and what disposition will be made of the returned goods.

It is usually best for the firm to develop their own proposed recall strategy, in writing, and submit it to the FDA for review and then make any needed modifications requested by the FDA. In fact, it is prudent to have a *basic* recall strategy in place at all times, which can then be quickly updated with specific details when needed. This basic document can also be a useful training tool.

Since time is of the essence, the firm should start implementing their strategy even before the FDA responds, making any necessary changes after being requested to do so by the agency.

The firm's recall strategy should identify the specific personnel responsible for recall decisions and how they can be contacted. The individuals authorized to communicate with the media should also be specified.

Care must be taken to properly identify what lot or lots are affected. For example, if the problem originated with a defective component, more than one lot or batch may be impacted. This may require checking to determine the lots in which the defective component was used. Unless this information can be readily accessed from computerized records, it may require tedious examination of paper-based records.

Communications from the firm to the trade regarding the recall must be reviewed by the FDA before issuance. The FDA is willing to work with firms on preparation of the most effective recall communications. Similarly, the FDA will advise the firm how returned goods should be handled. It must witness or otherwise verify the destruction or reconditioning of goods returned under a recall.

The firm will be asked to keep the FDA well informed as to the status of the recall and the quantity of recalled product returned or accounted for.

Once the FDA is convinced that all reasonable efforts have been made to follow the agreed-upon recall strategy, the firm will be notified that the recall has been terminated.

SUGGESTED READINGS

American Society for Quality. The Product Recall Planning Guide. 2nd ed. Milwaukee, WI: ASQ Quality Press, 1999.
FDA. Part 7, Enforcement Policy, 21 CFR. Subpart C, Recalls.
FDA. Product Recalls, Including Removals and Corrections—Industry Guidance, 2003.
FDA. Investigations Operations Manual, Chapter 7 Recall Activities, 2009.
Meadows M. The FDA and product recalls. FDA Consumer, May–June 2006.

24 Top management responsibility

BACKGROUND OF REASONS FOR MANAGEMENT'S ROLE
Although the dietary supplement GMPs are silent on the topic, it is well established that the FDA expects the top officers of a firm to take the initiative and responsibility for assuring full compliance with both the letter and the spirit of the regulations. It is both the expressed attitude toward earnestly striving for consistent excellence in quality and the company's stated policy toward full voluntary compliance that others within the organization need to understand as being what is required of them, and management should make these points clear.

One of the reasons for management's interest in the firm's quality system is that consumers expect and deserve to feel assured that the products they purchase contain the ingredients and perform in the manner claimed in the labeling. Since repeat sales depend on customers' and retailers' continued satisfaction with specific products and since competition is ever-present, from a strictly pragmatic business point of view, it is essential to ensure consistent quality. The consuming public is too sophisticated to spend money on products that have obvious defects or that fail to perform as expected. Long-term business is built through repeated sales, which simply do not materialize if merchandise is substandard. **Quality** is a key to business success, and for this reason (among others), it is important that top management should be genuinely and obviously interested and involved in all aspects of full GMP compliance.

When defective goods are introduced into interstate commerce, the possibility of a recall exists. As discussed in chapter 23, recalls are extremely costly and damaging to a firm's reputation. The resulting adverse publicity from a recall, although sometimes unfair, can impact future sales. The best way of avoiding recalls is careful and deliberate attention to manufacturing and quality control details as presented in the GMPs.

Defective products not infrequently result in product liability issues. Liability cases also are costly, and the cost of liability insurance tends to escalate when risks are high. The most effective way of avoiding such costs is strict compliance with the GMPs, which have been carefully crafted to help ensure consistent quality.

For these reasons, it is the responsibility of top management to ensure that proper planning and direction exist for quality control and for full GMP compliance. Whether the firm is large or small, it is up to management to check that the resources and infrastructure exist and that everyone in the organization understands management's commitment for doing things right each and every day.

Thus, top management must be actively involved, should demonstrate commitment, and should communicate this interest and concern regarding complete GMP compliance, both for regulatory reasons and for good business reasons. That is the ethical approach, in keeping with corporate social responsibility (CSR), even if not specifically mandated.

TOP MANAGEMENT DEFINED

The term "top management" comes from everyday language, not from the regulations. The term "senior management" is often used as a synonym. These terms refer to the persons who direct and control the organization, regardless of titles, although typical titles include chief executive officer (CEO), chairman, president, vice presidents, and others. These are the persons who implement the owners' business objectives, whether or not they are themselves part of the ownership. In a large multifacility company, the top management in any given site are the ones in charge there, whether or not they are in the top management of the parent company.

Subordinate managers and supervisors look to top management for direction and leadership. Thus, those in the roles of top management are the ones who set the company theme on quality and GMP compliance, and it is they who have the power and the duty to prevent violations. However, every employee, regardless of rank, also has responsibility for doing things correctly.

There is always top management in every firm, regardless of the size of the company. Even in very small firms (including "virtual" organizations), there is someone in overall charge who must ensure full GMP compliance.

Although not mandated, many firms publish mission statements and a corporate policy statement to inform employees, the trade, and consumers of their intent to ensure that all of their products are made to consistent high-quality standards and in full voluntary compliance with appropriate laws and regulations.

TOP MANAGEMENT'S LEGAL RESPONSIBILITY FOR GMP COMPLIANCE

The FD&C Act is a *strict liability* statute. This means that if a violation occurs, the FDA can request the Department of Justice to take action against the "persons" responsible for the violation(s) whether or not that "person" *intended*, or in some instances even *knew* about the violation. Under section 201(e) of the Act, a person can either be an individual, or a company. Therefore, not only can firms be subject to severe penalties, but individuals holding responsible positions within the company can also be subject to fines out of their own pockets and/or to jail terms.

This strict liability provision makes personnel, including but not limited to top management, vicariously liable for violations committed by others. This is different from most laws, which usually require intent or recklessness or negligence to incur punishment. However, the fact that the FD&C Act *is* a strict liability statute has been confirmed by two Supreme Court cases, *U.S. vs Dotterweich* and *U.S. vs Park*.

Warning Letters are typically addressed to a high-ranking officer of a firm, such as the president or CEO. Consent decrees and criminal prosecutions often also name specific individuals in addition to the firm. This is one reason why, during FDA establishment inspections, the investigator asks questions about the organizational structure of the company as a whole, and at that facility, so that the FDA would know against whom to bring charges in any regulatory action that might ensue from the inspection. It is stated in the FDA's *Investigations Operations Manual* (IOM) that it is the inspector's task to establish who had the duty and power to *detect, to prevent,* and *to correct* violations. Supervisory personnel at all levels should understand this.

AVAILABILITY OF RESOURCES

Among top management's duties is to ensure the availability of resources, since it is they who set budgets and priorities. In this instance, "resources" include the physical plant and grounds, the equipment and instruments, the building support systems such as HVAC and lighting, sufficient number of qualified personnel and supervisors to properly handle the GMP-required activities, ongoing training, proper maintenance, appropriate laboratory facilities (either on site, or contracted), and an adequate production and process control system.

ADDITIONAL MANAGEMENT DUTIES

It is also the responsibility of top management to see to it that a quality policy is established and clearly communicated to all employees as well as to ensure that processes, programs, and data are properly handled. It is further advisable that top management provide continuous feedback to employees on ideas and suggestions they may offer toward improvements in performance, cost reduction, and quality, even if these are not accepted and used.

Although not required by the regulations, it is advisable for top management to establish an organization chart and to ensure having the specific duties of key personnel set out in written job descriptions. Where appropriate, these job descriptions should also provide individuals with the necessary authority to carry out their tasks. Each person with specific GMP responsibilities should have an unambiguous understanding of what is expected of him or her. Similarly, it should be made clear to all concerned as to who does have the responsibility and authority to initiate action to prevent the occurrence of nonconformities or deviations that impact GMP compliance, for example, who has the authority to stop production if it becomes necessary to make corrections.

To fulfill these responsibilities, top management must understand the GMPs and must be kept informed whenever significant issues or problems arise. They must be certain that appropriate corrective actions are taken when needed. This is one reason for conducting internal periodic compliance audits, as discussed in chapter 30.

Some or all of the manufacturing and laboratory operations may be outsourced, as described in chapter 31. When this is done, it is advisable for top management to enter into a written contract with the firm supplying these services to establish lines of communication and to ensure GMP compliance on the part of the supplier. Management should conduct adequate *due diligence* to ensure that the supplier is capable of performing properly. This is another of top management's responsibilities.

SUMMARY

Although not mentioned per se in the regulations, it is clear that the FDA expects each firm's management to establish policy for and commitment to full GMP compliance, to establish the organizational structure to facilitate these goals and objectives, and to provide adequate resources including appropriate facilities and equipment and trained personnel. Moreover, management is expected to keep abreast of all significant issues related to quality and compliance with all applicable laws and regulations and to exert leadership in encouraging all employees to participate in ensuring that these objectives are met.

SUGGESTED READINGS

Food and Drug Administration. Section 5.3.6 Responsible Individuals, in Chapter 5, Investigations Operations Manual, 2011.

International Conference on Harmonization. Management Responsibility, in ICH Q10, Quality Systems, 2008.

International Organization for Standardization. Responsibilities of Top Management, Section 5 of ISO 9001.

Kowal SM. Personal responsibility and personal jeopardy: FDA's prosecution of individuals. Food and Drug Law Institute Update, May–June 2002:50–51.

Messplay GC, Heisey C. Management responsibilities. In FDA's Quality System Approach. Contract Pharma, March, 2007: 18–20.

United States v. Dotterweich, 320 U.S. 277, 1943.

United States v. Park, 421 U.S. 658, 1975.

Zaret EH. Management responsibility in modern quality systems. Pharmaceutical Formulation & Quality, December–January, 2005: 44–45.

25   Record keeping, documentation, and SOPs

In all business activities, including manufacturing and quality control, good *communication* between and among individuals, departments, and outside contacts is essential. Oral or spoken communication is of course the most commonly used method of sharing information, but this alone is often less than fully satisfactory, in part because the receiver(s) of the information must correctly interpret, understand, and at least for a period of time, remember what has been heard. This has been thoroughly studied in the field of communications theory. But many barriers exist in the interpretation of spoken messages, such as language problems, poor hearing and/or listening skills, distractions, lack of interest or involvement, emotional issues, and others.

Many of us remember from our childhood the game where a message is whispered to one person in a group who then whispers it to another person, and so on, throughout the rest of the group until the final person announces what he or she heard. The end result is usually a major departure from the initial message, often highly amusingly so. This clearly illustrates the hazards involved in using solely *spoken* messages to communicate information.

To avoid such problems, the use of *written* textual means of transmission of information is used, either alone or in addition to conveying the information in spoken form.

The word "document" can either be a noun or a verb. As a noun it refers to a written or printed paper (or now in the computer age, the equivalent in digital form) that conveys and records information. As a verb, "to document" means to prepare a document. Therefore, "documentation" implies the preparation of documents, including their use, retention, management, storage, and control.

Documentation is extremely important in manufacturing and quality control, as well as in many other facets of business, both to ensure correct and precise understanding of the information contained, and when needed, to create a traceable historical record of what was done, how it was done, when, and by whom.

In other words, documentation helps ensure that the GMPs are consistently followed and that records are retained to provide an effective trail that can be followed if circumstances require. This concept may seem to some to be a burdensome requirement, yet it is essential to ensure that each of the many steps is done, and is done in the correct way and sequence each and every time, to assure adherence to the firm's established procedural requirements. Most quality systems are based on documentation, including the GMPs for essentially every regulated health care products industry in all countries, the World Health Organization, and the systems established by the International Organization for Standardization (ISO).

The term documentation is frequently used in the preamble, but it seldom appears per se in the actual regulations. The word is used only occasionally, for example in §111.35(b), but instead other applicable terms, such as "written procedures" and "records," appear more frequently in the regulations. It is generally understood that these are indeed parts of documentation. The term further encompasses written specifications, formulae, test methods, laboratory records and analytical notebooks, log books, purchase orders, and a myriad of other such instances where written procedures and methods and/or records are necessary.

As emphasized in chapter 1, quality cannot be "tested into a product," it has to be *built-in,* and consequently the FDA places great stress on the importance of good documentation in doing so, which is an essential part of any quality system.

WRITTEN PROCEDURES

To ensure that every significant task called for in the dietary supplement GMPs gets properly carried out in a consistent manner, by whoever conducts each operation or *when* the operation is conducted, and at the required frequency, it is important to have clear, concise, and detailed written instructions, commonly called standard operating procedures (SOPs). These are useful not only for the persons conducting the tasks, but also for training, supervisory, and auditing purposes. SOPs are essentially the firm's "internal regulations" as to how things are to be done.

Although some specific SOPs are mandated, most firms have many more than those. It is not unusual for any given firm to have a hundred or more SOPs, albeit it is not necessary to have written procedures for every minute detail, leaving it to the judgment of the firm as to which are appropriate. Laboratory operations and quality control operations alone may require many separate SOPs. Each firm needs to determine what SOPs they consider necessary, and then to analyze their existing SOPs and cover any gaps through writing additional procedures if they see the need to do so. Correctly following well-written procedures helps ensure uniformity and quality at every step in the process. This also helps reduce the requirements for testing the finished products.

The SOPs specifically required in the regulations are the following:

- §111.8   Personnel
- §111.16   Cleaning the physical plant, and pest control
- §111.25   Equipment and utensils
- §111.103   Quality control operations
- §111.153   Components, packaging, and labels
- §111.303   Laboratory operations
- §111.353   Manufacturing operations
- §111.403   Packaging and labeling operations
- §111.453   Holding and distributing operations
- §111.503   Returned dietary supplements
- §111.553   Product complaints

The preamble to Part 111 correctly states that written procedures are necessary for the definition, operation, and documentation of a process control system, and without such procedures, it would be virtually impossible for any company, regardless of size, to consistently manufacture products meeting requirements for identity, purity, quality, strength, and composition. Written procedures contain the necessary instructions for employees to successfully execute their respective functions.

It is very important to note that in addition to *having* SOPs to fulfill the GMP requirements, it is necessary that these SOPs must always be properly *followed*. Failure to conduct operations as called for in the appropriate SOPs is a significant GMP violation, which may result in the product being considered to be legally adulterated even if there is nothing actually wrong with the product.

The concept is to *say* what you do and how you do it, and then actually *do* what you say. This is a topic looked for by the FDA during facility inspections, and failure to properly follow established SOPs is one of the most frequently cited GMP violations.

If an SOP needs to be changed, due to other changes that impact it, or because a better way has been found to conduct a step in an operation, it is advisable to have a formal change control procedure in place. SOPs must not be ignored, nor informally revised. Review and revision procedures should be clearly defined, but not made so burdensome as to preclude or delay conveniently making changes when such are needed.

Many firms handle this by providing a change request form that outlines the suggested change, citing the reasons why such a change is needed. This form then gets reviewed by the appropriate individuals or departments, to either confirm their agreement with the proposed revision or to suggest a modification to it as well as to evaluate potential impacts and related consequences in terms of product quality and GMP compliance.

There is no required format or layout for SOPs. This is left to the discretion of each firm. It is usually advisable to adopt a standard template for all of the firm's SOPs. It is usual (but not required) to start with a title that reflects the activities to be performed, an identification or control number, an effective date, the number of pages (e.g., page 1 of 8), followed by the stepwise procedure in detail. If any particular tools or supplies are needed, they are usually listed. If any records will be required, they are usually also detailed. References to other related documents may be included. Adding a glossary or definition of terminology used may be appropriate. The approval date and the appropriate approval signatures should be included.

Each SOP should clearly and succinctly spell out how each specific task is to be performed.

If many of the employees of the firm regularly speak a language other than English, it may be advisable to make translations of certain SOPs to ensure their comprehension by their users.

Generic SOPs exist, usually for purchase, but in practice, it often tends to be impractical to attempt to use these, except for possible suggestions they may contain. Firms' operations frequently differ significantly from those of others, often involving different equipment and methods, and therefore the generic SOPs may require such extensive changes as to preclude their usefulness.

Many firms keep the current version of each SOP in a central file, with copies distributed as appropriate, either in hardcopy or by computer. However, it is very important to devise a system to ensure that *only* the current approved version is in use, with obsolete versions destroyed (except for a master file copy retained for historical purposes).

SOPs should answer the traditional who, what, when, why, and how questions. It is usually preferable to have the initial draft of each SOP prepared by a person who has been involved in the task or procedure, or at least by someone quite familiar with it. These individuals may or may not be capable of clearly expressing their thoughts in writing, but if not they may need aid in preparing the draft. The draft is usually prepared using a word processing computer program. Such drafts should include each step, in sequential order, including any preparatory work that must be done prior to starting. Details of any records that must be prepared should be included. The initial draft should then be circulated for review by others, and changed or corrected if necessary. The final version of the SOP can then be prepared for formal approval and signatures by the designated persons.

SOPs should include notation of any safety precautions needed in the performance of the task.

In some circumstances, it may be advisable to include a *checklist* to be completed as the task is done, to ensure that no steps are overlooked. This is somewhat similar to the way airline pilots function to provide assurance that their aircraft is safe and properly configured for the particular flight.

The regulations require that quality control personnel approve *all* written procedures, including any changes to or revisions of SOPs.

RECORD KEEPING

The GMPs mandate that certain records and documentation are required, and failure to have a required record is a GMP violation that may lead to the product being considered adulterated. Moreover, all records required by Part 111 (or copies of them) must be readily available for inspection and copying by FDA when requested, according to §111.610(a).

A familiar phrase often used by FDA is "if it isn't written down, from our point of view, it didn't happen."

Deficiencies in records are one of the most frequently cited GMP violations, and in establishment inspections, it is usual for at least half of the time is spent in reviewing documentation including records.

The required records include those detailed in the following sections:

- §111.8 and §111.14(b)(1)
- §111.14(b)(2)
- §111.23(b)
- §111.23(c)
- §111.35(b)(1)
- §111.35(b)(2)
- §111.35(b)(3)
- §111.35(b)(4)
- §111.35(b)(5)
- §111.35(b)(6)
- §111.95(b)(1)
- §111.95(b)(2)
- §111.95(b)(3)
- §111.95(b)(4)
- §111.95(b)(5)
- §111.140(b)(1)
- §111.140(b)(2)
- §111.140(b)(3)

- §111.180(b)(1)
- §111.180(b)(2)
- §111.180(b)(3)
- §111.210
- §111.260
- §111.325(b)(1)
- §111.325(b)(2)
- §111.375(b)
- §111.430(b)
- §111.475(b)(1)
- §111.475(b)(2)
- §111.535(b)(1)
- §111.535(b)(2)
- §111.535(b)(3)
- §111.535(b)(4)
- §111.570(b)(1)
- §111.570(b)(2)

GOOD DOCUMENTATION PRACTICES

The term "good documentation practices" (GDP) is often used with reference to internal standards for creating documents that are clear, concise, accurate, unambiguous, legible, and traceable as to who prepared the document and when it was made. GDP is neither found in, nor specifically required by the GMPs, but is instead merely a prudent approach to follow. The FDA has not issued guidelines related to GDP.

If the firm elects to follow GDP, it is advisable that an internal procedure covering this is drafted, circulated, and approved so that each person involved with documentation understands the standardized methods that have been adopted. Such standards may include a list of allowable abbreviations or may state that abbreviations are inappropriate and all words should be entirely spelled out.

In GDP, it is usual to require any handwritten entries be made in permanent black or blue ink (never in pencil or by a flow pen). Correction fluid should never be used, but instead, if an entry needs correction, a single line should be drawn through the erroneous information (leaving the original incorrect version legible) and the correct wording entered above this, with the initials of the person who made the correction and the date the change was made.

The format for writing dates is usually specified.

Some firms insist on differentiating the capital letter "O" from the numeral zero by requiring a slash be drawn through each zero.

Data entries should be made directly in the record, not first onto a scrap of paper for later entry into the final document. The use of ditto marks for repetitive data is usually prohibited. Back-dating is also usually prohibited.

It is advisable to not leave blank spaces on forms empty, instead mark through them with a line or enter "N/A" for "not applicable" to prevent someone from later inserting data or other information in the blanks.

These are merely a few suggestions for consideration in drafting the firm's GDP standard. Many other such items may be applicable.

DOCUMENTATION MANAGEMENT AND CONTROL

As mentioned above, most firms have a wide range of documents related to GMP compliance, including not only SOPs and records, but also specifications, master manufacturing and batch production records, company policies, work instructions, log books, forms, analytical methods and related data, memos, agreements and contracts, and many others.

All of these require proper management and control. It is advisable to take the time to review and identify the various documents being generated and by whom, whether these are paper-based or computerized, or both, and then also review the existing procedures for approvals, control and distribution of documents, storage and the ability to retrieve documents, including updating documents when such is needed.

Proper documentation not only has GMP implications but is also vital to each firm's knowledge management scheme and therefore justifies careful attention.

It is advisable to have a plan to ensure the ability to retrieve important documents to maintain business continuity and GMP compliance after a disruption or a disaster.

Some important documents may exist in the form of e-mail messages and their attachments, and therefore may reside on various computers, servers, flash drives, and various databases.

Records and information management (RIM) has become a separate area of expertise, with associations such as ARMA (which was originally an acronym for the Association of Records Managers and Administrators). Some large firms now have document control departments.

Documents clearly need to be designed, prepared, reviewed, and distributed with care.

RECORDS RETENTION

According to §111.605(a), from the FDA point of view, required records must be kept one year past the shelf life date (if such is used) or for two years beyond the date of the distribution (not the date of manufacture) of the last batch of dietary supplements associated with those records. The shelf life date includes expiration dating or "best if used by" statements, if such are used.

Some firms elect to retain records for significantly longer periods since statutes of limitations on many civil actions (including product liability claims) may be longer, and certain records may prove useful in managing liability risk. However, this can be a "two-edged sword." It is prudent to establish actual records retention times with competent legal advice.

Paper-based records should be stored in an orderly way, for expedient retrieval, and under such conditions as to minimize deterioration with time. Care should also be taken to minimize the risk of damage of such documents by fire or water.

PART 11

When records are kept in electronic form, according to §111.605(c), they must comply with 21 CFR Part 11.

The full details of Part 11 are not contained in the GMPs per se and are thus beyond the scope of this book. However, since the use of electronic (as opposed to paper-based) records has significant advantages both to industry and to the FDA in that storage and retrieval of electronic information is far superior, plus digital information can be analyzed and used in ways not possible with paper records, the trend has been strongly toward the use of electronic records and forms. These are now essentially ubiquitous.

A basic problem with electronic records and electronic signatures is the verification of their accuracy and authenticity in that the data or other information contained can be deliberately or accidentally manipulated. From a regulatory point of view, it is essential to establish that this does not occur. The concept of a rule covering the handling of electronic information was initiated in the early 1990s by a consortium of pharmaceutical manufacturing industries trade associations. Many meetings with the FDA were held by industry groups in an effort to establish how paperless record systems could best be accommodated to the GMPs. The outcome was an Advanced Notice of Proposed Rulemaking (ANPR), which was published in the *Federal Register* in 1992 to solicit public views on a number of concerns that had been raised. A revised proposal was published in 1994, and the final rule was issued and became effective in 1997. This is available on the FDA's Web site. The actual Part 11 regulation consists of only three pages, but the preamble has 33 pages.

All of the industries regulated by the FDA that use paperless systems are required to be in compliance with Part 11, and the intricacies of it have been widely discussed. Voluminous literature exists regarding Part 11, and many seminars and webinars on the topic have been held.

No vendor can guaranty that software is fully compliant in that compliance is not limited simply to the technical details of the system, although software must be validated. Only specific individuals can have the authority to perform certain actions. Compliance also depends on training, SOPs, and other procedural details. The goal is to ensure the authenticity, reliability, and trustworthiness of all electronic records used, and this also involves the personnel using the systems.

Although Part 11 compliance is complex, the FDA does intend to take appropriate action to enforce its requirements during establishment inspections, and therefore this requires due attention.

SUGGESTED READINGS

DeSain C. Documentation Basics That Support Good Manufacturing Practices and Quality System Regulations. Duluth, MN: Tamarack Associates, 2001.

Dobbs JH. An overview of document management and document processing. Drug Inf J 1993; 27:417–423.

European Commission. The Rules Governing Medicinal Products in the European Union (EudraLex). Volume 4 GMP for Human and Veterinary Products, Part 1, Chapter 4, Documentation, Brussels, 2001.

FDA. Guide to Inspections of Pharmaceutical Quality Control Laboratories, Section D, Documentation, 1993.

FDA. Guidance for Industry: Part 11, Electronic Records; Electronic Signatures—Scope and Application, 2003.

FDA. 21 CFR Part 211, Current Good Manufacturing Practice for Finished Pharmaceuticals, Subpart J Records and Reports, Section 211.180.

GAMP 5: A Risk-Based Approach to Compliant GxP Computerized Systems. Tampa, FL: ISPE, 2010.

Good Documentation Practices: A Guide for the Dietary Supplement Industry.Silver Spring, MD: American Herbal Products Association, 2010.

Martin JM. Good automated manufacturing practices: the essential tool for maintaining 21 CFR part 11 compliance. Today's Chemist at Work. American Chemical Society, September 2004: 19–21.

Olson P. The challenge of data archiving. Contamination Control, Fall 2007: 27–29.

Vesper JL. Documentation Systems, Clear and Simple. Buffalo Grove, IL: Interpharm Press, 1998.

World Health Organization. A WHO Guide to Good Manufacturing Practice Requirements, Part 1, WHO/VSQ/97.01, Section 5, Documentation, Geneva, 1997.

26 Change control

In all of the product areas regulated by the FDA, adequate change control is stressed since change does necessarily occur in manufacturing operations and inadequate change control procedures can create a significant risk of unintended consequences, which may lead to noncompliance.

As stated in §111.105(a), quality control personnel must approve *all* processes, specifications, controls, and written procedures. This includes *revisions* to those already existing. Also, in §111.123(a)(1), quality control operations must review and approve all modifications to the master manufacturing records. Therefore, the need for change control is *implied* albeit not specifically mandated as related to dietary supplement manufacturing activities, and this is accomplished primarily through the assigned responsibilities of quality control personnel.

It is prudent to implement a robust change control system, including methods to maintain and track a history of changes made. This is widely recognized as a key component of any quality system, as is stressed in ISO-9001 *Quality Management Systems*, which although not specifically related to Part 111, is nevertheless an excellent guidance.

A formal Management of Change (MOC) program is clearly worthy of consideration. Such systems generally have the person who perceives the need for a change to submit a form that states the reason, rationale, and justification for the change, whether it is to be temporary or permanent, and other pertinent information. The completed form is then circulated to the appropriate parties for their consideration, and finally to designated quality control personnel for approval or rejection. If the change is approved, it is implemented with the documentation retained in a permanent file to record the history.

Although many firms handle change control using a paper-based system, software is commercially available to enhance the management of this function. Computerization of change control facilitates collecting useful data helpful in identifying trends that can lead to improvements.

When temporary changes are approved, it is prudent to set a time for reevaluating such changes to see whether they should remain in force or be eliminated. Temporary changes must sometimes be implemented quickly, in emergency situations, but such changes do require approval of quality control personnel and should be well documented.

A SOP should exist covering the details of this system, although some firms find it difficult to cover all aspects of change control in a single SOP. Instead, some have *separate* SOPs covering differing kinds of changes such as those involving documents (including specifications), another covering buildings and equipment changes, still another on process changes, another on matters relating to personnel and training, and others regarding computer systems, sampling plans, packaging materials, test procedures, etc.

Since change is inevitable and often needed, it is important that the change control system allow changes to be made easily, efficiently, and quickly, but

CHANGE CONTROL

correctly. If the system is overly complex and time consuming, there is the hazard of changes being made informally, sidestepping the controls.

If any SOPs become out-of-date and do not accurately reflect how a given task is actually performed, that is a significant noncompliance. That is why it is essential to ensure that all changes are properly made and carefully evaluated *before* they are implemented, but the goal should be to *control* not stifle change.

When changes are made, care should be taken to ensure that all documents affected by the changes are revised, and that the appropriate people are promptly informed, and if appropriate, adequate retraining instituted.

After any significant change has been made, it is advisable to monitor and evaluate the results of the change. This may, on occasion, necessitate further changes.

Change control is also related to deviations and corrective actions, as discussed in chapter 12.

Employees should be made aware of the importance of, and methodology for, change control. This should be incorporated into the training programs discussed in chapter 3.

It is important to be aware of changes made by component and packaging material suppliers that might affect the firm's quality and GMP compliance. It is prudent to insist that suppliers notify the firm of any significant changes they are making, as discussed in chapter 29, since meeting specifications does not necessarily mean that the supplied item has not undergone change. It is advisable to include details on the notification and approval of changes made by suppliers in the supply agreements or quality agreements, including a definition of what constitutes a change and the types of changes requiring notification. Similarly, suppliers should ensure that *their* suppliers also have adequate change control programs in place.

CONTINUOUS IMPROVEMENT

Although the regulations are silent on the topic, the preamble mentions that good records are needed to enhance improvement processes. This at least implies the importance of continuous or continual improvement (CI), which is a management philosophy that contends that most processes can be, and should be, improved. However, improvement relies on controlled change and is therefore a subset of change management.

Improvements may lead to reduced costs, improved yield, better uniformity of quality, reduced waste, better maintenance, lowering the possibility of contamination, and many other facets of the operations. It is of course important to "do things right the first time," but even so, there are always opportunities to do even better. Therefore, each company's quality system should be sufficiently flexible to allow for CI. Although this is not a regulatory mandate, it is good business sense to strive for improvement through small incremental steps.

CI is not a new concept. In Japan, where this is widely and successfully used, it is called "Kaizen," from "kai" meaning continuous change and "zen" meaning for the better.

Various versions of CI are also deliberately instilled into company cultures in many countries. Every firm has its own way of doing things, its own shared values, its own unique approaches to problem solving and decision-making, its own outlook on innovation, and its own viewpoint on change. To inculcate a positive attitude toward CI usually requires *adjusting* the firm's culture, which in turn, requires deliberate and planned action on the part of management. This is typically neither easy nor quickly accomplished, but is definitely worth the effort. Change is not always easy to accept, and to achieve full cooperation of all employees to regularly actively seek and apply small changes for the overall betterment of the company and the quality of its products requires effective leadership.

There are many "tools" that can be for CI, among which are just-in-time (JIT), lean manufacturing, Six Sigma, brainstorming, the Deming or Shewhart plan-do-check-act (PDCA) cycle, the reduction of variability mentioned in chapter 9, and others.

SUGGESTED READINGS

Buecker J, Tuttle J. Change management systems in the pharmaceutical industry. Pharm Eng 2002; 22(6):18–26.

Fish RC. Management of Change. Vol 1, No. 35. Rockville, MD: FDA News & Information, AAC Consulting Group, Inc., 2000.

Imai M. Kaizen, The Key to Japan's Competitive Success. New York: McGraw-Hill, Inc., 1986.

Jones DE. The training side of change control. J cGMP Compliance 2(4):32–37.

Monden Y. Toyota Production System, An Integrated Approach to Just-In-Time. 3rd ed. Norcross, GA: Engineering & Management Press, 1998.

Muchemu D. Change Control for FDA Regulated Industries. Bloomington, IN: AuthorHouse, 2007.

Scherkenbach WW. Deming's Road to Continual Improvement. Knoxville, TN: SPC Press, Inc., 1991.

Stephon DM. Considerations in effectively managing change control issues. J cGMP Compliance 1998; 2(4):51–55.

Turner SG, ed. Pharmaceutical Change Control. New York: Informa Healthcare, 1999.

27 Adverse event reporting and record keeping

In this context, an "adverse event" is any health-related negative event associated with the use of a dietary supplement product. Although reporting and record keeping with regard to adverse events are not covered in the GMP regulations per se, these are closely related to the handling of product complaints, which *is* covered in the regulations and discussed in chapter 21.

Moreover, adverse event reporting is specifically mandated by the Dietary Supplement and Nonprescription Drug Consumer Protection Act, Public Law 109-462, which took effect in 2007, and amended the FD&C Act at section 761.

Most adverse events are minor. However, the Act defines *serious* adverse events (SAEs) as those resulting in death, a life-threatening experience, inpatient hospitalization, persistent or significant disability or incapacity, congenital abnormality or birth defect, or one that requires medical or surgical intervention as based on reasonable medical judgment. It is *only* those SAEs that must be reported to the FDA, that is, they must be both "adverse" and "serious."

HOW AND BY WHOM SUCH REPORTS MUST BE FILED

The reports must be made on MedWatch Form 3500A by a manufacturer, packer, or distributor whose name appears on the label of a dietary supplement marketed in the United States. The entity that is required to submit the SAE report is referred to in the Act as being the "responsible person." The initial report must be filed within 15 business days of the receipt of information on the SAE. Follow-up reports of new medical information received within the first year after the initial report must also be submitted to the FDA within 15 business days. Prompt submission of the reports is important for public health reasons.

The MedWatch Form 3500A requires, at a minimum, the name of the injured person (the patient), the name of the initial reporter (the person who first notifies the responsible person about the SAE, whether this is the injured person or someone else), the identity and contact information for the responsible person, the suspect dietary supplement, and the serious adverse event. The responsible person should make diligent attempts to obtain and report complete information, including when appropriate getting the injured person's permission to contact the health care practitioner(s) familiar with the situation to obtain further information and relevant medical records, as needed. A copy of the label used on the suspect product should accompany the report.

The MedWatch Forms 3500A are available at FDA's Web site. These are *not* the same forms as used for the *voluntary* reporting, which are MedWatch Form 3500.

MedWatch Forms 3500A (with a copy of the label and any other attachments) should be mailed to

> FDA Center for Food Safety and Applied Nutrition, HFS-11
> Office of Food Defense, Communication and Emergency Response
> 5100 Paint Branch Parkway
> College Park, MD 20740

The forms must be mailed and cannot be submitted electronically or by fax.

RECORD KEEPING

Records on *all* adverse events (not *just* SAEs) must be maintained for six years, and must be available for FDA inspection for that period of time.

LABELING REQUIREMENTS

Product labels must include a full U.S. address or a domestic phone number (including the area code) through which the responsible person may receive a report of a serious adverse event. Although not required, FDA recommends that the label also include a clear and prominent statement informing consumers that they may use the indicated address or phone number for reporting serious adverse events associated with the use of the product.

WRITTEN PROCEDURES

Although there is no specific GMP requirement to have a SOP on adverse event reporting (AER), it is prudent to do so and to see to it that the responsible personnel receive adequate training to be able to properly handle this important topic.

THE SIGNIFICANCE OF AER REPORTING

The toxicology of many dietary supplement ingredients and combinations has not yet been thoroughly studied, nor has the possible interaction of such products with prescription or OTC drugs, nor are possible differences in tolerance among different age groups well known.

Although the FDA does get voluntary reports from consumers, health care providers (HCPs), pharmacists, poison control centers, and other sources, the AER data are extremely useful to the FDA.

It is important to keep in mind that AER reports do not necessarily mean that a given product caused an event with which it is associated in time. Instead, such reports signal that further investigation may be appropriate, and could in some instances produce an early warning of a problem. However, these reports, standing alone, do *not* establish a causal relationship between a product and an adverse event. Adverse events may occur for a variety of reasons unrelated to the product.

In general, the trade associations involved with dietary supplements have strongly supported the AER law as good both for the public as an additional and useful precautionary measure, and for the industry by helping secure more credibility for manufacturers of dietary supplement products.

AERs are considered "safety reports" under Section 756 of the FD&C Act, and are subject to the nondisclosure provisions of the Privacy Act and Health Insurance Portability and Accountability Act (HIPAA). Therefore, persons reporting are protected against personal information being released.

SUGGESTED READINGS

Food and Drug Administration. Guidance for Industry: Questions and Answers Regarding Adverse Event Reporting and Recordkeeping for Dietary Supplements as required by the Dietary Supplement and Nonprescription Drug Consumer Protection Act, October 2007, College Park, MD.

Food and Drug Administration. Guidance for Industry: Questions and Answers Regarding the Labeling of Dietary Supplements as Required by the Dietary Supplement and Nonprescription Drug Consumer Protection Act, September 2009, College Park, MD.

Frankos VH. Dietary Supplement and Nonprescription Drug Consumer Protection Act: Dietary Supplement 2008. Food and Drug Law Institute (FDLI) Conference, January 30, 2009.

Frankos VH. FDA regulation of dietary supplements and requirements regarding adverse event reporting. Clin Pharmacol Ther 2010; 87(2):239–244.

Public Law 109-462, 109th Congress, Dietary Supplement and Nonprescription Drug Consumer Protection Act, December 22, 2006.

Spangler DC. Dietary Supplement Adverse Event Reporting: an Industry Perspective. Food and Drug Law Institute (FDLI) Conference, January 30, 2009.

28 Adulteration and contamination

PROHIBITED ACTS
As mentioned in chapter 2, Section 301 of the FD&C Act specifies certain "prohibited acts," for which stiff penalties are provided if these sections are violated. One of these "acts" is the introduction into interstate commerce any product that is *adulterated*.

DEFINITIONS OF ADULTERATION
The definition of adulteration differs somewhat from one product class to another, but for dietary supplements it is defined in Section 402 of the FD&C Act in two separate ways.

Section 402(f) asserts that a dietary supplement or a dietary ingredient is deemed to be adulterated if it presents a significant or unreasonable risk of illness or injury under conditions of use recommended or suggested in labeling, or if no such recommendations or suggestions are made, under ordinary conditions of use. Also a *new* dietary ingredient is considered to be adulterated if there is inadequate information to provide reasonable assurance that it does not present a significant risk of illness or injury. Further, a dietary supplement or ingredient can be deemed adulterated if (under specified circumstances) it is declared to pose an imminent hazard to public health or safety. In addition, a product or ingredient may be considered legally adulterated if it is or contains an ingredient that is poisonous, insanitary, or contains a substance that may render it injurious to health. Certain other conditions may also render a dietary supplement or ingredient adulterated, including if it has been intentionally subjected to radiation.

The GMP regulations define *quality* to mean that a dietary supplement consistently meets its established specifications for identity, purity, strength, and composition, and limits on contaminants, and has been packaged, labeled, and held under conditions *to prevent adulteration* under Sections 402(a)(1), (a)(2), (a)(3), and (a)(4) of the Act. This definition stresses the importance of taking steps to ensure that adulteration is absent.

In addition to the definitions of adulteration contained in Section 402(f), another cause for a product being considered to be adulterated, under Section 402(g), is if it has been prepared, packed, or held under conditions that do not meet the *current good manufacturing practice regulations*. In other words, even if a product is technically satisfactory in all respects, if GMPs were not strictly followed in its production, it can be deemed to be legally adulterated. It is of obvious importance that this concept be clearly understood.

INTENTIONAL AND UNINTENTIONAL ADULTERATION
Most instances of adulteration arise from the accidental inclusion of contaminants as discussed below. However, occasionally unscrupulous manufacturers have been known to deliberately add undeclared and unlawful ingredients to

products, usually to enhance a specific pharmacologic effect. For example, the FDA discovered a number of tainted weight loss products sold as dietary supplements that contained potentially harmful prescription drug active ingredients. Similarly, other products labeled as dietary supplements have been found to contain drugs for erectile dysfunction, diabetes, and other conditions. There have also been controversial allegations of supplements used by athletes being adulterated with anabolic steroids and other compounds banned by various sports leagues. The FDA, in its role of protecting public health, considers all such instances of deliberate adulteration to be serious infractions, not infrequently resulting in criminal charges against the responsible parties. Moreover, the adverse publicity this generates tends to have a negative impact on the entire dietary supplement industry. The FDA has taken, and will continue to take, strong actions to prevent such tainted products from being on the market.

CONTAMINATION

The term *contamination* refers to the presence of small amounts of foreign, unintended, unwanted, undesirable impurities in a component or product that may be considered to be detrimental to the desired quality, and which may lead to adulteration.

In many places within the GMPs the need is stressed to ensure that *limits* are set and met on those types of contamination that may adulterate or lead to adulteration. See §111.365 and §111.75(c)(1) and (c)(3), §111.415, §111.15(e)(2), §111.410, §111.10(a), and §111.360 of the GMPs, among others, for examples.

This properly implies that *limiting contamination* is one key to avoiding adulteration.

The GMPs do not include a "laundry list" of the many forms of the undesirable contamination that may be encountered, nor on the practical limits allowable, but instead leave these decisions to each manufacturer since the firm is generally the best qualified to know the contaminants likely to occur in specific instances. Obviously, the FDA may or may not agree with the company, meaning that the manufacturer should be prepared to justify these decisions if called upon to do so. Such decisions should be made carefully, based on a combination of good science, experience, and common sense.

The FDA does not expect companies to set limits for, and to test for, *every potential* contaminant. Thus, the substances to be considered in determining whether to set limits for particular types of contamination would vary, depending on the source of the components (e.g., plant, animal, marine, microbial, manufactured chemical, etc.). However, §111.70(b) states the requirement to establish limits on *those types* of contamination that may adulterate or may lead to adulteration of the finished products, and similarly, §111.70(c) requires establishment of in-process specifications for any point or step where control is necessary for this purpose. Thus, the manufacturer must consider the kinds of contaminants *likely* to be encountered and set appropriate limits. This is, again, a judgment call on the part of each manufacturer.

Some have wondered why the FDA did *not* include a listing of possible contaminants and the allowable limits of each. But such a question is actually outside the concept of GMPs, which are not detailed "how to" regulations but instead are outlines of what must be accomplished as a minimum, leaving the details of methods to accomplish these goals to each firm to decide for themselves. The GMPs cover activities in manufacturing, packaging, labeling, and holding dietary supplements so that the products contain what the label purports, and are *not* contaminated with undesirable or harmful substances. The FDA expects firms to set the necessary and appropriate methods, procedures, and controls to accomplish these ends.

Clearly, the requirement to set *limits* on contamination implies that zero tolerance is *not* required. Some minor forms of contamination, while perhaps esthetically undesirable, do not lead to what is legally defined as adulteration. Not infrequently such insignificant contamination is essentially unavoidable.

ECONOMICALLY MOTIVATED CONTAMINATION

The widely publicized instances of the heparin contamination, melamine added to infant formula and pet food, diethylene glycol (DEG) in glycerin, and others, are all involved with deliberately adding low-cost materials to more expensive items of commerce in a manner that makes the spiked material conform to the usual quality tests for the genuine article. That is precisely why it is termed "economically motivated" since the perpetrators gain financially from their unethical activities. This is not new and has been going on for years. In fact, it was one of the reasons for the original Food and Drug Act of 1906. It is unlikely to stop, and will probably reappear in the future, for example, when there are shortages of supply of an ingredient but high demand of an item, where money can be made if criminals are able to devise a clever way to accomplish their goals. The globalization of the supply chain has made this even more likely to occur.

The FDA defines economically motivated adulteration (EMA) as "the fraudulent, intentional substitution or addition of a substance in a product for the purpose of increasing the apparent value of the product or reducing the cost of its production, i.e., for economic gain."

In the several instances of glycerin substituted by the less-costly DEG, it is understandable why this would not be immediately recognized without proper analytical testing. Both items are similar in appearance, being light-colored slightly viscous liquids at room temperature and having a sweet taste. However, glycerin (also called glycerol) is edible and safe, while DEG is poisonous. This points the need to deal *only* with reputable and trustworthy suppliers as discussed in chapter 29, and in conducting proper identification analyses as discussed in chapter 19.

FOREIGN MATERIAL CONTAMINATION

One form of contamination is foreign material (FM) getting into a product, such as metal, glass, or wood chips. This is usually not an ongoing problem but instead tends to be sporadic, which complicates the measures required for detection and control of such contaminants.

Unwanted particles can either arrive in the raw materials or may arise within the plant, for example, from fragments because of wear-and-tear of equipment, such as from the punches and dies used in tablet making, or from wire screens used in sifting powder or granular material.

Magnetic *separators* are useful in minimizing metallic particulate contamination. These typically consist of plate or grate-type magnets installed in chutes or over conveyor belts to extract ferrous metal particles that may be present.

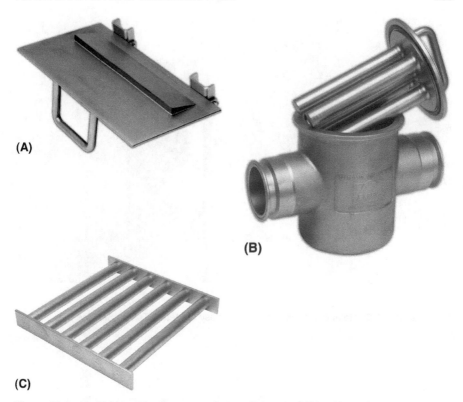

Figure 28.1 (A–C) Magnetic separators. *Source*: Courtesy of Eriez Magnetics © 2011.

These are used both to protect manufacturing equipment from damage and to help ensure freedom from product contamination that might lead to adulteration. Ferrous particles are attracted to magnets with a force proportional to the size of the particles, plus the strength of a magnet varies inversely with the square of the distance from the particles to the magnet. These factors need to be taken into consideration in selecting the size, strength, and location of magnetic separators.

Vendors of magnetic separators (Fig. 28.1) can be helpful in the selection, location, installation, and use of the optimum protective system, wherever the likelihood of metallic particulate contamination may exist.

The use of *metal detectors* (Fig. 28.2) at critical points in the manufacturing and packaging processes can significantly lower the probability of such problems, although just installing detectors does not necessarily eliminate this source of contamination. Detectors should be looked upon as one step in controlling this problem. Metal detection technology is applicable for dry, wet, and liquid products as well as specifically for tablets, capsules, and powders.

The sensitivity of metal detectors can be influenced by the kind, size, and shape of the metal contaminants, the physical characteristics and moisture content of the product being checked, the throughput speed, and environmental

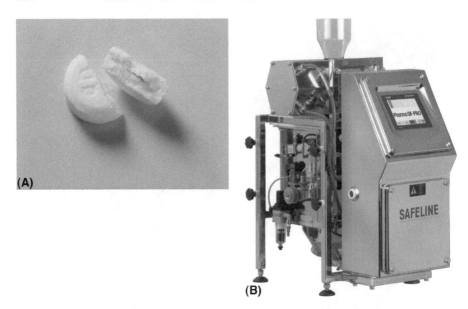

Figure 28.2 (A,B) Metal detectors. *Source*: Courtesy of Mettler Toledo Safeline Ltd.

conditions such as temperature and humidity. Supplying information to the vendor of the metal detectors about what to expect in the way of contamination and the nature of the products to be checked can be quite helpful in selecting the optimum equipment. Such units can be set up either for automatic rejection of contaminated product, or to sound an alarm or stop the product flow. One common arrangement is a fast-acting diverter valve, which "siphons-off" contaminated product so that an operator can visually examine and remove the foreign pieces.

Metal detectors are commercially available that are capable of detecting very small particles, including tiny fragments of stainless steel and metal flakes from punches and dies of tablet presses.

It is prudent to reduce the chances of glass particle contamination through the use of shatter-resistant lamps (light bulbs or fluorescent tubes) (Fig. 28.3). Moreover, if glass windows are used in areas where products or components are exposed, transparent shatter-resistant films exist to prevent shards of glass from contaminating in the event a window should break due to storms, attempted forced entry, or other such events.

Small foreign particles, often bits of stainless steel, can sometimes be seen in products with the naked eye, although it may be necessary to use a stereo-microscope to view them. In some instances it may be important to gain insight as to the *source* of such particulate contamination to prevent recurrence of the problem, through using the services of specialized microanalytical laboratories equipped to do forensic-like studies.

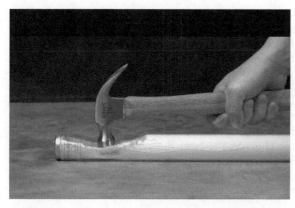

Figure 28.3 Shatter-proof light bulbs and fluorescent tubes. *Source*: Courtesy of Shat-r-Shield.

§111.365(i) calls for the use of measures against the inclusion of metal or other foreign material in components or dietary supplements, for example, through the use of filters, strainers, traps, magnets, or electronic metal detectors.

HEAVY METAL CONTAMINATION

Another possible contaminant is the so-called heavy metals, which generally refers to mercury, lead, arsenic, and cadmium, although the term is not clearly defined in the GMPs nor in the scientific literature. These are "heavy" in that their density is 5.0 or above. Interestingly, the human body needs trace amounts of these elements, but above certain limits they may accumulate in the tissues and become toxic. Even low levels of heavy metals in products can be troublesome when a dietary supplement product containing them is taken repeatedly over a prolonged period. Heavy metals may exist in botanical components, coming from the soil in which the plants were grown. Some raw materials derived from mineral sources may also contain heavy metals, as may synthetically produced fine-chemical ingredients, from contaminated starting materials, catalysts, or from the production equipment used. The USP establishes limits and test methods for heavy metals. Moreover, the American Herbal Products Association (AHPA) has published a guidance document on this. Also there are limits suggested in NSF/ANSI Standard 173. None of these sources of background information are mandated, leaving it up to each manufacturer of dietary supplement products to decide on the appropriate specifications.

NATURAL TOXINS

Particularly in the instance of botanical components, there are certain known natural toxins and known adulterants, which may be present and may need to be eliminated or controlled. One such natural toxicant that has received considerable negative publicity is aristolochic acid, associated with permanent kidney damage and cancer.

RESIDUAL SOLVENT CONTAMINATION

Small amounts of solvents may remain in some dietary ingredients and excipients, coming from the manufacturing processes used in their production. These are often termed organic volatile impurities (OVIs). Some solvents can be toxic and therefore lead to adulteration and/or can cause undesirable taste or odor in products. Residual solvents can also cause manufacturing problems such as changes in the "flowability" of powders or changes in crystal structure. The determination of the presence and amounts of such solvents in components is typically conducted by chromatographic analytical techniques such as those described in USP Chapter 467. This has also been addressed by the International Conference on Harmonization (ICH) in document Q3C. However, such testing is not necessary if the supplier can give satisfactory assurance that there is no potential for undesirable residual solvents in their items.

PESTICIDE RESIDUE CONTAMINATION

Botanical materials may be contaminated with pesticides, used to eliminate insects and other pests in agricultural settings. Many of the countries from which botanical items are imported have different regulatory controls over allowable pesticides than in the United States, which opens the possibility that sources of disallowed pesticide contamination may exist. It is the responsibility of dietary supplement manufacturers to be aware of this, and to set appropriate specifications and test methods to ensure that pesticide residue problems are avoided. The USP Chapter 561, while not mandated by the GMPs, gives useful information on this topic, and further information is available from the Pesticide Data Program of the U.S. Department of Agriculture and from the Pesticide Residue Program Laboratory of the California Department of Food and Agriculture.

MICROBIOLOGICAL CONTAMINATION

Dietary supplement products are not sterile (totally free of microorganisms), nor is there any need for them to be. However, the nature of the so-called bioburden (the numbers and types of microorganisms present) *is* of importance and requires consideration.

The finished products should be free from *objectionable* organisms, meaning those that may endanger the health of consumers or that may adversely impact product stability. However, there is no clear definition of *which* organisms are objectionable.

Throughout Part 111, emphasis is placed on limiting the *sources* of potential microbiological contamination. For example, as discussed in chapter 3 and §111.10, personnel hygienic practices and gowning address this. Requirements concerning the grounds, physical plant facilities, pest control, plumbing, sewage, and trash disposal, etc., as discussed in chapter 4, as well as selection and care and cleaning of equipment, as discussed in chapters 5 and 6, are directly related to minimizing the potential for such contamination.

The microbiological quality of the water, both when used as a component and for cleaning and sanitizing, is stressed. The regulations also require holding and distribution of dietary supplements under conditions that will protect against contamination. Cleaning and sanitizing all filling and packaging equipment (as appropriate) is mandated by §111.415 as a further step to minimize possible microbiological contamination of finished products.

§111.70 states that the manufacturer is responsible for establishing specifications for incoming components, including limits on those types of contaminants that may adulterate or lead to adulteration of the finished product, while §111.73 requires that such specifications be met, and §111.75 details how it must be determined that they *are* met. The FDA recognizes that the limits for particular types of contaminants (including microbiological contaminants) will vary depending on many factors.

Useful guidance for this somewhat confusing topic can be found in USP <2022> titled "Microbiological Procedures for Absence of Specified Microorganisms—Nutritional and Dietary Supplements," and USP <2023> titled "Microbiological Attributes of Nonsterile Nutritional and Dietary Supplements." Also, USP <62> titled "Microbiological Examination of Nonsterile Products: Tests for Specified Microorganisms" is of interest. Moreover, NSF/ANSI Standard 173 also contains helpful suggestions for microbiological control.

MYCOTOXINS AND AFLATOXINS

Mycotoxins present still a further issue in the microbiological purity of dietary supplements. Mycotoxins (including aflatoxins, a group of toxic compounds from certain strains of *Aspergillus* fungi) are toxic compounds produced by fungi or molds and yeasts. Botanical items, particularly if improperly stored at high humidity and elevated temperatures, may become contaminated with these molds (particularly species of *Aspergillus, Penicillum, Mucor,* and *Fusarium*), which in turn may produce the toxic chemicals, which can present health-significant safety issues. Analytical methods exist for determining the presence, identity, and amounts of mycotoxins and aflatoxins present.

For herbal ingredients and finished products, the AHPA has a useful publication titled *Guidance on Microbiology & Mycotoxins.*

Similarly, endotoxins may possibly present another health hazard if present, although these are typically of significance in sterile pharmaceuticals as opposed to dietary supplements. Endotoxins are formed by certain gram-negative bacteria. A sensitive test method for detecting the presence of endotoxins is the Limulus Amebocyte Lysate (LAL) technique, which utilizes blood from the horseshoe crab to produce a color change if endotoxins are present.

TESTING FOR MICROORGANISMS AS CONTAMINANTS

It is obviously possible that various types of microorganisms may contaminate dietary supplement products, which could lead to adulteration. This is in particular true of botanicals, but could also apply to other types of products as well. Therefore, it was originally suggested that *all* finished products be tested for objectionable microorganisms. However, the FDA decided that such testing of *all* finished products is *not* necessary, and it is instead up to each manufacturer to evaluate which products require testing. Such a decision should be based on the characteristics of each product, the nature and source of the components, the specifications for absence of microbial contaminants in the components, and whether this is properly addressed in the certificates of analysis obtained from the suppliers, the firm's in-process specifications, and the nature of the manufacturing process.

§111.365(d) requires microbiological testing *as necessary*, leaving it up to the manufacturer to determine *if*, *when*, and *how* such testing should be done. To repeat, there is *no* specific requirement to test each lot of each component, or each batch of finished product, for microbiological purity.

WATER ACTIVITY

Water activity is mentioned in §111.365(e) as one way to prevent the proliferation of microorganisms. This useful concept is often not well understood. It has been applied to good advantage in foods since 1950s, and more recently also in pharmaceuticals.

Water activity, a_w, is *not* the same as total moisture content as commonly measured by gravimetric loss-on-drying procedures, or by the Karl Fischer method or other techniques. The basic concept is that some of the water present is chemically or physically "bound" in a crystal lattice or trapped in small capillaries, while other water is present in "free" or "unbound" form. In this view, a_w refers specifically to the unbound water. The importance of a_w is that each species of microorganisms has a limiting a_w value below which it cannot grow. Therefore, controlling a_w (and pH) through product formulation constitutes an effective way of preventing microbial growth, and also in prolonging the physical stability of components and dietary supplement products. A number of easy-to-use commercial instruments facilitate the measurement of water activity.

AIRBORNE CONTAMINATION

As previously noted, §111.20(d)(1)(iii) requires proper ventilation systems to help prevent airborne contamination. Most such systems recirculate a portion of the air in each area, but bring in some fresh air from outdoors. Such outdoor air may contain significant quantities of particulate and microbiological impurities. Therefore, adequate filtration of such air is important, requiring careful attention to the design and maintenance of air filters, the cleanliness and integrity of ducts, proper airflow patterns, and pressure differentials between areas of the plant.

In handling dry materials, powders, and granules, care should be taken to minimize the generation and dissemination of dust. The use of dust-capturing hoods, and dust collecting systems, should be considered where appropriate.

In addition to possible contaminants from outdoor air, personnel add to the situation both from the normal constant shedding of skin cells and from droplets of saliva emitted when people talk, cough, or sneeze. Added to this is the dirt and microorganisms carried on people's hair, clothing, and shoes. The larger the number of people in a given area, and the greater their degree of activity, the greater they add to airborne contamination. Moreover, manufacturing operations may generate dust particles from handling, mixing or conveying, or filling granular products, or droplets from liquids processing.

Equipment, ledges, walls, ceilings, floors, and other surfaces that are not kept clean can also be significant sources of airborne contamination, as can drains. Also, the waste put down drains is typically rich in nutrients, and the warm, moist, and dark conditions favor the growth of microorganisms. When floors and equipment are washed, floor drains are suddenly flooded with water resulting in the entrapped air being forced out with a highly turbulent motion, carrying droplets and microorganisms into the air.

The degree and nature of the movement of indoor air depends on many factors, including the dimensions and configuration of the area, the movement and activity of personnel and equipment, the opening of doors and windows, convection currents from temperature gradients, the agitating effect of equipment in use, and others. The end result is that the room air is usually turbulent, causing dust particles and microorganisms to become airborne. Smaller particles and droplets may remain suspended for prolonged periods, while the larger particles and droplets settle onto the floor and other surfaces. This can result in the contamination of surfaces, components, and products that are exposed.

Floors tend to become heavily contaminated with dust and microorganisms due to the settling from the air, plus persons entering from outdoors can introduce more dirt and microorganisms on their shoes. This type of contamination can easily be tracked from area to area within the plant, on shoes as well as by forklift trucks, and moving equipment, pallets, shipping cases, drums, etc., from one area to another, resulting in potential cross-contamination. These factors require careful attention.

THE IMPORTANCE OF CLEAN HANDS IN AVOIDING CONTAMINATION

Another major source of possible contamination is from the hands of workers. Whenever employees touch a product, components, inside surfaces of packaging, or other contact surfaces, the possibility of microbiological contamination is real. Therefore, frequent and proper handwashing is important. In some instances, the use of disposable plastic gloves may be advisable. White cotton gloves, while presenting an appearance of being sanitary, actually tend to retain soil and microorganisms and may contribute to the problem rather than to its solution.

DEFECT ACTION LEVELS

Some food products, even when produced under GMPs, contain natural or unavoidable defects at low levels that are not hazardous to health. It is often impractical to grow, harvest, or process raw products completely free from defects. The FDA has established maximum acceptable levels of many of such defects, for use in deciding when regulatory action is appropriate for foods, called *defect action levels* (DALs). At the time the dietary supplement GMPs were being developed, consideration was given to establishing DALs for some dietary ingredients (particularly botanicals). However, the FDA decided *not* to do so since there were insufficient data available to identify appropriate DALs for many dietary ingredients.

CONCLUSIONS

It is frequently repeated in Part 111 that it is essential to establish limits on the types of contamination that may lead to adulteration. It is the responsibility of the firm to determine specifically what must be done to accomplish these goals. Each company's line of products, sources of components, facilities, equipment, processes, and personnel are of course unique, and therefore it is not possible to state explicitly what steps are required to meet the regulatory requirements for contamination and adulteration avoidance. However, this topic needs to be a top priority of management, with reasoned input from all concerned.

SUGGESTED READINGS

Barbosa-Canovas GV. Water Activity in Foods: Fundamentals and Applications. Hoboken, NJ: Wiley-Blackwell Publishers, 2007.

Bugno A. Occurrence of toxigenic fungi in herbal drugs. Braz J Microbiol 2006; 37:47–51.

Cole MR, Fetrow CW. Adulteration of dietary supplements. Am J Health Syst Pharm 2003; 60(15):1576–1580.

Ernst E. Toxic heavy metals and undeclared drugs in Asian herbal medicines. Trends Pharm Sci 2002; 23(3):138–139.

Forcinio H. Preventing metal contamination. Pharmaceutical Technology, July 2003:36–40.

Fung F, Clark RF. Health effects of mycotoxins: a toxicological overview. Clin Toxicol 2004; 42(2):217–234.

Geyer M, Parr MK. Nutritional supplements cross-contaminated and faked with doping substances. J Mass Spectrom 2008; 43(9):802–902.

Huggett DB. Organochlorine pesticides and metals in botanical dietary supplements. Bull Environ Contam Toxicol 2001; 66(2):150–155.

Jeong ML, Zahn M. Pesticide analysis of a dietary supplement. J AOAC 2008; 91(3):630–636.

Jimenez L, ed. Microbial Contamination Control in the Pharmaceutical Industry. New York: Informa Healthcare, 2004.

Kemsley J. Improving metal detection in drugs. Chemical & Engineering News, December 8, 2008:32–34.

Liva R. Facing the problem of dietary supplement heavy metal contamination. Integr Med 2007; 6(3):36–38.

Maragos CM, Busman M. Rapid and advanced tools for mycotoxin analysis: a review. Food Addit Contam 2010; 27(5):688–700.

Maughan RJ. Contamination of dietary supplements and positive drug tests in sports. J Sports Sci 2005; 9:883–889.

Mindak WR. Lead in women's and children's vitamins. J Agric Food Chem 2008; 36:6892–6896.

Prince R. Microbiology in Pharmaceutical Manufacturing. 2nd ed. Bethesda, MD: PDA Books, publisher, 2008.

Raman P. Evaluation of metal and microbial contamination in botanical supplements. J Agric Food Chem 2004; 52(26):7822–7827.

Sifman NR, Obermeyer WR. Contamination of botanical supplements by *digitalis ianata*. N Engl J Med 1998; 339(12):806–811.

St. Jeor VL. Identifying foreign material contamination. Microscopy Today, May 2008:10–15.

Stellmack M. Microanalytical methods for sleuthing contaminants. Contract Pharma, January–February 2009:78–81.

Trucksess MW, Whitaker TB. Sampling Plans of Mycotoxins in Foods and Dietary Supplements. In: Mycotoxin Prevention and Control in Agriculture, ACS Symposium Series. Vol 1031. New York: Oxford University Press; 2010:207–221.

Veeramuthu G. Sort out best methods for foreign object detection. Food Quality, October–November 2007:50–55.

Williams KL. Endotoxins: Pyrogens, LAL Testing, and Depyrogenation. 3rd ed. New York: Informa Healthcare, 2007.

29 Supply chain integrity

It is a truism that to consistently manufacture quality products it is necessary to employ only high-quality raw materials. Since most dietary supplement manufacturers *purchase* components and packaging materials from other manufacturers (as opposed to making these items themselves), the evaluation, selection, and control of the sources of supply are therefore critical considerations. Similarly, for goods and services obtained from contractors, it is equally important that these sources also be selected and monitored with care.

The network of vendors, suppliers, and contractors used by a firm is frequently referred to as the "supply chain," whereas the choice of, and control of, this network is typically known as "supply chain management." This usually involves input from whoever is in charge of the purchasing function in cooperation with manufacturing, quality control, and other entities within the firm. Another facet of the supply chain is the logistics involved, for example, getting the goods from the supplier to the place where they are needed in the most expedient, safest, secure, and least costly way. If materials are imported, this includes proper documentation for customs clearance.

The term "integrity" in this instance refers to the system performing as intended, in a proper and ethical manner, without unauthorized manipulations or other wrongdoings. It implies visibility, physical security, trust, and good relationships. This is of particular importance in this age of globalization, where sources of supply are often geographically far away and sometimes from countries with less regulatory oversight than we are used to in the United States. The achievement of supply chain integrity requires careful attention, defined roles, and good communications.

The terms "vendor" and "supplier" are usually considered to be synonymous.

Not infrequently the vendor or supplier is *not* the firm that *produces* the materials, but is instead a distributor or broker. This often involves repackaging and/or relabeling, in some instances multiple times, with the attendant risk of possible errors, misunderstandings, or contamination during repackaging as well as the possibility of mislabeling. This also can introduce difficulty with traceability. It is obviously important to know with surety where the items were made and by whom, not just who is selling them.

It is also important to keep in mind that the suppliers of components and packaging materials used in making dietary supplement products are *not* bound by Part 111, although they may be required to be in compliance with the *food* GMPs (Part 110), as explained in detail in Comment 29 in the preamble to Part 111. But this means that adequate quality systems may or may not be in place with suppliers, which in turn implies the practical necessity of determining what control systems each supplier does in fact use. The ultimate party legally responsible for the quality of finished dietary supplement products is of course the firm named in the labeling, and therefore supply chain integrity is of significant importance to dietary supplement firms.

SELECTION OF SUPPLIERS

Since consistent quality of the finished products is in large part dependent on the quality of the components and packaging materials used, it is important to carefully choose each supplier. Similarly, firms supplying bulk or finished products on a contract manufacturing basis must also be selected with due care. Cost alone should not be the sole criterion in making such decisions.

Identifying appropriate suppliers to work with first involves gathering information about potential candidates, by making inquiries within the industry, and/or by sending questionnaires such as the Standardized Information on Dietary Ingredients (SIDI) method discussed in chapter 30. The information gleaned may include, for example, the country of origin, the proposed supplier's experience and regulatory history, and details of the candidate's existing manufacturing and quality systems. It may also be helpful to obtain the names of some of the customers the vendor has served.

It is advisable to furnish proposed suppliers with the detailed specifications of what is required, asking if the vendor can, indeed, regularly meet those specifications. Moreover, it is important to ascertain if the vendor can consistently deliver the required material in the quantities foreseen in a timely manner. Pricing is of course a factor, but as mentioned above, it should not be the *main* consideration. These steps are usually accomplished by discussions between the parties, preferably with a record made of the discussions. If the response is positive, the usual next step would be obtaining samples of the items the proposed vendor would supply (usually from three or more different batches), and then checking the samples against the specifications.

In short, it is important to ensure that ongoing quality is part of the due diligence process in selecting suppliers. Moreover, it is also important to do business only with trustworthy sources.

RISK MANAGEMENT IN THE SELECTION OF SUPPLIERS

There are many places in quality systems where risk management is an important consideration, vendor selection being one of them. In this approach, it is necessary to consider the possible hazards involved and the probable consequences of each, then reaching decisions on the best ways of minimizing the negative impacts that may result. The overall goal is to balance risks, costs, and quality while maintaining regulatory compliance.

Among the risks to consider are the consequences of experiencing an interruption in supply, the advantages and disadvantages of having more than one supplier of each item, the difficulty in identifying alternative suppliers, the possibility of geopolitical problems arising when using an overseas source, the amount and quality of the information available to enable making reasonable judgments of potential vendors' quality systems, the ease or difficulty and costs involved in arranging for an audit of the proposed vendors' operations, the available knowledge about the regulatory system in the country of manufacture, the likelihood of logistical or customs clearance problems arising including necessary dependence on third-party logistics suppliers, and many other such factors.

The possibility of economically motivated adulteration (as discussed in chapter 28) should also be considered.

Such factors, properly balanced, facilitate reaching informed decisions on the optimum selection of vendors for each item needed in the supply chain.

Because of limited resources, a risk-based approach is the only feasible way to qualify suppliers.

QUALIFICATION OF SUPPLIERS

After deciding which suppliers to use, the next step is documenting *qualification* of each. The FDA often uses the phrase "know your supplier," and the qualification process is an important step toward this.

Qualification is *required* by §111.75(a)(2)(ii) *if* it is intended to rely on certificates of analysis (CofAs) from a component supplier. Although not specially stated in the regulations, the FDA expects firms to qualify *all* suppliers of components used in manufacturing dietary supplement products, and preferably also vendors of key packaging components. Unfortunately, the regulations are essentially silent on explicitly *how* suppliers should be qualified, aside from establishing the reliability of the supplier's CofAs.

A CofA containing specified information on numerous aspects of the component's qualities can be supplied with each delivered lot, which can be relied on to confirm the identity of any component that is *not* a dietary ingredient, and can also be accepted to determine whether other specifications are met for *any* component. This requires establishing the *reliability* of the supplier's CofAs through confirmation of the results of the supplier's tests or examinations. This raises the question as to how closely the firm's test results need to match those on the vendor's CofA. Although the regulations are silent on this point, a reasonable limit would be an acceptance range of ±2%, although this may differ depending on the type of test being performed.

The testing for CofA reliability must be repeated periodically to reconfirm the supplier's CofAs, although the regulations are silent on how frequently this must be done.

CofAs must include a description of the test methods used, limits of the tests or examinations, and must contain the actual results for the lot received.

Some manufacturers choose *not* to rely on CofAs, but instead conduct their own appropriate tests or examinations to evaluate compliance with all specifications, in which case there is no regulatory requirement to formally qualify suppliers, although doing so remains a good business decision and is strongly recommended, since full testing alone may not always be adequate to ensure the desired level of quality.

The various dietary supplement trade associations, jointly working through the Joint SIDI^{TM} Working Group, have prepared a useful document titled *Certificate of Analysis for Dietary Supplement Components: A Voluntary Guideline*, which is available on the Web sites of the trade associations.

In considering supplier qualification from a practical (as opposed to regulatory) point of view, it is worthwhile taking into account whatever is known about the supplier, for example, 483s or Warning Letters from FDA inspections if any, informal information from others in the industry, any third-party certifications, the supplier's change management program and documentation and record keeping systems, to what extent the vendor qualifies its own suppliers, etc.

Where vendor qualification is required, quality control personnel must review and approve the documentation setting forth the qualification (and any requalification) of any supplier, as stated in §111.75(a)(2)(E). It is prudent to also have the same review and approval process conducted on the nonmandated vendor qualifications, since the FDA expects firms to qualify *all* suppliers of components.

Audit reports (if any) may be part of the qualification records, and if so, they should be specific and complete, not merely brief opinion letters. Vendor audits, as discussed in chapter 30, are quite useful in supplier selection and qualification, albeit they are not mandated by the regulations.

Although not specifically mandated, it is advisable to have SOPs covering vendor selection, evaluation, and qualification. It is similarly advisable to maintain an up-to-date listing of approved suppliers.

As discussed in chapter 28, there have been several incidents of ingredient contamination and adulteration from materials furnished by unscrupulous vendors, with disastrous outcomes. Avoiding such situations must have a high priority.

SUPPLIER CONTROLS

It is important to know who the *actual* manufacturer is, as well as any repackers or relabelers involved, for all dietary ingredients and excipients, as well as for contract manufacturing operations. Moreover, it is important to monitor and keep records on the performance of each vendor and to follow trends.

Particularly with the growing complexity in the supply chain due to increased global sourcing, good purchasing controls and solid relationships with suppliers must become the norm. There needs to be a comprehensive mutual understanding of roles and responsibilities coupled with good two-way communication. Not all suppliers thoroughly understand the GMP concept, and may need assistance in that regard.

If a significant problem occurs that is definitely traceable to a specific supplier, an appropriate action is to issue a Supplier Corrective Action Request (SCAR). This is a way of formally requesting the supplier to do an investigation to determine the root cause of the problem and to take the appropriate corrective action. It is usual to state a date by which the vendor should complete the investigation and respond with a report on the findings and the corrective action taken. If the vendor's response is considered inadequate, the vendor may be designated a "Restricted Supplier" until and unless more sufficient corrective action is taken.

Many firms also use what are termed Supplier Rating Systems (SRS) that track each vendor's performance regarding quality and delivery, and other factors; the results of which can impact the amount of business a firm decides to do with each approved vendor. Both SCARS and SRS are fairly standard purchasing tools widely used by many industries including dietary supplement manufacturers.

If a supplier fails to live up to its commitments and is unwilling or unable to take the necessary corrective actions, they should be disqualified. Care must be taken to be sure that once disqualified, a vendor should not be requalified without making real and substantial changes.

The degree of supplier control necessary tends to vary with the type and significance of the product or service purchased, but it is advisable to define the extent of control to be exercised over each supplier.

QUALITY AGREEMENTS

It is advisable to enter into quality agreements with the major vendors in the supply chain. Such agreements are formal documents that establish mutual understandings of the quality management system to be applied to the supplied items. This should preferably be separate from any commercial agreements that involve such details as quantities, prices, delivery points, etc.

A quality agreement establishes relationship expectations and communication mechanisms. It typically covers specifications, manufacturing documentation, labeling and packaging, sampling, testing, CofAs (if applicable), release and shipment, what constitutes a "lot" and lot numbering, the right to visit and/or audit the vendor's facility, change control procedures, notification of FDA inspections or actions, and other such details. There is usually a provision included for the termination of, or amendments to, the quality agreement.

Although firms often have a standard format or template for such agreements, they typically need to be tailored to the specific circumstances. Quality agreements can be quite helpful in clarifying goals, expectations, and responsibilities. Nondisclosure and confidentiality can be included either in a quality agreement or in a separate document. It is also usual to include a provision for dispute resolution.

It is advisable to establish a SOP indicating the types of vendors and contract service organizations for which quality agreements are required and the details that should be included in such agreements.

SHOW AND SHADOW

An interesting fact uncovered by the FDA in their investigation of the contaminated heparin case was that the plant that was successfully audited was not the true manufacturing site. The one audited was called the "show" facility that met expectations, whereas the defective product was actually made at an entirely different location, now called the "shadow" facility. In conducting plant audits, this incident implies the necessity of confirming that the plant being inspected is actually the one supplying the purchased items. Heparin is of course a drug, not a dietary supplement, but the incident illustrates the extremes to which supply chain integrity issues can reach and can be quite difficult to detect if the vendor plans to deliberately commit fraud. Fortunately, such extreme cases are rare, but the rule of *caveat emptor* (buyer beware) clearly does apply.

It is advisable to verify that each item actually comes from approved suppliers or manufacturers through expected routes and were not diverted. It is also advisable to implement a comprehensive approach to monitoring supply chain integrity and to examine all containers upon receipt for damage or evidence of tampering.

SUGGESTED READINGS

Carter WD. Excipient quality agreements: exploring the IPEC quality agreement guide and template. Contract Pharma, September 2010.
Chopra S, Meindl P. Supply Chain Management: Strategy, Planning and Operations. Upper Saddle River, NJ: Prentice Hall, 2000.
Joint SIDI™ Working Group. Certificate of Analysis for Dietary Supplement Components—A Voluntary Guideline, April, 2010.
Ortiz B. Increased FDA scrutiny of purchasing/supplier controls. Contract Pharma 2010; 12(8):100–103.
Rhoades B. Global supply chain quality management. Contract Pharma, January/February 2011:68–73.
Supply Chain Security: A Comprehensive and Practical Approach, White Paper. Tampa, FL: ISPE, 2010.
Tremblay J-F. Sourcing from China. Chem Eng News 2008; 86(20):15–20.

30 Audits

REASONS FOR AUDITS
Although not specifically mandated by the GMPs, the basic reason for conducting audits is as control mechanism to identify potential problems before they result in nonconformances that could have either business or regulatory consequences. Audits help identify and reduce risks, as well as aid in performance improvement.

TYPES OF AUDITS
Aside from other areas of the business realm, in manufacturing and quality control there are many types and kinds of audits commonly used. These are commonly classified into three general categories, first-party or self-inspection audits, second-party or vendor/supplier audits, and third-party external audits conducted by outside organizations usually for registration or certification purposes. However, these classical definitions are not always clear-cut.

SELF-INSPECTIONS
Rather than waiting for FDA inspections to detect violations and possible weaknesses in GMP compliance methods, it is prudent and sensible to monitor the situation via periodic deliberate reviews of policies, systems, methods, and procedures. Even well-meaning, well-motivated firms may miss significant items in the total compliance scheme. Therefore, self-inspection is an effective way to identify and correct potential problem areas, both to avoid regulatory issues and to ensure consistent product quality to help protect consumer confidence for pragmatic and for ethical reasons.

Self-inspections can also lead to improvements in costs, yields, efficiency, and reductions in rejects, reworking, and scrap.

Large firms may have departments staffed by specialists to conduct internal audits, whereas smaller firms typically either assign such duties to existing personnel in addition to their other functions or they engage outside organizations for such purposes. Hired outside firms or consultants qualify for the term "self-inspectors" if they have been engaged specifically to help management understand and evaluate the firm's compliance status and to detect any shortcomings and recommend needed corrective actions.

It is wise to have the inspection team directed by an individual designated by top management, to assure getting the full attention of subordinate management, and to be certain that necessary changes are made when weaknesses are uncovered. It is also good to have various operating departments represented on the inspection team, all of whom should be quite familiar with the GMP regulations and the company's operations.

The self-inspection team should meet at prescribed intervals to discuss their goals and strategy, and to assess the progress in improving the status of their firm's regulatory compliance. Their activities should cut across organizational lines, and should encompass a review of each function, which could conceivably affect product quality, including but not necessarily limited to purchasing, receiving, warehousing, quarantine methods, manufacturing or processing, packaging, shipping, maintenance, quality control, and documentation.

The team should also review the firm's handling of customer complaints, sampling plans and methods, adverse event reporting, the recall system, and the handling and disposition of returned goods.

The inspection team should have the power to get needed changes and corrective actions taken where necessary, but should not be in a position to punish or take disciplinary action against individuals.

The technique of inspection may either be to visit one department or area after another in sequence, or to follow components through all of the steps from receipt until the finished product is shipped. Still another approach is to essentially follow FDA's Compliance Program 7321.008, in effect conducting a mock FDA inspection.

The inspection team should inform supervisory personnel in the facility of their objectives, scope, and the functions of the inspection. Supervisors in turn should be sure that all employees in their departments clearly understand the reasons for, and the techniques of, the inspection process. It is important that everyone comprehend the concepts and advantages of self-inspection, to minimize possible resentment or morale problems and to maximize voluntary input to the team. Where unions are involved, the officers, stewards, and business agents should also have the concept explained to them prior to the start of the inspection.

Companies initially embarking on a self-inspection program may find that the law of diminishing returns applies, that is, the first inspections may detect many deficiencies and potential sources of problems, but once these are unearthed and corrected, subsequent self-inspections yield less requirements for corrective action. This may indicate that the frequency of such inspections can be reduced.

Self-inspection programs tend to be evolutionary, starting on a modest scale but growing in complexity and sophistication as the company and the inspection team become more familiar with the techniques and the advantages gained.

In order for a self-inspection program to be successful, it is essential to have the full support of management at all levels, stemming from the top. Auditors must be given the authority to do their jobs and retain status within the firm's hierarchy. It is advisable to have a written statement issued by top management, establishing and endorsing the concept.

Self-audits may be either surprise inspections on the grounds that they prefer to see the existing conditions without advance notice of the arrival of the inspection team, although experience tends to show that it is usually more productive to openly schedule and announce audits in advance. One possible impact of announced and scheduled audits is the tendency for department heads to do their own reviews and take corrective actions *before* the arrival of the inspection team.

In preparation for self-audits, the inspection team needs to do their homework. Each member of the team should become familiar with the details of each area, including the processes, systems, equipment, and applicable records, documents, and logs. A study of the relevant SOPs can be a useful starting point. Previous audit records, if they exist, and correspondence on corrective actions taken should be reviewed. This should include information on any previous FDA inspections and any follow-up to them.

A useful tool in self-inspections is to prepare a checklist of the items to be considered, to avoid the possibility of inadvertently overlooking significant details. However, checklists, if followed too rigidly, can hamper the scope of the inspection. The team should act like detectives, following leads and clues that occur, without limiting their scope by a checklist.

A study of records and logs is typically the most important and time-consuming step in the audit.

What is actually being done should be compared with what SOPs and manufacturing directions call for. Watching operators in the performance of their duties can be a useful step.

At the conclusion of the audit, the team should meet with the supervisors of each area for a discussion and critique of the findings and conclusions, and of what will be reported to top management. Effort should be made to make this discussion frank, open, and two-way, to clear up any misunderstandings. If deficiencies brought to light during the audit have been corrected, this should be noted in the team's final report.

As soon as possible once the internal audit has been completed, a written report should be prepared to inform all concerned with the findings and recommendations. This should include both deficiencies and favorable details, including praise when deserved. Serious problems needing urgent attention should be mentioned. Target dates for corrections should be suggested.

Since the final report may be lengthy, it is advisable to preface it with a brief, clear, concise executive summary.

If any supervisory personnel should disagree with any of the findings or recommendations in the final report, they should be given the opportunity to express their viewpoints, and the matter resolved.

Thereafter, provision should be made for follow-ups, and reports to top management, on the status and timing of the suggested corrective actions.

As an alternative to an "all out" overall self-inspection, some firms prefer to handle this on a section-by-section basis at different times. Examples of this approach would be audits of packaging and labeling operations, laboratory audits, sanitation audits, maintenance audits, warehousing audits, etc. Taking self-inspection in smaller "bites" can cause less disruption and can provide the inspection team more time for in-depth preparation.

FDA'S ENTITLEMENT TO SEE INTERNAL AUDIT REPORTS

It has been FDA's policy to *not* routinely ask to see and review internal audit reports, since this might cause firms to carefully edit what is contained in the reports to avoid self-incrimination, thereby giving top management a limited or distorted view of the actual compliance picture.

Even if during establishment inspections where they do not elect to request to review self-inspection reports, it is common for investigators to ask if there *is* such a self-inspection program in use. In the event of litigation, the FDA may ask for access to the internal audit reports, and the courts would likely grant them such access.

The agency's policy on the review of firms' quality audits is stated in Compliance Policy Guide (CPG) 7151.02, subchapter 130.300.

FDA's policy of generally *not* requesting review of internal audit reports gives the firm assurance that such reports will not be made available to competitors nor to plaintiffs' attorneys under Freedom of Information, since the FDA will usually not even *have* such information. Still, it is important for the firm to remember that while audit reports serve an essential function, they are also potentially dangerous documents. It is advisable to control the distribution of audit reports to ensure they are not coupled to other document files that are available to, and may be copied by, FDA inspectors.

The dietary supplement GMPs do *not* mandate a self-inspection process, although some other industry areas regulated by the FDA do require them.

GAP ANALYSIS AND NEEDS ASSESSMENTS

The terms "gap analysis" and "needs assessments" refer to voluntary studies firms make to establish steps they may need to take to identify their areas of noncompliance, where appropriate corrective action should be devised, or when other GMP issues should be given further attention. Such studies, although useful, are not mandated.

SECOND-PARTY AUDITS

As discussed in more detail in chapter 29, it is imperative to ensure that vendors and suppliers of dietary ingredients, excipients, packaging materials, outside contract manufacturing services, and contract laboratories are properly selected and qualified. The evaluation and qualification of suppliers is mentioned in §111.75(a), albeit not much detail is provided. Although not specifically mandated in Part 111, supplier audits are quite useful in this regard, however, both from regulatory and good business perspectives. This is particularly true since many items are sourced from overseas vendors.

Accepting the fact that vendor audits are desirable even if not required, the question becomes when and how to conduct such audits, and how far to go. The answers to these questions should be based on considerations of the importance or criticality of the items being sourced from each supplier, what quantities of goods are involved, the complexity and sophistication of each supplier's operations, and the known quality history of each vendor. These issues are subjective and require seasoned judgments. It is prudent to have a SOP that establishes the rationale for such decision-making.

Some vendor audits are triggered as part of the due diligence prior to doing business with a new supplier, some are "for cause" brought about by a worrisome problem or deviation, whereas still others are part of a scheme of regular routine checks that some firms incorporate into their quality system.

Particularly with the trend toward global sourcing, it is increasingly important to be certain that suppliers are reputable and trustworthy and have the appropriate manufacturing and quality systems in place. Prior to being put on the approved supplier's list, and at periodic intervals thereafter, each supplier should be evaluated in some effective manner. The idea is of course to conduct periodic audits by visiting the manufacturing facility from which the goods supplied originate, essentially following the same steps as outlined above for a self-inspection, or the steps similar to those used by FDA inspectors as discussed in chapter 33. However, on-site visits for audits are sometimes simply not feasible due to the necessary travel involved and in some instances also because of language barriers.

One step in such evaluations is to gather significant data regarding each supplier. This can be achieved by asking each to fill out a questionnaire. However, it has proved difficult for the suppliers to fill out the many different forms sent to them by companies needing the information, in part because the specific questions asked vary from company to company, and in part because it is often difficult for the suppliers to interpret the intent of the questions asked because of the phrasing used. To facilitate this for all parties concerned, the four leading dietary supplement trade associations in the United States have jointly developed a voluntary guideline to improve the process of gathering such information in an efficient and orderly way, called the *Standardized Information on Dietary Ingredients* (SIDI^{TM}). This simplifies the task for the vendors, and also reduces the occurrence of mistakes or inaccurate information. It also facilitates updating the information when changes occur. The forms used cover where and specifically how the product is manufactured, quality control steps used, information on the chemical and physical properties of the product, mention of known or potential sources of impurities, and/or contaminants including but not limited to residual solvents and possible microbiological issues, and many other such details that help facilitate the evaluation of dietary ingredient and excipient suppliers.

SIDI is applicable to the full range of dietary supplement ingredients, including vitamins, minerals, and botanicals as well as to excipients. Although extremely useful, this type of evaluation should not be looked upon as a *replacement* for audits. However, when audits are not feasible, or as a "pre-audit" step, SIDI is quite useful as one tool in the overall component supplier evaluation.

Further information and copies of the SIDI forms may be obtained by contacting any of the trade associations involved, the Council for Responsible Nutrition, the Consumer Products Healthcare Association, the American Herbal Products Association, or the Natural Products Association. It is not necessary to be a member of these associations to get the information and the forms.

Another approach is to hire outside firms or consultants to conduct audits of existing or potential future suppliers. However, the costs involved tend to limit this practice.

Still another evolving alternative is a not-for-profit international membership-based consortium called Rx-360, where audits can be shared by various firms that use the same suppliers. This is attractive both to the firms needing audit information and to the suppliers themselves that spend considerable time and effort hosting multiple audits, many of which are duplicative. Originally focused on the pharmaceutical industry, this is now also available to dietary supplement manufacturers, both as cost-effective tool for qualifying suppliers and to help ensure having a secure and reliable supply chain. The consortium is also in the process of adoption of standards and guidelines appropriate for dietary ingredients, excipients, and products from contract manufacturing organizations. Further details are available on Rx-360's Web site or by email to info@Rx-360.org.

THIRD-PARTY AUDITS

Before the final Part 111 was finalized, many firms wanted to let their customers know that they were quite serious about the quality of their products, so they engaged outside certification firms to come in and inspect them, and in some instances also teach them the basics of GMP. At the successful conclusion, the firm could then claim that they had been certified as being compliant. Examples include the certifications offered by the NSF International, the National Nutritional Foods Association, the U. S. Pharmacopoeia, and others. The seals of approval on labels, from a respectable trade group, were useful, both in improving the image of the dietary supplement industry in general, and as a marketing tool for specific companies.

Many firms still find advantages to being certified in this way, so such third-party audits and certifications continue to exist. FDA's point of view is that although they do not recognize or endorse *any* third-party certifications, if they are helpful in preparing firms for compliance with Part 111, that is good. Third-party certifications are now in less demand since the "playing field has been leveled" by the fact that *all* manufacturers must be in compliance with the GMPs. However, *ingredient suppliers* that are not covered by the dietary supplement GMPs have continued to show considerable interest in third-party certifications for business reasons, as have some contract manufacturing companies.

On occasion, firms that have been third-party certified have subsequently received Warning Letters from the FDA. This implies that such certifications are not necessarily guarantees that the certified firm is, in fact, *always* in full compliance with the GMPs. The term "being in full compliance" is somewhat nebulous and is arguably rarely achieved. Even after a thorough FDA inspection that does not result in a 483, it is important to keep in mind that *any* inspection or audit is merely "a snapshot" of conditions found by the investigators at a specific point in time, but cannot guaranty that at another time the situation would necessarily be the same.

ANSI STANDARD 173

The American National Standards Institute (ANSI) works with other organizations on the establishment of voluntary consensus standards. One such standard was prepared in cooperation with NSF International in connection with their NSF Dietary Supplements Certification Program, as NSF/ANSI Standard 173. This is the first (and so far, the only) American National Standard for dietary supplements and for the ingredients used in dietary supplements. This is a useful document with information on possible methods of compliance with Part 111, as well as with the adverse event reporting requirements. It provides suggested criteria and evaluation methods to help ensure that dietary supplement products contain the identity and quantity of ingredients listed on the product labels and that the products are free of any unacceptable levels of undesired contaminants. It also covers suggested ways to analyze products to ensure they do not contain undeclared ingredients, as well as suggestions as to ways of meeting other GMP requirements.

In drafting the final GMPs, the FDA considered the possibility of adopting the entire ANSI Standard 173, but it was decided that this would be impractical. Instead, the agency decided to give industry the flexibility of choosing their own ways of achieving the GMP goals as opposed to requiring the particular details and testing methods included in the standard. However, the agency supports the use of the testing methods in ANSI Standard 173, where appropriate, in complying with the GMP requirements.

SUGGESTED READINGS

Arter DR. Quality Audits for Improved Performance. 3rd ed. Milwaukee, WI: ASQ Quality Press, 2003.

Burr JT. Overcoming resistance to audits. Quality Progress, January 1987:15–18.

Gombas KL. Auditing mechanics 101: a guide for auditors and auditees. Food Safety Magazine, April/May 2005; 11(2):40–49.

Kausek J. Sharpen your auditing skills. Quality Progress, February 2008.

Mead WJ. Auditing Pharmaceutical Processing. In: Swarbrick J, Boylan JC, eds. Encyclopedia of Pharmaceutical Technology. Vol 1. New York: Marcel Dekker, Inc., 1988.

Reisch MS. Signed, sealed, and delivered: a seal of approval for dietary supplements and ingredients can mean different things. C&E News, October 7, 2002:14–15.

Russell JP. The ASQ Auditing Handbook. 3rd ed. Milwaukee, WI: ASQ Quality Press, 2005.

Wells J. Brace for impact—internal audits. Quality Progress, October 2010.

31 Outsourcing

It is acceptable, and often advantageous, to have outside companies handle specific (or even all) functions for a firm. Obtaining services from an outside source is often called "outsourcing" or "contracting out." As discussed in detail below, it is usual to spell out in detail the functions, duties, responsibilities, and fees for such services in a formal contract, which is where the terms "contractor" and "contract services" originate.

In general, contractors provide services or information, as opposed to tangible physical components or products.

It is, of course, important to establish how compliance with the regulatory issues must be properly attended to in contractual matters, as discussed below. However, in the United States, the FDA's dietary supplement GMP regulations are silent on the topic of the contractual documents per se.

Part 111 does make clear as to which parties are responsible for compliance with the GMPs where more than one party is involved in the manufacture, packaging, labeling, or holding of dietary supplements. In short, each of the parties is *directly responsible* for the specific operations that they perform and must comply with the *applicable portions* of the regulations, but the firm that owns the product and distributes it is ultimately responsible.

The firm (product owner) may or may not actually conduct the manufacturing steps (some or all of which may be contracted to others), but the firm retains the obligation to know what and how such activities are performed to make decisions as to whether the final product conforms to the established specifications to approve and release the product for distribution.

The nomenclature regarding "who's who" in outsourcing relationships tends to be somewhat confusing. In this book, the term "firm" is used to mean the entity that owns the product, whose name appears on the product's label. This is equivalent to the term "contract giver" or "sponsor." The term contractor is used to designate the name of the organization that performs certain tasks for the firm for a fee, also sometimes called the "contract acceptor."

Contracting is often used for specific operational steps such as maintenance repair and operations, calibration, cleaning and sanitation, pest control, information technology, and others. As discussed below, contracting is sometimes also used for *all* of the steps and operations in manufacturing as well as for laboratory work. However, the FDA has made it clear that when using a contractor, it is still the *firm* that has the final responsibility for the work done, *not* the contractor.

WHY FIRMS ELECT TO OUTSOURCE

The most common reason to outsource is that a firm may elect to concentrate its time and efforts on its core competencies, contracting out many or all other activities. Another principal reason is to avoid the need to invest in equipment and/or trained personnel to perform specific functions. For example, it may be easier and less costly to outsource pest control or calibration or packaging and labeling than to handle those functions in-house. Similarly, if sales outpace production capacity, it may be less costly or more expedient to outsource all or part of the needs. Also, if a new product concept seems to have potential, it may be faster and easier to get the item to market through outsourcing.

CONTRACT MANUFACTURING OPERATIONS

Not infrequently, studying the cost of goods will show the possibility that having products made by a contract manufacturing operation (CMO) may yield substantial savings as compared to the firm's manufacturing them. This is in part because the contractor typically has excellent equipment and processes, in part because of economies of scale for the contractor in the purchase of components and supplies, and in part because of efficiencies and expertise of the contractor's personnel who generally handle large volumes of many products day in and day out, in addition contractors may have good in-house laboratories for testing components and finished products. The "make or buy" decision is critical.

The choice of the optimum CMO is, of course, dependent on the kinds of dosage forms involved. Some CMOs specialize in specific dosage forms, whereas others are more general and make many types of products.

In selecting a CMO, it is usual to start with a request for proposal (RFP), which is an invitation for potential contractors to submit information to enable the firm to narrow the number of organizations to be considered. RFPs should encompass more than just pricing, and instead should provide the potential CMO with as much information on the firm's needs and expectations as feasible, to enable the CMO to explain how it can best serve its prospective client. Cost should only be one factor in choosing a contractor. On the basis of the responses received, the appropriate persons within the firm can reach a decision on which of the CMOs contacted should be further explored.

The usual next step is a visit to each facility to assess the building and equipment, the expertise and experience of their personnel, any environmental concerns, their maintenance program, and the general housekeeping of the plant. This is also an opportunity to discuss any issues requiring further clarification, which is likely to cover topics such as the their SOPs, their status with the FDA regarding regulatory compliance, and their view of what constitutes "quality." This is also the time to obtain assurance that there are no conflicting activities or practices, and to make a subjective evaluation of the probable ease of working with the CMO's management team.

VIRTUAL COMPANIES

Some firms use outsourcing for *all* of their activities, not only manufacturing and quality control, but also marketing, advertising, physical distribution, and all other functions. Such firms usually operate with just a small office staff for coordination and to ensure that regulatory issues are properly handled. Such firms are called *"virtual companies."*

In some virtual companies, the owners and the small staff (if any) work from their homes, usually communicating primarily via e-mail and text-messaging. This fragmented form of communication sometimes tends to make problem-solving difficult. However, not requiring a physical office obviously yields a substantial cost saving.

The FDA is aware of, and not particularly concerned by, the existence of virtual firms, provided the products fully comply with Part 111. The firm owns the products and the contractors merely perform a service, but both entities must comply with those sections of the GMPs that apply to them. Clearly, the virtual firm has less direct control, but is still ultimately responsible for the quality of products released for distribution.

Virtual firms are subject to FDA inspections, and should always be prepared for them, although such inspections may be daunting due to the complexities involved.

OFFSHORE OUTSOURCING

In this age of globalization, there are many CMOs in other countries. These are frequently excellent, and may offer better pricing than their domestic competitors, even considering transportation and customs duty. This tends to be particularly true in labor-intensive operations since workers in many counties have lower wages and fewer benefits than in the United States.

Obviously, distance complicates monitoring overseas producers, and the cost of travel to visit such operations should be considered in the decision-making involved with selecting an offshore CMO.

Lead times and logistical issues can pose problems in using offshore CMOs, which may require additional and more complex production planning. This often results in the decision to hold larger quantities of finished goods in the United States to ensure being able to meet any unexpected upturns in sales.

Language problems may impede working with, and communicating with, some overseas CMOs. Cultural differences may also cause difficulties in the interpretation of instructions.

Risks should be considered in deciding whether to use offshore CMOs. Some of the factors may include the possibility of wars or political upheavals, possible problems in dealing with local officials (some of whom may not operate on the same ethical standards as in the United States), erratic electric power supplies in some countries, protection of intellectual property, holidays and vacation times in various countries, the possibility of natural disasters in the area, and other factors.

CMOs in other countries that are involved with the manufacturing of dietary supplement products intended for sale in the United States *must* comply with the applicable portions of Part 111 and are subject to FDA inspection.

SUPPLY AGREEMENTS

Although not specifically required by the GMPs, generally two different contracts are used with CMOs, a supply agreement (sometimes called the business agreement) and a quality agreement. These may be combined into a single contract, although they are most frequently separate documents.

The supply agreement is a legal document that names the two parties, and typically defines certain words used such as "Product," "Services," "Components," "Packaging Materials," "Specifications," "current Good Manufacturing Practice," "FDA," "FD&C Act," "Master Batch Record," etc. An effective date of the agreement is usually specified often also with a provision for termination of the agreement. Typically such agreements include details of public and product liability insurance. Pricing and payments are stipulated, including how price increases will be handled. There is usually a clause enabling the firm to send personnel to the CMO's facility for monitoring at reasonable times, and whenever the FDA or other governmental agencies are conducting an inspection. Production planning forecasts and production schedules are usually covered. The purchasing and ownership of goods and materials is usually defined. Provision is often made for the CMO to provide a listing of current inventories upon request, including an annual physical inventory. Storage and shipment of goods is usually covered. The fact that the CMO should not subcontract with other companies or use third-party storage without authorization is usually stated.

There is at the end a page of signatures accepting and mutually agreeing to the terms of the contract. All such contracts are usually drawn by competent attorneys familiar with contract law, but with input from manufacturing and technical personnel from both companies involved. Such contracts form the basis for an important working relationship between the firm and the CMO.

Either in the supply contract or in a separate document there is usually a provision covering the ownership and confidentiality of certain information.

Some firms prefer to avoid the use of formal supply agreements, depending instead on using purchase orders, either written or digital, which often contain preprinted "boiler plate" guaranty statements.

QUALITY AGREEMENTS

Quality agreements (sometimes called "technical agreements") clarify and formalize the firm's expectations and requirements regarding the quality and regulatory aspects of outsourced manufacturing and/or laboratory operations. The use of written agreements of this type helps avoid misunderstandings and enhances communications.

Manufacturing quality agreements typically specify the respective quality roles and responsibilities of each party, including how products will be released and how quality matters are to be communicated between the parties (often listing specific contact persons). They also usually clearly state that the GMPs must be followed. Moreover, they typically will provide for documentation (including SOPs and appropriate records) and whether the firm wants full copies of all such documents or only the right to access them. The agreements usually detail what samples are to be taken and what is to be done with them. Such agreements often also spell out what must be done regarding any errors or deviations that may occur, and it may be advantageous to define how significant these must be to require action. Moreover, it is advisable to establish time limitations on investigations of deviations and failures.

Since change control is of significant importance, such agreements may require the contractor to inform the firm of any changes in the facility, equipment, in the manufacturing process, or in component suppliers (if the contractor is purchasing the components). The handling of complaints and adverse event reporting is usually clarified. It is not unusual to include a clause on how recalls or product withdrawals would be handled, if the need should arise. It is usual to include how on-site visits and audits will be handled, and the role of each party in the event of an FDA inspection.

In the event that significant disagreements may occur between the firm and the contractor, it is advisable to include a mutually satisfactory dispute resolution procedure.

Pricing, forecasting, delivery terms, and other such business-related topics should not be included in a quality agreement, but should be limited to the supply agreement, assuming these are handled as separate documents, as is usually the preferred format.

As with all contracts, quality agreements are usually prepared by lawyers, but quality agreements also need significant input from the appropriate technical personnel.

If many such agreements are expected to be required, it may be advisable to prepare a template with standardized clauses and boiler plates, which can be used to simplify and expedite future preparation of other such agreements. Having such formats can also be useful in negotiating contractual arrangements.

Although the FDA does not *require* quality agreements between product owners and CMOs at various industry meetings, the FDA representatives have stated that they encourage having such contracts in place with the responsibilities of the contract facility clearly defined. They have further stated that it is prudent to closely monitor the work of contractors through open communications and frequent in-plant visits or audits to know what is actually being done at the contract facility, which is in effect an extension of the firm's own organization.

FDA'S VIEWS ON OUTSOURCING

From FDA's point of view, whether a firm does testing or manufacturing at its own facility or at a contract facility, the expectations are the same. The contract facility must follow the applicable portions of the GMPs and must generate good data. Although everyone involved in the production or testing of dietary supplement products is responsible for ensuring compliance, the firm (the owner or sponsor) has the ultimate responsibility for the products.

If a problem occurs at a contractor's facility, which results in a GMP violation, it is still up to the owner of the product to take action against the product in the marketplace, initiating a recall if appropriate. If need be, the FDA may issue a Warning Letter or take other regulatory action against a contractor, but that does not relieve the product owner from responsibility.

CONTRACT RESEARCH ORGANIZATIONS

Although some CMOs also offer services for developing new products for their clients, there are also contract *research* organizations (CROs) that specialize in dietary supplement product formula development.

Since *extensive* research and development tends to be costly, when this is needed, it may be economically beneficial to have this done offshore. The drug industry, being much more research-intensive than the dietary supplement industry, now does much of their basic product development work overseas to hold down costs, resulting in the existence of many well-staffed offshore CROs.

CONTRACT ANALYTICAL LABORATORIES

As detailed in chapter 19, the dietary supplement GMPs *do* require component identification and other analytical testing to ensure components, in-process items, and finished products meet their specifications. Particularly small and medium-sized firms may lack adequate laboratory facilities and trained personnel, and therefore may elect to use the services of outside contract analytical laboratories (CALs) rather than going to the expense of providing these in-house.

In some instances, even with having an in-house laboratory, it may be desirable to use a CAL for certain specific purposes, for example, for work the in-house laboratory cannot handle due to equipment limitations, or for the development of new analytical methods.

It is advisable for the firm to have a SOP or at least an internal guidance document on the selection and use of CALs, if such use is contemplated. This would typically include the initial steps to be taken in selecting a CAL, such as clearly identifying the needs, doing due diligence, including visiting the laboratory to judge the facility and available equipment, checking with others who have used or are currently using the CAL, checking any appropriate and available FDA EIRs or 483s or Warning Letters concerning the CAL, determining how long the CAL has been in business, looking into the educational and experience backgrounds of the personnel involved, and any accreditations that may exist. Once a determination has been made to use a specific CAL, the SOP or guidance document should call for entering into both a business agreement and a quality agreement as discussed above.

After the selection and entering into contracts, the document should cover the working details, such as the transmittal and logging-in of samples, analytical methods to be employed, calibration and reference materials to be used, information to be included in the written report, and how the reports are to be delivered to the firm.

In the beginning of the relationship with a CAL it may be prudent to have identical samples analyzed in two or more different laboratories, and then to compare the results.

As is true in all outsourcing arrangements, good communication with CALs is an essential element.

CONSULTANTS

Consultants are considered to be a form of outsourcing, providing information and advice. Consultants can be of significant assistance in many areas, but obviously must be competent, although assessing this can be difficult. Therefore, the firm should be certain that consultants used do have sufficient and appropriate education, training, and experience (or a combination of these) to offer advice on the topic for which they are retained. Records should be maintained on the qualification of consultants used, the type of service they provided, and the dates they were used.

SUGGESTED READINGS

Anon. Early drafting of separate quality agreements with contractors urged. The Gold Sheet 2002; 36(11).

Brooks K. CRO industry update: leveraging expertise for innovation. Contract Pharma Magazine, May 2011.

Damon E. Manufacturer or retailer: who's responsible for a private label brand? Private Label Magazine, September–October 2002:61.

Dentali S. Choosing an analytical lab. Natural Products Industry Insider, September 15, 2003:84–86.

Guzman R. A regulatory approach to contract manufacturing. Pharmaceutical Outsourcing Magazine, March–April 2010:16–20.

Handel T. Quality agreements: how to clarify what happens after the contract is signed. Contract Pharma Magazine, October 2002:74–80.
Piepoli M. The changing role of contract laboratories. Pharmaceutical Technology, November 1999:38–40.
Trankler K. All in a day's work: contact labs handle problems your in-house lab can't. Food Quality, April/May 2008:38–42.
Vogel C. A contract laboratory questionnaire: the right questions can ensure that a contract lab is a good fit. Food Quality, April/May 2008:44–45.
Wyrick ML. A win-win guide to contracting services. Pharmaceutical Technology Sourcing and Management, October 2006:e10–e20. Available at: http://electronic.pharmtech.com.

32 The Food and Drug Administration

The Food and Drug Administration (FDA) is the oldest federal regulatory agency in the United States and is responsible for helping to ensure the safety and quality of a vast array of health care products, including dietary supplements. It carries out its mission in many ways, one of which is through enforcing compliance with its Good Manufacturing Practice (GMP) regulations. It is therefore of importance for manufacturers of dietary supplements to be familiar with the agency and its *modus operandi*.

Through enforcement activities and endeavoring to teach industry the means of (and the advantages of) full GMP compliance, the FDA emphasizes the role of *preventing* actual or legally adulterated dietary supplements from reaching the marketplace. Prevention is a fundamental principle in public health practice.

The FDA is a regulatory and science-based public health agency that has jurisdiction over most food products (other than meat and poultry), including dietary supplements. It also has jurisdiction over human and animal drugs, therapeutic agents of biological origin, medical devices, cosmetics, animal feed, and tobacco products. The agency has more than 12,000 employees, a large team of well-qualified individuals who work to ensure that the products they regulate are in compliance with the law.

Recent advancements in science and technology impacting public health have resulted in what is termed *regulatory science*, the science of developing new tools, standards, and approaches to assess the safety, efficacy, quality, and performance of FDA-regulated products. The FDA is actively pursuing this new concept of regulatory science.

With the changes that have occurred over time, including globalization, FDA's job is now far more complex than it was even a few years ago. This has been recognized by Congress and additional funding has been provided to help meet the challenges.

FDA ADVISORY COMMITTEES

To assist its mission, the FDA uses a large number of groups of outside experts, including knowledgeable consumer and industry personnel, to provide independent advice to the agency on a range of scientific, technical, and policy issues. Although the FDA is *not required* to follow the advice provided, and it makes the final decisions, the recommendations are in fact usually followed. Advisory committees are a valuable resource for the FDA, and they make significant contributions to the agency's decision-making processes. Often, a series of questions is proposed to the committees both in advance and during the discussion at the meetings. In some (but not all) advisory committee deliberations votes are taken. Transparency and public participation are critical features of this process.

Recognizing the importance of public's role in developing effective policies, other agencies in the executive branch of the federal government also use advisory committees. Therefore, in 1972, Congress enacted the Federal Advisory Committee Act (FACA) to ensure that the advice and guidance provided by such committees are objective and accessible to the public. Moreover, the FDA has established procedures for the use of advisory committees in 21 CFR Part 14.

One specific and important advisory committee is FDA's Science Board, which provides advice to the commissioner and other appropriate officials on specific and complex technical issues as well as emerging issues within industry and academia. The board also aids the agency in keeping pace with evolutions in the field of regulatory science, including upgrading scientific and research facilities to keep pace with the changes occurring. This board consists of 21 members.

Still another committee of interest is the Food Advisory Committee, which comprises experts in applicable fields such as toxicology, chemistry, environmental health, neuroscience, and others. This committee reviews and evaluates data, and makes recommendations on broad scientific and technical topics on emerging health-related issues of interest to the Center for Food Safety and Applied Nutrition, which on occasion may involve dietary supplement matters. The Food Advisory Committee may also be asked to provide suggestions on ways of communicating potential risks to the public.

A BRIEF OVERVIEW OF GMP HISTORY

In the early 20th century, after years of discussion about the miserable conditions existing in some meat processing facilities, it became clear both to the public and to President Theodore Roosevelt that steps needed to be taken to correct the situation. This was enhanced by a book titled *The Jungle*, written by Upton Sinclair, which resulted in public pressure for corrective action. This led to Congress passing the Pure Food and Drug Act of 1906, the forerunner of the food and drug legislation we have today. This act was first administered by the Bureau of Chemistry of the Department of Agriculture, then headed by Harvey W. Wiley, MD, who served in that position until 1912.

The agency's name was changed in 1927 to the Food, Drug, and Insecticide Administration, and again changed in 1930 to the present name of Food and Drug Administration. The FDA remained under the Department of Agriculture until 1940, and then through a series of changes is now part of the Department of Health and Human Services (DHHS).

In 1937, a crisis occurred from the inadvertent use of diethylene glycol as the solvent in a liquid product called Elixir of Sulfanilamide, an antibacterial drug. An "elixir" is usually a sweetened mixture of alcohol and water containing a relatively small amount of a drug substance. In other words, it is a liquid dosage form. The manufacturer of this product chose to use sweet-tasting diethylene glycol rather than alcohol and water, not realizing that diethylene glycol is toxic. This resulted in many deaths. At that point in time, premarket safety testing was not required. However, Congress thereafter quickly passed the 1938 Federal Food, Drug, and Cosmetic (FD&C) Act that *did* require proof of *safety* of every new drug product. The 1938 act has been extensively amended several times, but is still the basic (albeit not the only) law covering FDA's regulatory authority.

In 1962, still another tragic occurrence resulted in further amendments to the FD&C Act. A sedative, painkiller, and tranquilizer drug called *thalidomide* was sold in a number of countries in the late 1950s, and an application to the FDA for its introduction into the United States market was pending. A physician and pharmacologist on the FDA's technical staff, Francis Oldham Kelsey, refused to approve the pending application on the grounds that more studies were needed. It soon came to light that in Europe, many babies were born deformed due to the mothers' use of thalidomide during pregnancy. So, the thalidomide disaster was avoided in the United States. However, this near miss caused Congress to again amend the act, this time requiring not only safety but also *effectiveness* be established by adequate and well-controlled studies before approval of a new drug application. These amendments are called the Kefauver–Harris Amendments of 1962. Included in these amendments, what is now Section 501(a)(2)(B) of the act, is the following requirement:

> A drug shall be deemed to be adulterated if the methods used in, or the facilities or controls used for, its manufacture, processing, packing, or holding do not conform to or are not operated or administered in conformity with current good manufacturing practice to assure that such drug meets the requirements of this Act as to the safety and has the identity and strength, and meets the quality and purity characteristics which it purports or is represented to posses.

At that time, in 1962, this was the first-ever mention of Good Manufacturing Practice in *any* U.S. law or regulation, and neither the FDA nor the industry was precisely certain just what Congress *meant* by the term. The FDA therefore sent representatives to a number of the appropriate trade associations to discuss this in detail. From these discussions arose the first version of the drug GMP regulations. These regulations have of course been revised and updated with the passage of time. The drug GMP concept quickly was picked up around the world, and was also expanded to other fields covered by the FDA (and equivalent agencies in other countries) such as foods, medical devices, biological products, and of course now also dietary supplements.

GMPs were sometimes aptly referred to as "manufacturing controls," although this term is now seldom used.

The Dietary Supplement Health and Education Act (DSHEA) amended the FD&C Act, by adding Section 402(g), which authorized the FDA to promulgate GMPs for dietary supplements. This also stipulates that such regulations must be modeled after those for foods as opposed to those for drugs. The FDA issued an Advance Notice of Proposed Rulemaking in February 1997, based on an industry-submitted draft. Then, in March 2003, the FDA published the proposed dietary supplement GMPs, and eventually in June 2007, the final version was published with compliance dates ranging from June 2008 to June 2010, depending on the size of the firm. These regulations now have the full force of law. The constitutionality of the GMP regulations has been challenged, with claims that the FDA exceeded their statutory authority in promulgating them, and they are vague since they contain nonexplicit terms such as "adequate," "suitable," and "qualified." However, the courts have reiterated that the FDA does indeed have broad authority to regulate the steps in the manufacturing, packaging, labeling, and holding of dietary supplements. In short, FDA's interpretation of the statute was appropriate.

FDA's GMP regulations for foods (21 CFR 101.110) were adopted in 1986. In addition, other manufacturing or quality regulations have been adopted for specific categories of food, including low-acid canned foods, bottled water, and infant formula. Another more rigorous approach to the evaluation and control of food manufacturing practices was developed in the 1960s as part of the NASA effort to ensure the quality and safely of foods prepared for use by astronauts. This is known as HACCP (Hazard Analysis and Critical Control Points) and involves the systematic analysis of those points in a food production process where hazards may exist or may be introduced and where control is needed. Necessary controls, including methods of analysis to determine whether control was achieved, must be specified, and records demonstrating successful control must be kept. HACCP procedures have been voluntarily adopted by many food manufacturers and have been required by the FDA or USDA for some categories of food, including seafood, juices, and meat and poultry.

ORGANIZATIONAL STRUCTURE OF FDA

As stated earlier, the FDA is part of the DHSS. It consists of the Office of the Commissioner and the Office of Regulatory Affairs, and the following six centers:

- Center for Food Safety and Applied Nutrition (CFSAN)
- Center for Drug Evaluation and Research (CDER)
- Center for Devices and Radiological Health (CDRH)
- Center for Biologics Evaluation and Research (CBER)
- Center for Veterinary Medicine (CVM)
- Center for Tobacco Products (CTP)

Dietary supplements fall under the purview of the Division of Dietary Supplement Programs (DDSP), within the Center for Food Safety and Applied Nutrition, located at 5100 Paint Branch Parkway, College Park, MD 20740.

The Office of the Commissioner (OC) is located on the White Oak Federal Research Center Campus in Silver Spring, Maryland. The commissioner is the head of the FDA, appointed by the U.S. president after approval from the Senate. There are also deputy commissioners who oversee the efficient and effective implementation of FDA's mission. The OC includes offices of Administration, the Chief Counsel, Chief Scientist, External Affairs, Foods, International Programs, Medical Programs, Policy, Legislation, Crisis Management, Combination Products, Orphan Products, Pediatrics, Therapeutics, Counter Terrorism and Emerging Threats, Public Affairs, Planning and Budget, Women's Health, and Equal Employment Opportunity and Diversity Management.

The Office of Regulatory Affairs (ORA) is of particular interest to dietary supplement firms in that it is the lead office for all FDA field activities, including establishment inspections. It also develops FDA policy on compliance and enforcement. Although the headquarters of the ORA is at the same location as the OC, the ORA also operates in five regional offices and 20 district offices, as well as numerous laboratories, resident posts, and border stations. The head of the ORA is an associate commissioner. The five ORA regions are Northeast, Southeast, Central, Southwest, and Pacific.

The ORA maintains an Office of Enforcement, and an Office of Criminal Investigations (OCI) with special agents that focus on violations of the FD&C Act and on the Federal Anti-tampering Act. These agents operate much like other federal law enforcement organizations, gathering facts that are then presented to the U.S. Attorney's Office, which determines whether a case will be prosecuted. There is also an ORA Forensic Chemistry Center that performs appropriate tests on suspect products involved in OCI investigations.

The ORA also collaborates with appropriate state and local regulatory agencies, and with equivalent agencies in other countries. The ORA personnel conduct inspections in foreign countries that export products for distribution in the United States.

As discussed in chapter 33, ORA personnel conduct inspections of establishments involved with the manufacture, processing, or holding of dietary supplement products to determine the establishments' compliance with all of the regulations administered by the FDA, including the GMPs. If objectionable conditions are observed, the senior personnel of the establishment are furnished with a document called FDA Form 483 detailing such observations. If no enforcement action is contemplated, or after enforcement action is concluded, the FDA provides the inspected establishment with a final report, called an Establishment Inspection Report (EIR).

ORA's inspections are prioritized based on the agency's perception of the magnitude of the public health issues at stake. Those firms making products that may present health risks or that have previous records of poor performance get the highest priority. This is one of the many places where FDA employs risk-based decision-making, using criteria based on definition of the of the decision options, followed by an estimation of the public health consequences of each potential action, to enable selection of the optimum option. This helps maximize FDA's performance. Clearly, there are instances where it is difficult to accurately predict the consequences of each option, but in many of the decisions the FDA deals with, this approach is very useful. Moreover, the FDA has experts who are trained in and comfortable with decision analysis, risk assessment, risk management, and the assessment of uncertainties. The basic concept of risk-based decision-making is generally understood and used where appropriate by the agency.

The FDA is obviously very interested in ensuring that dietary supplement manufacturers *are* following the GMP regulations, so within ORA's manpower and time limits, such firms *are* being (and will continue to be) regularly and thoroughly inspected.

The majority of the establishments diligently attempt to fully comply with the GMPs, whereas those that do not are viewed by others in the industry as taking an unfair commercial competitive advantage. The playing field must be level, with all firms endeavoring to be fully compliant with all applicable laws and regulations.

The FDA personnel frequently participate in industry meetings and seminars to help inform and educate participants regarding compliance topics.

The FDA properly expects and encourages voluntary compliance with the several laws it enforces. However, when failure to comply is discovered, the agency is fully prepared to take strong enforcement actions.

FDA GUIDANCE DOCUMENTS

In addition to the actual regulations, the FDA also issues informal guidance documents to the industry. These are useful in that they represent FDA's current thinking on various topics, and it is prudent (but *not* mandatory) to follow the suggestions contained in such documents. Guidance documents are not binding on either the agency or the industry.

The FDA has established *good guidance practices* (GGPs) that set forth the agency's policies and procedures for developing, issuing, and using guidance documents. These are formalized in §10.115. This sets forth two levels of such documents. Level 1 guidance documents cover FDA's initial interpretations of statutory or regulatory requirements and any significant changes in interpretation or policy. It also covers highly controversial issues. Level 2 guidance documents set forth existing practices or minor changes in interpretation or policy. All guidance documents that are not Level 1 are Level 2. The Office of Management and Budget (OMB) is responsible for promoting good management practices and for overseeing and coordinating the administration's regulatory policy for all of the agencies, including the FDA. The OMB recognizes the enormous value of well-designed guidance documents.

CONFLICT RESOLUTION

Disputes may occasionally arise between firms or individuals and agency personnel.

The FDA has mechanisms for handling disputes, and complaints about responses from FDA offices, claims of unfair or unequal treatment, and similar topics. Some matters of this type can often be resolved through meetings with personnel at the appropriate ORA District Office, which is usually the best place to start. Moreover, there is an Office of the Ombudsman available to provide guidance and assistance in resolving problems and disagreements.

Disputes are sometimes triggered by differences in interpretation of the regulations and guidance documents. However, some disputes may also arise from differences in opinion or jurisdiction between ORA offices and specific personnel. It is not always entirely clear as to who is the decision-maker, which needs to be clearly defined in each instance. In any event, good communication is a necessary component of resolving disputes.

People from industry sometimes tend to fear the possibility of retaliation when FDA decisions are challenged, which implies that industry personnel should use care and discretion in raising argumentative issues. It is prudent to be reasonably certain of the proper interpretation of regulatory issues, and on technical matters to have scientifically strong background information available to support positions.

When they do occur, disputes should be resolved at the earliest feasible time, and at the lowest possible level. Informal resolution processes often work well, although this is largely dependent on the people involved.

FDA'S WEB SITE

The FDA maintains a truly excellent Web site at www.fda.gov, containing much useful information and links to many other appropriate sources. It also has a search engine, allowing access to a great variety of FDA matters. The home page is user-friendly, covering news and events, regulatory information, recall data, plus the ability to subscribe to free e-mail notices and alerts on many pertinent topics. Since this site is updated frequently, it is prudent to visit it often.

FREEDOM OF INFORMATION

The Freedom of Information Act (FOIA) is a federal law, passed in 1966, to facilitate making information from executive branch government agencies available to the public. This has been amended several times, both formally and by executive orders. The FDA is one of the agencies covered by this mandate.

Since requests for specific documents from FDA's FOIA service are typically very slow in coming and can be costly, this should be a tool of last resort. It is advisable to first do research to see if the desired information may be more readily available elsewhere.

Specific FOIA requests must be made in writing, and should contain the requestor's name, address, phone number, and a brief but clear description of the specific records desired that are releasable to the public. It is inadvisable to request "all available" information on a topic. FDA's Division of Freedom of Information does not accept requests sent by e-mail, although the requests may be sent by fax to 301/827-9267. If using mail, the address is

Food and Drug Administration
Division of Freedom of Information
Office of Public Information and Library Services
12420 Parklawn Building, ELEM-1029
Rockville, MD 20857

The phone number of the FOI Offices is 301/796-3900.

It is possible to visit FDA's FOI Public Reading Room at the above address, to search available documents.

Separate requests should be made for each product, firm, or other such entity about which records are desired.

Each written request should also contain a statement concerning the willingness to pay applicable fees, including any limitations.

Some information in FDA records is protected from public disclosure under the FOIA. This includes certain confidential proprietary information. Such information is redacted (blacked out) or deleted before making the records available to the public. In some instances, such as in long documents, it can take considerable time for the FDA to remove the nonreleasable portions, and the FDA charges the requestor for such time. FDA's regulations governing disclosures are in 21 CFR Part 20.

There are commercial services that can obtain FOI records much more quickly than by the direct written request route. Although costly, this is a fast and useful way of obtaining releasable records.

ELECTRONIC READING ROOMS

The ORA maintains an electronic reading room where some records are made readily available to the public, at ORA's discretion, as called for in the Electronic Freedom of Information Act Amendments of 1996. Among other documents, all Warning Letters (WLs) are on display.

FDA TRANSPARENCY INITATIVE

In keeping with the other executive branch agencies, the DHHS (including the FDA) is making *transparency* a priority. In this context, the term "transparency" refers to having well-defined regulations, policies, and procedures that are open to public scrutiny. This implies letting the public have easy access to agency information and decision-making processes. Keeping the public informed is the basis for accountability and public confidence in the agency. Using this principle, the public can easily have input and participate in policymaking. Moreover, this leads to enhanced collaboration among agencies and with outside nonprofit organizations, businesses, academia, and individuals.

FDA'S ENFORCEMENT POWERS

The FDA has many actions that can be used to help ensure compliance with the laws and regulations it enforces.

These include the issuance of WLs to firms notifying them of violations, but allowing them time to voluntarily comply with the law. The FDA also uses what they call "Untitled Letters" regarding more minor infractions, but these suggest and request corrective action as opposed to demanding it. The agency's position is that WLs are issued only for violations that are of regulatory significance and may lead to further enforcement action if not promptly and adequately corrected. This is FDA's principal means of achieving prompt voluntary compliance for significant violations. These are advisory, communicating the agency's position on a matter. A WL is *not* a prerequisite to the agency taking enforcement action, which in some instances may be taken *without* necessarily first issuing a WL. As a general rule, a WL is not issued if the agency concludes that a firm's corrective actions have been adequate. WLs are often the result of conditions found during establishment inspections, but may also be based on other sources of information. A key goal, however, is to get individuals and/or firms to take prompt and adequate corrective actions. WLs are generally sent to the highest known official in the firm involved, such as the CEO, president, or another senior official. A few typical comments in WLs relating to dietary supplements include the following:

- You failed to conduct appropriate tests to verify that your finished batch met specifications in §111.75(c)(2)
- You failed to establish specifications as required by §111.70(b)(2)
- You failed to determine that specifications were met as required by §111.75 (a)(2)
- You failed to establish in-process specifications in the Master Manufacturing Record (MMR) as required in §111.70(c)(1)
- You failed to prepare a MMR for each unique formulation and batch size as required by §111.205(a)
- Your quality control personnel failed to conduct a material review and make a disposition decision as required by §111.113(a)(1)
- Your BPR failed to include reconciliation of discrepancies between the issuance and use of labels as required by §111.260(k)(1)

Clearly, many more examples could be cited, but the above indicates the *types* of items frequently included in WLs.

It is useful to read copies of the WLs in the online electronic reading room mentioned above, since they provide useful insight into the findings and concerns of FDA inspectors.

When WLs fail to bring about the necessary corrective actions, the FDA can resort to seizures or injunctions, which are civil actions initiated through the U.S. Attorney's Office and the U.S. District Courts.

Seizures result in U.S. Marshals physically taking possession of the goods. This is usually *not* FDA's preferred enforcement option due to the time and costs involved, but is sometimes used when WLs have failed to resolve the matters. However, there is no legal requirement that the FDA first issue a WL before proceeding to other enforcement actions. The FDA uses the terms "mass" and "open-ended" to distinguish these from "lot-specific" seizures. A mass seizure involves *all* FDA-regulated products at an establishment or facility, for example, for product held in a filthy warehouse or produced in nonconformance with GMPs. An open-ended seizure involves all units of a specific product(s) regardless of lot or batch numbers. Open-ended seizures may be conducted when the violation extends to all lots or batches of a product, but not to all of the products in the firm.

An *injunction* is a judicial process initiated to stop or prevent actions that are in violation of the law, for example, to halt the flow of violative goods in interstate commerce. Injunctions can be considered for any out-of-compliance circumstance, but particularly when a health hazard or gross consumer deception is involved, requiring immediate action and a seizure is impractical. When requesting an injunction, the FDA usually tells the court that it has made a strong effort to get the objectionable products or practices corrected without court involvement but to no avail. The FDA usually prefers to have the individuals named defendants who had the authority and responsibility to prevent or correct the violations to help prevent future violations.

Seizures and injunctions are discussed in detail in FDA's *Regulatory Procedures Manual*.

Seizures and injunctions are frequently resolved by what is called a *consent decree*, which is an agreement (approved by the court) by a defendant to cease the illegal activities alleged by the FDA, in return for an end to the charges. A consent decree is a legally binding settlement that the defendant is *required* to abide by, and if the firm or individuals involved fail to do so, the court may impose other penalties, including stiff fines for each violation. In some instances, such decrees require the use of qualified independent consultants to review the facilities and procedures involved, and submit reports of their findings to the FDA. Fines may be imposed for failure to meet the prescribed schedules. Moreover, some consent decrees also call for *disgorgement of profits*, a legal term meaning the forced giving up of profits made through illegal or unethical ways, as a part of the settlement.

Still another powerful tool in forcing compliance is the criminal prosecution of persons responsible for noncompliance. Investigation and prosecution of individuals for violative conduct is an important component of FDA's enforcement options. This may lead to fines, incarceration, and/or debarment (which relates to persons convicted of felony or misdemeanor offenses being disallowed to continue to work in an FDA-regulated firm for a specified period of time, which can cause significant career disruptions).

The FD&C Act is one of a very few federal laws that applies the concept of "strict liability," meaning that individuals may be held accountable for the misconduct of *others*, even if they had no knowledge of or involvement in the offense. In other words, actual knowledge or personal participation in an offense is unnecessary to secure a conviction. This can make executives *vicariously* liable for the wrongdoings of employees working for them. The basis for this is that an executive with the responsibility and authority to *prevent* or *correct* a violation can be held fully accountable. It is enough that company employees *committed* a violation and the executive had the *authority* to prevent or correct it. Executives are therefore in serious jeopardy for the failures that may occur.

The Supreme Court has twice confirmed the validity of this important feature of the FD&C Act, which is now called the "Park Doctrine" after one of the court's decisions. This is also sometimes called the "Responsible Corporate Officer (RCO) Doctrine."

It is obviously of great importance for executives of dietary supplement manufacturing firms to be aware of this unusual fact of law. Although comprehensive in scope, covering *all* of the so-called prohibited acts (including misbranding), this has been and will continue to be used also for significant or repeated GMP violations.

In the decision in the Park case, the Supreme Court stated that "The FD&C Act imposes a positive duty to seek out and remedy violations when they occur, but also and primarily, a duty to implement measures to insure that violations will *not* occur." This relates to the responsibilities of management as covered in chapter 24.

In most instances when the Park Doctrine is used the person accused is allowed what is called a "305" hearing, which refers to Section 305 of the FD&C Act. This gives the person an opportunity to argue why the FDA and the Department of Justice should not proceed with a criminal prosecution. However, the FDA is not *required* to hold such hearings.

SUGGESTED READINGS

Anon. Medicine: the thalidomide disaster. Time, August 10, 1962; 80(6).
Ballentine C. Taste of raspberries, taste of death: the 1937 Elixir Sulfanilamide incident. FDA Consumer Magazine, June 1981.
Barkan ID. Industry invites regulation: the passage of the pure food and drug act of 1906. Am J Public Health 1985; 75(1):18–26.
Davis JB. New use of old tools: career-ending OIG exclusion and FDA debarment. FDLI Update, September/October 2010:56–58.
Eglovitch JS. FDA resurrects park doctrine in enforcement of pharmaceutical GMPs. The Gold Sheet, April 2011:15–19.
Fleder JR. The park criminal liability doctrine: is it dead or is it awakening? FDLI Update, September/October 2009:48–52.
Hilts PJ. The FDA at work: cutting-edge science promoting public health. FDA Consumer Magazine, January–February 2006:28–41.
Kowal SM. Personal responsibility and personal jeopardy: FDA's prosecution of individuals. FDLI Update, May/June 2002:50–52.
Nordenberg T. Inside FDA: barring people from the industry. FDA Consumer Magazine, March 1997.
Rados C. FDA law enforcement: critical to product safety. FDA Consumer Magazine, January–February 2006.
Zaret EH. Management responsibility in modern quality systems. Pharmaceutical Formulation & Quality, December/January 2005:44–45.
Ziporyn T. The food and drug administration: how those regulations came to be. J Am Med Assoc 1985; 254(15):2037–2039, 2043–2046.

33 FDA inspections

One of FDA's primary missions is to ensure that industry is, and remains, in full compliance with the FD&C Act and with the regulations the agency has promulgated. This of course includes compliance with the dietary supplement GMPs. One of the agency's most effective tools in doing this is to periodically inspect establishments that manufacture, process, pack, and/or hold such products.

Most inspections are unannounced. Therefore, it is important to always be prepared, which includes having employees trained in their roles during an inspection. It is advisable (but not required) to have a company SOP covering FDA inspections.

Under §704 of the FD&C Act, the FDA has the authority to inspect at reasonable times, reasonable limits, and within a reasonable manner. If the plant operates on more than one shift, inspectors may, therefore, arrive at times other than normal daytime hours. Inspections may either be for routine surveillance or "for cause" (meaning that the agency has a specific reason for conducting the inspection). It is appropriate to ask the reason for the inspection, although the inspector may or may not say.

QUALIFICATION OF FDA INVESTIGATORS

Investigators are also called Consumer Safety Officers (CSOs). CSOs must, as a minimum requirement, hold a bachelor's or higher degree from an accredited college or university, including at least 30 semester hours of appropriate science courses. They must have the physical ability to work long hours when necessary. The FDA provides them with both classroom and on-the-job training not only at the beginning of their careers with the agency but also on an ongoing basis, so those with a reasonable amount of experience are generally highly competent.

THE OPENING STAGE

On arrival, each investigator will introduce himself/herself and will present photo identification. An FDA Form 482 (called a "Notice of Inspection") will then be given to the most responsible person of the firm who is present. The 482 does not specifically state the reason for the inspection, or what the inspectors expect to find.

It is inadvisable to make investigators wait long when they arrive.

While sometimes only one investigator is involved, it is not unusual for a team of specialists from the FDA (e.g., chemists, microbiologists, computer specialists, etc.) to participate in the inspection, although they may not necessarily all arrive or be present at the same time.

It is highly advisable for the company to designate at least one person (and a backup, in case a designated person is unavailable) to receive and accompany the investigator(s), both to be of assistance and to take comprehensive notes on what transpires. These individuals should be well trained on the firm's policies regarding the issues likely to arise during an inspection, as well as regarding both the FDA's and the firm's rights. Moreover, they should be knowledgeable about the firm's manufacturing and quality control procedures to the extent of enabling them to either answer the investigators' questions or at least to know to whom such questions should best be directed.

As a general rule, it is unwise to let the investigators just wander alone through the facility, or to strike up conversations in the absence of one of the designated individuals. However, the firm has very little ability to control to whom an investigator speaks, as investigators are entitled to speak to any employee, including outside the plant.

The FDA does *not* require a warrant from a judge or magistrate to authorize an inspection, although warrants may be used under some circumstances.

Under §301(f) of the FD&C Act, "refusal to permit entry or inspection" is a criminal offense. Therefore, there is no choice but to let the inspection begin when the FDA personnel arrive.

It is common practice (but not a requirement) for the firm to provide the agency personnel with an office (or at least a desk) for use during their stay. It is inadvisable to offer to provide food, or to offer any kind of gifts (even company products) to investigators.

CONDUCT OF THE INSPECTION

It is usual for the investigator(s) to want a preliminary tour of the facility, called an "inspection walk through." This gives a visual overview of the manufacturing site, including the products, equipment, and processes involved. It also enables the investigators to plan their inspection strategy. The walk through is also helpful in establishing the necessary depth of the inspection, and identifies potential areas of concern. Moreover, it yields information on the environmental conditions inside and outside the plant, including a general view of the housekeeping. It may also indicate possible sources of contamination.

The depth of the inspection is typically determined in part by the purpose of the inspection, and in part by the investigators' knowledge of the manufacturing and quality control aspects of the dietary supplement industry. More important is the firm's regulatory history, and the conditions found as the inspection progresses.

The investigators will have done their homework thoroughly before their arrival at the plant, by studying the agency's files on the firm (including records on the outcome of previous inspections, if any). They will also receive information and instructions from appropriate ORA personnel.

Investigators tend to differ in their approaches to conducting inspections, although all follow the ORA's instructions. Inexperienced new CSOs tend to be less likely to accept industry's explanations of situations previously unknown to them, which may impact their concept of what is and what is not acceptable.

PREPARING TO BE INSPECTED

It is important to be ready to be inspected. Everyone in the firm needs to understand both the importance of inspections as well as what to do (and not do) while an inspection is being conducted. Preparation and training can help facilitate a favorable outcome of the experience.

It is prudent to have a SOP on this topic. Conducting internal audits to help ensure being in compliance is a useful step, as discussed in chapter 30. Holding "mock" inspections can also be helpful both for training purposes and to detect any conditions that may need correction or improvement.

It is advisable to tell the inspector that *all* questions and requests for records, files, SOPs, etc., are to be directed *only* to a designated representative. Many firms use an escort team, as opposed to only one individual, which facilitates quickly getting any documents requested.

If a person does not know the answer to a question the inspector asks, he/she should refer it to an individual who does know. It is prudent and proper for a person to say they do not know if, in fact, they do not or are unsure. It is important to listen carefully to each question and answer only after the question is fully understood. It is highly advisable to ask for clarification if necessary. Questions must be answered directly and honestly, without providing any misinformation. Having answered a question, it is wise to stop speaking and not volunteer information beyond what was specifically requested.

It is important that no employees of the firm sign or initial *any* affidavits or other documents without first consulting competent legal advice. This needs to be clearly understood by all.

A designated person should accompany the investigator at all times and should take copious notes on everything the inspector asks, says, or does.

After the inspection has been completed, it is advisable for the appropriate company personnel to hold a meeting to discuss the details. If feasible, the firm's legal counsel may be a useful part of this discussion.

The investigators will follow *Compliance Program 7321.008* (titled "Dietary Supplements—Import and Domestic"). Moreover, general details on how inspections are done can be found in Chapter 5 (titled "Establishment Inspections") of the current edition of the *Investigations Operations Manual* (IOM). Both of these publications are readily available to the public and are worthy of study as aids in preparation for being inspected.

INFORMATION TO WHICH FDA IS NOT ENTITLED

CSOs know that they are not to ask for the firm's financial data or pricing or sales data, other than sufficient information to establish the conduct of interstate commerce. Apart from confirming the qualifications of individuals performing certain specific functions, as called for in Part 111, personal information on personnel is not available to the agency. However, refusing a specific request from an inspector is risky and could lead to enforcement actions. If unsure in borderline instances, it is prudent to check with the firm's legal counsel to avoid the possibility of being charged with a partial refusal, the term used for refusing to give information to which the FDA is entitled under the law.

The other side of this coin is that employees of the firm should not try to get proprietary information about other companies from the investigators. Investigators typically know a considerable amount about other firms in the industry, but they are not permitted to disclose how other companies handle their scientific, technical, engineering, and business details.

On occasion, investigators may make suggestions on ways to improve certain aspects of a firm's operations. When this occurs, they are merely suggestions, not mandates.

It is best to keep in mind that investigators conduct inspections to *detect violations* and to gather evidence when appropriate. Inspections are not for the purpose of teaching or helping the firm to improve. The FDA does frequently participate in (and sometimes sponsors) seminars and other types of forums to help inform the industry on details of the GMPs and possible means of compliance, but that is not what the inspection process is about.

PHOTOGRAPHS, A CONTESTED ISSUE

As mentioned above, one of the objectives of facility inspections is to collect evidence to support regulatory actions if the investigator perceives that violations have occurred. Obviously, in some instances, photographs provide excellent visualization of violations, useful in court cases. Photographic evidence can be very damaging to the firm.

Investigators are taught the basics of photography and are provided with cameras. This is discussed in detail in Section 5.3.4 "Photographs," of the FDA's *Investigations Operations Manual*. Among other pertinent statements in this, one that is particularly controversial is the following:

> Do not request permission from management to take photographs during an inspection. Take your camera into the firm and use it as necessary, just as you would any other inspectional equipment. If management objects to taking photographs, explain that photos are an integral part of an inspection and present an accurate picture of plant conditions. Advise management that the U.S. Courts have held that photographs may lawfully be taken as part of an inspection.

The FD&C Act does not specifically refer to or authorize in-plant photography.

A firm's refusal to permit photography might be considered as a partial refusal to permit an inspection, which can be a criminal offense. The court cases cited by the FDA in support of their right to take photos are arguable. There has long been ongoing debate over this issue, which remains unsettled. Many firms continue to refuse to consent to in-plant photography during an inspection. In establishing a firm's policy on this topic, it is advisable to consult competent legal advice.

FACTORY SAMPLES

On occasion, the FDA may decide to collect physical samples during an inspection, and they have the right to do so. The investigators must provide the company with a form FDA 484 covering such samples, but they are unlikely to comment on the type of examination or analysis that is expected, nor are they likely to promise a report on the findings. It is prudent for the firm to observe what was sampled and how the samples were taken, and then take their own equivalent samples to hold for future use in case the need arises.

INSPECTION OF DOCUMENTATION

As discussed in chapter 25, it is essential to have certain SOPs in place and that they are properly followed.

Timely, concise, and accurate records are also of great importance, whether they are written or digital. In most establishment inspections, the checking of proper documentation is the most time-consuming activity, often taking roughly half of the time of the total inspection. In other words, the FDA looks not only on what is being done at the time of the inspection but also what has been done in the past as reflected in records.

Investigators usually ask to see and discuss the firm's organization chart, as well as complaint files, batch records, training records, and logs covering maintenance, calibration, cleaning, and sanitation. Laboratory records are usually examined in detail, as discussed in chapter 19.

INSPECTIONAL OBSERVATIONS

As the inspection proceeds, the investigators make note of what they perceive to be objectionable conditions and practices. These are their *observations* and opinions, but it is not up to the investigators to decide whether or not these are *violations*. That decision is ultimately made by the agency, after considering all of the facts and circumstances.

Inspectional observations are passed along to the firm's management as the inspection proceeds, usually on a daily basis, to minimize any surprises, errors, and misunderstandings at the end of the inspection. This allows discussion and clarification during the inspection, and gives the firm the opportunity to inform the inspection team what corrections have been or will be made.

When the inspection has been completed, but before the investigators leave the facility, they prepare a form (either handwritten or digitally typed) listing all of the *significant* deviations from the GMPs that they have observed. Obviously the term "significant" is subjective and open to interpretation. The form used for this is called FDA 483, more commonly just referred to as "483." This form was created in 1953 by an amendment of the FD&C Act adding Section 704(b).

The 483 listing the inspectors' objectionable observations is also discussed with the firm's management in a final wrap-up meeting.

The printed 483 form used to record observations contains this disclaimer:

> This document lists observations made by the FDA representatives during the inspection of your facility. They are inspectional observations, and do not represent a final Agency determination regarding your compliance.

At the conclusion of the inspection, a closeout meeting is held with the firm's management, at which time the 483 (if any) is presented and discussed. Other items, not of sufficient significance as to be on the 483, may also be discussed. If the investigator has placed any incorrect observations on the 483, this should be pointed out. Corrective actions accomplished *during* the inspection should be annotated. The 483 becomes part of the firm's regulatory history, and is available to the public under Freedom of Information.

The original 483 is signed in pen and ink by the agency representatives, and is left with the individual that received the FDA 482 Notice of Inspection, if that person is present and qualifies as being the "most responsible." If that person is not available, or is outranked by someone else, it is given to the most senior executive available. Legible copies of the original 483 are then distributed.

If no significant deviations from GMPs were observed by the inspection team, no 483 is issued. This is of course a highly desired outcome, and does occur on occasion, albeit in most instances there is a 483 at the conclusion of an inspection.

RESPONDING TO 483s

There is no legal mandate to respond to a 483, but it is prudent to do so. A well-reasoned, well-written response, sent promptly, can head off any further action. The FDA's *Regulatory Procedures Manual* says that generally a Warning Letter (WL) is not issued if the agency concludes that the firm's corrective actions are adequate.

If the firm decides to respond, it is prudent to make it clear that the company does understand what the law requires and further understands the 483 observations. Moreover, it is appropriate to state that the company does intend to take corrective action as rapidly as feasible, preferably outlining what those actions will be and a reasonable timetable as to when they can be accomplished. It is inadvisable to dispute the violations in the response letter, since, if appropriate, that should have been done in the wrap-up meeting with the inspector.

In August 2009, the FDA announced a policy on the time frame for firms' submission of responses to 483s. Before issuing a WL, the FDA will generally allow firms 15 business days to provide their response. If FDA gets the response within that time, they will conduct a detailed review of it before determining whether to issue a WL. Therefore, it is of importance to promptly submit responses to 483s.

The response should also demonstrate that a logical and comprehensive approach was taken to determining *why* the violations occurred, and the steps that will be taken to prevent recurrence.

If specific corrective actions are stated in the response (as they should be), it is certain that the FDA *will* eventually follow-up to see if those actions were, in fact, properly taken. Failure to keep commitments is an almost certain sure way to invite strong regulatory actions.

ESTABLISHMENT INSPECTION REPORTS

In addition to preparing a 483 (if needed) after the inspection, the investigator(s) are also required to prepare an Establishment Inspection Report (EIR). This is a lengthy narrative document that details the inspection process and inspectional findings. Being comprehensive, it takes time to write an EIR, often several days or weeks. This is an important record of the inspection.

EIRs typically start with a summary stating the name and location of the establishment, what prompted the inspection, the dates of the inspection, the names of the CSOs involved, and any special instructions from the ORA or CFSAN regarding the conduct of the inspection, together with a brief overview of the conclusions.

Another section gives a brief description and history of the firm, and names the most responsible individual and other responsible persons. A brief description of the results of previous inspections (if any) may be included. A statement is included describing how it was confirmed that the products from the facility enter interstate commerce.

If various people were interviewed or provided information during the inspection, they are named, together with the name of the person to whom the FDA 482 (Notice of Inspection) was given.

A section describing the operations and equipment is often included, as well as a section on any samples collected. If a 483 was issued, the details of it and any discussions of it with management may be included. The EIR may also include comments on training, cleaning and sanitation, pest control, sampling, handling of returned dietary supplements, customer complaints, laboratory testing, packaging and labeling operations, a review of the firm's documentation, whether SOPs are properly followed, and other such pertinent GMP details covered in Part 111.

Details of the closing discussion among the investigators and the firm's management are included.

EIRs must be concise, factual, and free from unsupportable conclusions. There are often several attachments and exhibits.

CLASSIFICATION OF EIRS

The agency reviews each EIR, and makes a determination of how to classify it, based in part on the investigators' recommendations and in part on following current agency policies. Following are the most usual (but not the only) classifications:

NAI (no action indicated), signifying that no significantly objectionable conditions or practices were found that justify further FDA action.

VAI (voluntary action indicated), meaning some objectionable conditions were found but they do not reach the threshold of requiring regulatory action, and the firm has agreed to voluntarily make the necessary corrections.

OAI (official action indicated), meaning that significant objectionable conditions were found warranting regulatory action such as the issuance of a WL, requesting a recall, or consideration of criminal charges.

If OAI, the ORA district's Compliance Branch conducts a further review and decides on the appropriate action to be taken.

The EIR classification process yields reasonable uniformity in decision-making, following an agency program called the Field Accomplishments and Compliance Tracking System (FACTS).

When the FDA deems the inspection to be closed, a copy of the EIR is sent to the establishment that was inspected, and also makes a redacted copy available to FDA's Freedom of Information Office for public distribution as requests are received.

TURBO EIR

In 2002, the FDA implemented a software program called "TurboEIR," intended to assist CSOs in creating and printing both 483s and EIRs. Most investigators have this on their laptop computers, and they are encouraged by the agency's management to use TurboEIR to help them write accurate, consistent, and complete documents.

WARNING LETTERS

As discussed in chapter 32, WLs are typically the first step if and when FDA regulatory action is needed to "push" dietary supplement firms to get into full

GMP compliance after objectionable conditions are discovered during establishment inspections.

CONFIDENTIAL INFORMANTS

On occasion, individuals may want to provide information to the FDA, but remain anonymous. Provision for handling such instances is included in the IOM, including the necessary steps to keep the identity of the source confidential. The preferred method of handling this is on a face-to-face interview as opposed to a telephone call, to enable assessment of the person's demeanor, body language, and truthfulness. When feasible, such contacts involve two investigators, one to conduct the interview and the other to take notes and to serve as a witness to the interview. If the informant makes allegations, the investigators explore how he/she knows them to be true, whether the informant personally saw, heard, or read about the incidents, and whether they can provide any proof. They also ask the informant *why* he/she is divulging this information, to establish the motivation.

SUGGESTED READINGS

Branding FH. Underdeveloped: FDA's authority to take photographs during an FDA establishment inspection. Food Drug Law J 2003; 58(1):9–17.

Cooper RM, Fleder JR. Responding to a form 483 or warning letter, a practical guide. Food Drug Law J 2005; 60(4):479–493.

FDA, Field Management Directive 145 (FMD 145). Procedure for Release of Establishment Inspection Reports, April 1997.

FDA, Dietary Supplements, Import and Domestic. Program 7321.008, Compliance Program Guidance Manual, 2010.

FDA, Investigation Operations Manual, Chapter 5 Establishment Inspection, Subchapter 523 Photographs, and Subchapter 590 Establishment Inspection Reports (EIRs), 2010.

FDA, ORA Laboratory Manual, ORA Division of Field Science Document III-5, Section 5.6, TurboEIR and Turbo483s, May 2011.

Fortin ND. Is a picture worth more than 1,000 words? The 4th amendment and FDA's authority to take photographs under the Federal Food, Drug, and Cosmetic Act. J Food Law Policy 2005; 1:239–266.

Shuren J. (FDA), Review of post-inspection responses. Fed Regist 2009; 74(153):40211–40213.

Terzotis K. Basics of FDA inspection preparation. Contract Pharma, November–December 2004:58–61.

Appendix FDA dietary supplement GMP regulations

These are the actual GMP regulations as published in the *Federal Register* of June 25, 2007 (72 FR 34968), and as they now appear in 21 CFR Part 111.

PART 111—CURRENT GOOD MANUFACTURING PRACTICE IN MANUFACTURING, PACKAGING, LABELING, OR HOLDING OPERATIONS FOR DIETARY SUPPLEMENTS
Subpart A—General Provisions
§ 111.1 Who is subject to this part?

(a) Except as provided by paragraph (b) of this section, you are subject to this part if you manufacture, package, label, or hold a dietary supplement, including:
  (1) A dietary supplement you manufacture but that is packaged or labeled by another person; and
  (2) A dietary supplement imported or offered for import in any State or territory of the United States, the District of Columbia, or the Commonwealth of Puerto Rico.
(b) The requirements pertaining to holding dietary supplements do not apply to you if you are holding those dietary supplements at a retail establishment for the sole purpose of direct retail sale to individual consumers. A retail establishment does not include a warehouse or other storage facility for a retailer or a warehouse or other storage facility that sells directly to individual consumers.

§ 111.3 What definitions apply to this part?
The definitions and interpretations of terms in section 201 of the Federal Food, Drug, and Cosmetic Act (the act) apply to such terms when used in this part. For the purpose of this part, the following definitions also apply:

Actual yield means the quantity that is actually produced at any appropriate step of manufacture or packaging of a particular dietary supplement.

Batch means a specific quantity of a dietary supplement that is uniform, that is intended to meet specifications for identity, purity, strength, and composition, and that is produced during a specified time period according to a single manufacturing record during the same cycle of manufacture.

Batch number, lot number, or control number means any distinctive group of letters, numbers, or symbols, or any combination of them, from which the complete history of the manufacturing, packaging, labeling, and/or holding of a batch or lot of dietary supplements can be determined.

Component means any substance intended for use in the manufacture of a dietary supplement, including those that may not appear in the finished batch of the dietary supplement. Component includes dietary ingredients (as described in section 201(ff) of the act) and other ingredients.

Contact surface means any surface that contacts a component or dietary supplement, and those surfaces from which drainage onto the component or dietary supplement, or onto surfaces that contact the component or dietary supplement, occurs during the normal course of operations. Examples of contact surfaces include containers, utensils, tables, contact surfaces of equipment, and packaging. *Ingredient* means any substance that is used in the manufacture of a dietary supplement and that is intended to be present in the finished batch of the dietary supplement. An ingredient includes, but is not necessarily limited to, a dietary ingredient as defined in section 201(ff) of the act.

In-process material means any material that is fabricated, compounded, blended, ground, extracted, sifted, sterilized, derived by chemical reaction, or processed in any other way for use in the manufacture of a dietary supplement.

Lot means a batch, or a specific identified portion of a batch, that is uniform and that is intended to meet specifications for identity, purity, strength, and composition; or, in the case of a dietary supplement produced by continuous process, a specific identified amount produced in a specified unit of time or quantity in a manner that is uniform and that is intended to meet specifications for identity, purity, strength, and composition.

Microorganisms means yeasts, molds, bacteria, viruses, and other similar microscopic organisms having public health or sanitary concern. This definition includes species that:
(1) May have public health significance;
(2) May cause a component or dietary supplement to decompose;
(3) Indicate that the component or dietary supplement is contaminated with filth;
   or
(4) Otherwise may cause the component or dietary supplement to be adulterated.

Must is used to state a requirement.

Pest means any objectionable insect or other animal including birds, rodents, flies, mites, and larvae.

Physical plant means all or any part of a building or facility used for or in connection with manufacturing, packaging, labeling, or holding a dietary supplement.

Product complaint means any communication that contains any allegation, written, electronic, or oral, expressing concern, for any reason, with the quality of a dietary supplement, that could be related to current good manufacturing practice. Examples of product complaints are: Foul odor, off taste, illness or injury, disintegration time, color variation, tablet size or size variation, under-filled container, foreign material in a dietary supplement container, improper packaging, mislabeling, or dietary supplements that are superpotent, subpotent, or contain the wrong ingredient, or contain a drug or other contaminant (e.g., bacteria, pesticide, mycotoxin, glass, lead).

Quality means that the dietary supplement consistently meets the established specifications for identity, purity, strength, and composition, and limits on contaminants, and has been manufactured, packaged, labeled, and held under conditions to prevent adulteration under section 402(a)(1), (a)(2), (a)(3), and (a)(4) of the act.

Quality control means a planned and systematic operation or procedure for ensuring the quality of a dietary supplement.

Quality control personnel means any person, persons, or group, within or outside of your organization, who you designate to be responsible for your quality control operations.

Representative sample means a sample that consists of an adequate number of units that are drawn based on rational criteria, such as random sampling, and that are intended to ensure that the sample accurately portrays the material being sampled.

Reprocessing means using, in the manufacture of a dietary supplement, clean, uncontaminated components or dietary supplements that have been previously removed from manufacturing and that have been made suitable for use in the manufacture of a dietary supplement.

Reserve sample means a representative sample of product that is held for a designated period of time.

Sanitize means to adequately treat cleaned equipment, containers, utensils, or any other cleaned contact surface by a process that is effective in destroying vegetative cells of microorganisms of public health significance, and in substantially reducing numbers of other microorganisms, but without adversely affecting the product or its safety for the consumer.

Theoretical yield means the quantity that would be produced at any appropriate step of manufacture or packaging of a particular dietary supplement, based upon the quantity of components or packaging to be used, in the absence of any loss or error in actual production.

Water activity (aw) is a measure of the free moisture in a component or dietary supplement and is the quotient of the water vapor pressure of the substance divided by the vapor pressure of pure water at the same temperature.

We means the U.S. Food and Drug Administration (FDA).

You means a person who manufactures, packages, labels, or holds dietary supplements.

§ 111.5 Do other statutory provisions and regulations apply?

In addition to this part, you must comply with other applicable statutory provisions and regulations under the act related to dietary supplements.

Subpart B—Personnel

§ 111.8 What are the requirements under this subpart B for written procedures?

You must establish and follow written procedures for fulfilling the requirements of this subpart.

§ 111.10 What requirements apply for preventing microbial contamination from sick or infected personnel and for hygienic practices?

(a) *Preventing microbial contamination.* You must take measures to exclude from any operations any person who might be a source of microbial

contamination, due to a health condition, where such contamination may occur, of any material, including components, dietary supplements, and contact surfaces used in the manufacture, packaging, labeling, or holding of a dietary supplement. Such measures include the following:
   (1) Excluding from working in any operations that may result in contamination any person who, by medical examination, the person's acknowledgement, or supervisory observation, is shown to have, or appears to have, an illness, infection, open lesion, or any other abnormal source of microbial contamination, that could result in microbial contamination of components, dietary supplements, or contact surfaces, until the health condition no longer exists; and
   (2) Instructing your employees to notify their supervisor(s) if they have or if there is a reasonable possibility that they have a health condition described in paragraph (a)(1) of this section that could result in microbial contamination of any components, dietary supplements, or any contact surface.
(b) *Hygienic practices.* If you work in an operation during which adulteration of the component, dietary supplement, or contact surface could occur, you must use hygienic practices to the extent necessary to protect against such contamination of components, dietary supplements, or contact surfaces. These hygienic practices include the following:
   (1) Wearing outer garments in a manner that protects against the contamination of components, dietary supplements, or any contact surface;
   (2) Maintaining adequate personal cleanliness;
   (3) Washing hands thoroughly (and sanitizing if necessary to protect against contamination with microorganisms) in an adequate handwashing facility:
      (i) Before starting work; and
      (ii) At any time when the hands may have become soiled or contaminated;
   (4) Removing all unsecured jewelry and other objects that might fall into components, dietary supplements, equipment, or packaging, and removing hand jewelry that cannot be adequately sanitized during periods in which components or dietary supplements are manipulated by hand. If hand jewelry cannot be removed, it must be covered by material that is maintained in an intact, clean, and sanitary condition and that effectively protects against contamination of components, dietary supplements, or contact surfaces;
   (5) Maintaining gloves used in handling components or dietary supplements in an intact, clean, and sanitary condition. The gloves must be of an impermeable material;
   (6) Wearing, where appropriate, in an effective manner, hair nets, caps, beard covers, or other effective hair restraints;
   (7) Not storing clothing or other personal belongings in areas where components, dietary supplements, or any contact surfaces are exposed or where contact surfaces are washed;
   (8) Not eating food, chewing gum, drinking beverages, or using tobacco products in areas where components, dietary supplements, or any contact surfaces are exposed, or where contact surfaces are washed; and

(9) Taking any other precautions necessary to protect against the contamination of components, dietary supplements, or contact surfaces with microorganisms, filth, or any other extraneous materials, including perspiration, hair, cosmetics, tobacco, chemicals, and medicines applied to the skin.

§ 111.12 What personnel qualification requirements apply?
(a) You must have qualified employees who manufacture, package, label, or hold dietary supplements.
(b) You must identify who is responsible for your quality control operations. Each person who is identified to perform quality control operations must be qualified to do so and have distinct and separate responsibilities related to performing such operations from those responsibilities that the person otherwise has when not performing such operations.
(c) Each person engaged in manufacturing, packaging, labeling, or holding, or in performing any quality control operations, must have the education, training, or experience to perform the person's assigned functions.

§ 111.13 What supervisor requirements apply?
(a) You must assign qualified personnel to supervise the manufacturing, packaging, labeling, or holding of dietary supplements.
(b) Each supervisor whom you use must be qualified by education, training, or experience to supervise.

§ 111.14 Under this subpart B, what records must you make and keep?
(a) You must make and keep records required under this subpart B in accordance with subpart P of this part.
(b) You must make and keep the following records:
  (1) Written procedures for fulfilling the requirements of this subpart B; and
  (2) Documentation of training, including the date of the training, the type of training, and the person(s) trained.

Subpart C—Physical Plant and Grounds
§ 111.15 What sanitation requirements apply to your physical plant and grounds?

(a) *Grounds.* You must keep the grounds of your physical plant in a condition that protects against the contamination of components, dietary supplements, or contact surfaces. The methods for adequate ground maintenance include:
  (1) Properly storing equipment, removing litter and waste, and cutting weeds or grass within the immediate vicinity of the physical plant so that it does not attract pests, harbor pests, or provide pests a place for breeding;
  (2) Maintaining roads, yards, and parking lots so that they do not constitute a source of contamination in areas where components, dietary supplements, or contact surfaces are exposed;

(3) Adequately draining areas that may contribute to the contamination of components, dietary supplements, or contact surfaces by seepage, filth or any other extraneous materials, or by providing a breeding place for pests;
(4) Adequately operating systems for waste treatment and disposal so that they do not constitute a source of contamination in areas where components, dietary supplements, or contact surfaces are exposed; and
(5) If your plant grounds are bordered by grounds not under your control, and if those other grounds are not maintained in the manner described in this section, you must exercise care in the plant by inspection, extermination, or other means to exclude pests, dirt, and filth or any other extraneous materials that may be a source of contamination.

(b) *Physical plant facilities.*
(1) You must maintain your physical plant in a clean and sanitary condition; and
(2) You must maintain your physical plant in repair sufficient to prevent components, dietary supplements, or contact surfaces from becoming contaminated.

(c) *Cleaning compounds, sanitizing agents, pesticides, and other toxic materials.*
(1) You must use cleaning compounds and sanitizing agents that are free from microorganisms of public health significance and that are safe and adequate under the conditions of use.
(2) You must not use or hold toxic materials in a physical plant in which components, dietary supplements, or contact surfaces are manufactured or exposed, unless those materials are necessary as follows:
   (i) To maintain clean and sanitary conditions;
   (ii) For use in laboratory testing procedures;
   (iii) For maintaining or operating the physical plant or equipment; or
   (iv) For use in the plant's operations.
(3) You must identify and hold cleaning compounds, sanitizing agents, pesticides, pesticide chemicals, and other toxic materials in a manner that protects against contamination of components, dietary supplements, or contact surfaces.

(d) *Pest control.*
(1) You must not allow animals or pests in any area of your physical plant. Guard or guide dogs are allowed in some areas of your physical plant if the presence of the dogs will not result in contamination of components, dietary supplements, or contact surfaces;
(2) You must take effective measures to exclude pests from the physical plant and to protect against contamination of components, dietary supplements, and contact surfaces on the premises by pests; and
(3) You must not use insecticides, fumigants, fungicides, or rodenticides, unless you take precautions to protect against the contamination of components, dietary supplements, or contact surfaces.

(e) *Water supply.*
(1) You must provide water that is safe and sanitary, at suitable temperatures, and under pressure as needed, for all uses where water does not become a component of the dietary supplement.

APPENDIX FDA DIETARY SUPPLEMENT GMP REGULATIONS

    (2) Water that is used in a manner such that the water may become a component of the dietary supplement, e.g., when such water contacts components, dietary supplements, or any contact surface, must, at a minimum, comply with applicable Federal, State, and local requirements and not contaminate the dietary supplement.

(f) *Plumbing.* The plumbing in your physical plant must be of an adequate size and design and be adequately installed and maintained to:
    (1) Carry sufficient amounts of water to required locations throughout the physical plant;
    (2) Properly convey sewage and liquid disposable waste from your physical plant;
    (3) Avoid being a source of contamination to components, dietary supplements, water supplies, or any contact surface, or creating an unsanitary condition;
    (4) Provide adequate floor drainage in all areas where floors are subject to flooding-type cleaning or where normal operations release or discharge water or other liquid waste on the floor; and
    (5) Not allow backflow from, or cross connection between, piping systems that discharge waste water or sewage and piping systems that carry water used for manufacturing dietary supplements, for cleaning contact surfaces, or for use in bathrooms or hand-washing facilities.

(g) *Sewage disposal.* You must dispose of sewage into an adequate sewage system or through other adequate means.

(h) *Bathrooms.* You must provide your employees with adequate, readily accessible bathrooms. The bathrooms must be kept clean and must not be a potential source of contamination to components, dietary supplements, or contact surfaces.

(i) *Hand-washing facilities.* You must provide hand-washing facilities that are designed to ensure that an employee's hands are not a source of contamination of components, dietary supplements, or any contact surface, by providing facilities that are adequate, convenient, and furnish running water at a suitable temperature.

(j) *Trash disposal.* You must convey, store, and dispose of trash to:
    (1) Minimize the development of odors;
    (2) Minimize the potential for the trash to attract, harbor, or become a breeding place for pests;
    (3) Protect against contamination of components, dietary supplements, any contact surface, water supplies, and grounds surrounding your physical plant; and
    (4) Control hazardous waste to prevent contamination of components, dietary supplements, and contact surfaces.

(k) *Sanitation supervisors.* You must assign one or more employees to supervise overall sanitation. Each of these supervisors must be qualified by education, training, or experience to develop and supervise sanitation procedures.

§ 111.16 What are the requirements under this subpart C for written procedures?

You must establish and follow written procedures for cleaning the physical plant and for pest control.

§ 111.20 What design and construction requirements apply to your physical plant?

Any physical plant you use in the manufacture, packaging, labeling, or holding of dietary supplements must:

(a) Be suitable in size, construction, and design to facilitate maintenance, cleaning, and sanitizing operations;
(b) Have adequate space for the orderly placement of equipment and holding of materials as is necessary for maintenance, cleaning, and sanitizing operations and to prevent contamination and mixups of components and dietary supplements during manufacturing, packaging, labeling, or holding;
(c) Permit the use of proper precautions to reduce the potential for mixups or contamination of components, dietary supplements, or contact surfaces, with microorganisms, chemicals, filth, or other extraneous material. Your physical plant must have, and you must use, separate or defined areas of adequate size or other control systems, such as computerized inventory controls or automated systems of separation, to prevent contamination and mixups of components and dietary supplements during the following operations:
   (1) Receiving, identifying, holding, and withholding from use, components, dietary supplements, packaging, and labels that will be used in or during the manufacturing, packaging, labeling, or holding of dietary supplements;
   (2) Separating, as necessary, components, dietary supplements, packaging, and labels that are to be used in manufacturing from components, dietary supplements, packaging, or labels that are awaiting material review and disposition decision, reprocessing, or are awaiting disposal after rejection;
   (3) Separating the manufacturing, packaging, labeling, and holding of different product types including different types of dietary supplements and other foods, cosmetics, and pharmaceutical products;
   (4) Performing laboratory analyses and holding laboratory supplies and samples;
   (5) Cleaning and sanitizing contact surfaces;
   (6) Packaging and label operations; and
   (7) Holding components or dietary supplements.
(d) Be designed and constructed in a manner that prevents contamination of components, dietary supplements, or contact surfaces.
   (1) The design and construction must include:
      (i) Floors, walls, and ceilings that can be adequately cleaned and kept clean and in good repair;
      (ii) Fixtures, ducts, and pipes that do not contaminate components, dietary supplements, or contact surfaces by dripping or other leakage, or condensate;
      (iii) Adequate ventilation or environmental control equipment such as airflow systems, including filters, fans, and other air-blowing equipment, that minimize odors and vapors (including steam and noxious fumes) in areas where they may contaminate components, dietary supplements, or contact surfaces;
      (iv) Equipment that controls temperature and humidity, when such equipment is necessary to ensure the quality of the dietary supplement; and

APPENDIX FDA DIETARY SUPPLEMENT GMP REGULATIONS

(v) Aisles or working spaces between equipment and walls that are adequately unobstructed and of adequate width to permit all persons to perform their duties and to protect against contamination of components, dietary supplements, or contact surfaces with clothing or personal contact.

(2) When fans and other air-blowing equipment are used, such fans and equipment must be located and operated in a manner that minimizes the potential for microorganisms and particulate matter to contaminate components, dietary supplements, or contact surfaces;

(e) Provide adequate light in:
   (1) All areas where components or dietary supplements are examined, processed, or held;
   (2) All areas where contact surfaces are cleaned; and
   (3) Hand-washing areas, dressing and locker rooms, and bathrooms.
(f) Use safety-type light bulbs, fixtures, skylights, or other glass or glass-like materials when the light bulbs, fixtures, skylights or other glass or glass-like materials are suspended over exposed components or dietary supplements in any step of preparation, unless your physical plant is otherwise constructed in a manner that will protect against contamination of components or dietary supplements in case of breakage of glass or glass-like materials.
(g) Provide effective protection against contamination of components and dietary supplements in bulk fermentation vessels, by, for example:
   (1) Use of protective coverings;
   (2) Placement in areas where you can eliminate harborages for pests over and around the vessels;
   (3) Placement in areas where you can check regularly for pests, pest infestation, filth or any other extraneous materials; and
   (4) Use of skimming equipment.
(h) Use adequate screening or other protection against pests, where necessary.

§ 111.23 Under this subpart C, what records must you make and keep?

(a) You must make and keep records required under this subpart C in accordance with subpart P of this part.
(b) You must make and keep records of the written procedures for cleaning the physical plant and for pest control.
(c) You must make and keep records that show that water, when used in a manner such that the water may become a component of the dietary supplement, meets the requirements of § 111.15(e)(2).

Subpart D—Equipment and Utensils
§ 111.25 What are the requirements under this subpart D for written procedures?

You must establish and follow written procedures for fulfilling the requirements of this subpart D, including written procedures for:

(a) Calibrating instruments and controls that you use in manufacturing or testing a component or dietary supplement;

(b) Calibrating, inspecting, and checking automated, mechanical, and electronic equipment; and
(c) Maintaining, cleaning, and sanitizing, as necessary, all equipment, utensils, and any other contact surfaces that are used to manufacture, package, label, or hold components or dietary supplements.

§ 111.27 What requirements apply to the equipment and utensils that you use?

(a) You must use equipment and utensils that are of appropriate design, construction, and workmanship to enable them to be suitable for their intended use and to be adequately cleaned and properly maintained.
   (1) Equipment and utensils include the following:
      (i) Equipment used to hold or convey;
      (ii) Equipment used to measure;
      (iii) Equipment using compressed air or gas;
      (iv) Equipment used to carry out processes in closed pipes and vessels; and
      (v) Equipment used in automated, mechanical, or electronic systems.
   (2) You must use equipment and utensils of appropriate design and construction so that use will not result in the contamination of components or dietary supplements with:
      (i) Lubricants;
      (ii) Fuel;
      (iii) Coolants;
      (iv) Metal or glass fragments;
      (v) Filth or any other extraneous material;
      (vi) Contaminated water; or
      (vii) Any other contaminants.
   (3) All equipment and utensils you use must be:
      (i) Installed and maintained to facilitate cleaning the equipment, utensils, and all adjacent spaces;
      (ii) Corrosion-resistant if the equipment or utensils contact components or dietary supplements;
      (iii) Made of nontoxic materials;
      (iv) Designed and constructed to withstand the environment in which they are used, the action of components or dietary supplements, and, if applicable, cleaning compounds and sanitizing agents; and
      (v) Maintained to protect components and dietary supplements from being contaminated by any source.
   (4) Equipment and utensils you use must have seams that are smoothly bonded or maintained to minimize accumulation of dirt, filth, organic material, particles of components or dietary supplements, or any other extraneous materials or contaminants.
   (5) Each freezer, refrigerator, and other cold storage compartment you use to hold components or dietary supplements:
      (i) Must be fitted with an indicating thermometer, temperature-measuring device, or temperature-recording device that indicates

and records, or allows for recording by hand, the temperature accurately within the compartment; and
　(ii) Must have an automated device for regulating temperature or an automated alarm system to indicate a significant temperature change in a manual operation.
(6) Instruments or controls used in the manufacturing, packaging, labeling, or holding of a dietary supplement, and instruments or controls that you use to measure, regulate, or record temperatures, hydrogen-ion concentration (pH), water activity, or other conditions, to control or prevent the growth of microorganisms or other contamination must be:
　(i) Accurate and precise;
　(ii) Adequately maintained; and
　(iii) Adequate in number for their designated uses.
(7) Compressed air or other gases you introduce mechanically into or onto a component, dietary supplement, or contact surface or that you use to clean any contact surface must be treated in such a way that the component, dietary supplement, or contact surface is not contaminated.

(b) You must calibrate instruments and controls you use in manufacturing or testing a component or dietary supplement. You must calibrate:
(1) Before first use;
(2) At the frequency specified in writing by the manufacturer of the instrument and control; or
(3) At routine intervals or as otherwise necessary to ensure the accuracy and precision of the instrument and control.

(c) You must repair or replace instruments or controls that cannot be adjusted to agree with the reference standard.

(d) You must maintain, clean, and sanitize, as necessary, all equipment, utensils, and any other contact surfaces used to manufacture, package, label, or hold components or dietary supplements.
(1) Equipment and utensils must be taken apart as necessary for thorough maintenance, cleaning, and sanitizing.
(2) You must ensure that all contact surfaces, used for manufacturing or holding low-moisture components or dietary supplements, are in a dry and sanitary condition when in use. When the surfaces are wet-cleaned, they must be sanitized, when necessary, and thoroughly dried before subsequent use.
(3) If you use wet processing during manufacturing, you must clean and sanitize all contact surfaces, as necessary, to protect against the introduction of microorganisms into components or dietary supplements. When cleaning and sanitizing is necessary, you must clean and sanitize all contact surfaces before use and after any interruption during which the contact surface may have become contaminated. If you use contact surfaces in a continuous production operation or in consecutive operations involving different batches of the same dietary supplement, you must adequately clean and sanitize the contact surfaces, as necessary.

(4) You must clean surfaces that do not come into direct contact with components or dietary supplements as frequently as necessary to protect against contaminating components or dietary supplements.
(5) Single-service articles (such as utensils intended for one-time use, paper cups, and paper towels) must be:
   (i) Stored in appropriate containers; and
   (ii) Handled, dispensed, used, and disposed of in a manner that protects against contamination of components, dietary supplements, or any contact surface.
(6) Cleaning compounds and sanitizing agents must be adequate for their intended use and safe under their conditions of use;
(7) You must store cleaned and sanitized portable equipment and utensils that have contact surfaces in a location and manner that protects them from contamination.

§ 111.30 What requirements apply to automated, mechanical, or electronic equipment?

For any automated, mechanical, or electronic equipment that you use to manufacture, package, label, or hold a dietary supplement, you must:

(a) Design or select equipment to ensure that dietary supplement specifications are consistently met;
(b) Determine the suitability of the equipment by ensuring that your equipment is capable of operating satisfactorily within the operating limits required by the process;
(c) Routinely calibrate, inspect, or check the equipment to ensure proper performance. Your quality control personnel must periodically review these calibrations, inspections, or checks;
(d) Establish and use appropriate controls for automated, mechanical, and electronic equipment (including software for a computer controlled process) to ensure that any changes to the manufacturing, packaging, labeling, holding, or other operations are approved by quality control personnel and instituted only by authorized personnel; and
(e) Establish and use appropriate controls to ensure that the equipment functions in accordance with its intended use. These controls must be approved by quality control personnel.

§ 111.35 Under this subpart D, what records must you make and keep?

(a) You must make and keep records required under this subpart D in accordance with subpart P of this part.
(b) You must make and keep the following records:
   (1) Written procedures for fulfilling the requirements of this subpart, including written procedures for:
      (i) Calibrating instruments and controls that you use in manufacturing or testing a component or dietary supplement;
      (ii) Calibrating, inspecting, and checking automated, mechanical, and electronic equipment; and

(iii) Maintaining, cleaning, and sanitizing, as necessary, all equipment, utensils, and any other contact surfaces that are used to manufacture, package, label, or hold components or dietary supplements;
(2) Documentation, in individual equipment logs, of the date of the use, maintenance, cleaning, and sanitizing of equipment, unless such documentation is kept with the batch record;
(3) Documentation of any calibration, each time the calibration is performed, for instruments and controls that you use in manufacturing or testing a component or dietary supplement. In your documentation, you must:
    (i) Identify the instrument or control calibrated;
    (ii) Provide the date of calibration;
    (iii) Identify the reference standard used including the certification of accuracy of the known reference standard and a history of recertification of accuracy;
    (iv) Identify the calibration method used, including appropriate limits for accuracy and precision of instruments and controls when calibrating;
    (v) Provide the calibration reading or readings found;
    (vi) Identify the recalibration method used, and reading or readings found, if accuracy or precision or both accuracy and precision limits for instruments and controls were not met; and
    (vii) Include the initials of the person who performed the calibration and any recalibration.
(4) Written records of calibrations, inspections, and checks of automated, mechanical, and electronic equipment;
(5) Backup file(s) of current software programs (and of outdated software that is necessary to retrieve records that you are required to keep in accordance with subpart P of this part, when current software is not able to retrieve such records) and of data entered into computer systems that you use to manufacture, package, label, or hold dietary supplements.
    (i) Your backup file (e.g., a hard copy of data you have entered, diskettes, tapes, microfilm, or compact disks) must be an exact and complete record of the data you entered.
    (ii) You must keep your backup software programs and data secure from alterations, inadvertent erasures, or loss; and
(6) Documentation of the controls that you use to ensure that equipment functions in accordance with its intended use.

Subpart E—Requirement to Establish a Production and Process Control System
§ 111.55 What are the requirements to implement a production and process control system?

You must implement a system of production and process controls that covers all stages of manufacturing, packaging, labeling, and holding of the dietary supplement to ensure the quality of the dietary supplement and that the dietary supplement is packaged and labeled as specified in the master manufacturing record.

§ 111.60 What are the design requirements for the production and process control system?

(a) Your production and in-process control system must be designed to ensure that the dietary supplement is manufactured, packaged, labeled, and held in a manner that will ensure the quality of the dietary supplement and that the dietary supplement is packaged and labeled as specified in the master manufacturing record; and

(b) The production and in-process control system must include all requirements of subparts E through L of this part and must be reviewed and approved by quality control personnel.

§ 111.65 What are the requirements for quality control operations?

You must implement quality control operations in your manufacturing, packaging, labeling, and holding operations for producing the dietary supplement to ensure the quality of the dietary supplement and that the dietary supplement is packaged and labeled as specified in the master manufacturing record.

§ 111.70 What specifications must you establish?

(a) You must establish a specification for any point, step, or stage in the manufacturing process where control is necessary to ensure the quality of the dietary supplement and that the dietary supplement is packaged and labeled as specified in the master manufacturing record.

(b) For each component that you use in the manufacture of a dietary supplement, you must establish component specifications as follows:
   (1) You must establish an identity specification;
   (2) You must establish component specifications that are necessary to ensure that specifications for the purity, strength and composition of dietary supplements manufactured using the components are met; and
   (3) You must establish limits on those types of contamination that may adulterate or may lead to adulteration of the finished batch of the dietary supplement to ensure the quality of the dietary supplement.

(c) For the in-process production:
   (1) You must establish in-process specifications for any point, step, or stage in the master manufacturing record where control is necessary to help ensure that specifications are met for the identity, purity, strength, and composition of the dietary supplements and, as necessary, for limits on those types of contamination that may adulterate or may lead to adulteration of the finished batch of the dietary supplement;
   (2) You must provide adequate documentation of your basis for why meeting the in-process specifications, in combination with meeting component specifications, will help ensure that the specifications are met for the identity, purity, strength, and composition of the dietary supplements and for limits on those types of contamination that may adulterate or may lead to adulteration of the finished batch of the dietary supplement; and

(3) Quality control personnel must review and approve the documentation that you provide under paragraph (c)(2) of this section.
(d) You must establish specifications for dietary supplement labels (label specifications) and for packaging that may come in contact with dietary supplements (packaging specifications). Packaging that may come into contact with dietary supplements must be safe and suitable for its intended use and must not be reactive or absorptive or otherwise affect the safety or quality of the dietary supplement.
(e) For each dietary supplement that you manufacture you must establish product specifications for the identity, purity, strength, and composition of the finished batch of the dietary supplement, and for limits on those types of contamination that may adulterate, or that may lead to adulteration of, the finished batch of the dietary supplement to ensure the quality of the dietary supplement.
(f) If you receive a product from a supplier for packaging or labeling as a dietary supplement (and for distribution rather than for return to the supplier), you must establish specifications to provide sufficient assurance that the product you receive is adequately identified and is consistent with your purchase order.
(g) You must establish specifications for the packaging and labeling of the finished packaged and labeled dietary supplements, including specifications that ensure that you used the specified packaging and that you applied the specified label.

§ 111.73 What is your responsibility for determining whether established specifications are met?

You must determine whether the specifications you establish under § 111.70 are met.

§ 111.75 What must you do to determine whether specifications are met?

(a) Before you use a component, you must:
  (1) Conduct at least one appropriate test or examination to verify the identity of any component that is a dietary ingredient; and
  (2) Confirm the identity of other components and determine whether other applicable component specifications established in accordance with § 111.70(b) are met. To do so, you must either:
    (i) Conduct appropriate tests or examinations; or
    (ii) Rely on a certificate of analysis from the supplier of the component that you receive, provided that:
      (A) You first qualify the supplier by establishing the reliability of the supplier's certificate of analysis through confirmation of the results of the supplier's tests or examinations;
      (B) The certificate of analysis includes a description of the test or examination method(s) used, limits of the test or examinations, and actual results of the tests or examinations;
      (C) You maintain documentation of how you qualified the supplier;

(D) You periodically re-confirm the supplier's certificate of analysis; and
(E) Your quality control personnel review and approve the documentation setting forth the basis for qualification (and re-qualification) of any supplier.

(b) You must monitor the in-process points, steps, or stages where control is necessary to ensure the quality of the finished batch of dietary supplement to:
(1) Determine whether the in-process specifications are met; and
(2) Detect any deviation or unanticipated occurrence that may result in a failure to meet specifications.

(c) For a subset of finished dietary supplement batches that you identify through a sound statistical sampling plan (or for every finished batch), you must verify that your finished batch of the dietary supplement meets product specifications for identity, purity, strength, composition, and for limits on those types of contamination that may adulterate or that may lead to adulteration of the finished batch of the dietary supplement. To do so:
(1) You must select one or more established specifications for identity, purity, strength, composition, and the limits on those types of contamination that may adulterate or that may lead to adulteration of the dietary supplement that, if tested or examined on the finished batches of the dietary supplement, would verify that the production and process control system is producing a dietary supplement that meets all product specifications (or only those product specifications not otherwise exempted from this provision by quality control personnel under paragraph (d) of this section);
(2) You must conduct appropriate tests or examinations to determine compliance with the specifications selected in paragraph (c)(1) of this section;
(3) You must provide adequate documentation of your basis for determining compliance with the specification(s) selected under paragraph (c)(1) of this section, through the use of appropriate tests or examinations conducted under paragraph (c)(2) of this section, will ensure that your finished batch of the dietary supplement meets all product specifications for identity, purity, strength, and composition, and the limits on those types of contamination that may adulterate, or that may lead to the adulteration of, the dietary supplement; and
(4) Your quality control personnel must review and approve the documentation that you provide under paragraph (c)(3) of this section.

(d) (1) You may exempt one or more product specifications from verification requirements in paragraph (c)(1) of this section if you determine and document that the specifications you select under paragraph (c)(1) of this section for determination of compliance with specifications are not able to verify that the production and process control system is producing a dietary supplement that meets the exempted product specification and there is no scientifically valid method for testing or examining such exempted product specification at the finished batch stage. In such a case, you must document why, for example, any component and in-process testing, examination, or monitoring, and any other information, will ensure that such exempted product specification is met without verification through periodic testing of the finished batch; and

(2) Your quality control personnel must review and approve the documentation that you provide under paragraph (d)(1) of this section.

(e) Before you package or label a product that you receive for packaging or labeling as a dietary supplement (and for distribution rather than for return to the supplier), you must visually examine the product and have documentation to determine whether the specifications that you established under § 111.70 (f) are met.

(f) (1) Before you use packaging, you must, at a minimum, conduct a visual identification of the containers and closures and review the supplier's invoice, guarantee, or certification to determine whether the packaging specifications are met; and

(2) Before you use labels, you must, at a minimum, conduct a visual examination of the label and review the supplier's invoice, guarantee, or certification to determine whether label specifications are met.

(g) You must, at a minimum, conduct a visual examination of the packaging and labeling of the finished packaged and labeled dietary supplements to determine whether you used the specified packaging and applied the specified label.

(h) (1) You must ensure that the tests and examinations that you use to determine whether the specifications are met are appropriate, scientifically valid methods.

(2) The tests and examinations that you use must include at least one of the following:
 (i) Gross organoleptic analysis;
 (ii) Macroscopic analysis;
 (iii) Microscopic analysis;
 (iv) Chemical analysis; or
 (v) Other scientifically valid methods.
 (vi) You must establish corrective action plans for use when an established specification is not met.

§ 111.77 What must you do if established specifications are not met?

(a) For specifications established under § 111.70(a), (b)(2), (b)(3), (c), (d), (e), and (g) that you do not meet, quality control personnel, in accordance with the requirements in subpart F of this part, must reject the component, dietary supplement, package or label unless such personnel approve a treatment, an in-process adjustment, or reprocessing that will ensure the quality of the finished dietary supplement and that the dietary supplement is packaged and labeled as specified in the master manufacturing record. No finished batch of dietary supplements may be released for distribution unless it complies with § 111.123(b).

(b) For specifications established under § 111.70(b)(1) that you do not meet, quality control personnel must reject the component and the component must not be used in manufacturing the dietary supplement. (c) For specifications established under § 111.70(f) that you do not meet, quality control personnel must reject the product and the product may not be packaged or labeled for distribution as a dietary supplement.

§ 111.80 What representative samples must you collect?

The representative samples that you must collect include:

(a) Representative samples of each unique lot of components, packaging, and labels that you use to determine whether the components, packaging, and labels meet specifications established in accordance with § 111.70(b) and (d), and as applicable, § 111.70(a) (and, when you receive components, packaging, or labels from a supplier, representative samples of each unique shipment, and of each unique lot within each unique shipment);
(b) Representative samples of in-process materials for each manufactured batch at points, steps, or stages, in the manufacturing process as specified in the master manufacturing record where control is necessary to ensure the identity, purity, strength, and composition of dietary supplements to determine whether the in-process materials meet specifications established in accordance with § 111.70(c), and as applicable, § 111.70(a);
(c) Representative samples of a subset of finished batches of each dietary supplement that you manufacture, which you identify through a sound statistical sampling plan (or otherwise every finished batch), before releasing for distribution to verify that the finished batch of dietary supplement meets product specifications established in accordance with § 111.70(e), and as applicable, § 111.70(a);
(d) Representative samples of each unique shipment, and of each unique lot within each unique shipment, of product that you receive for packaging or labeling as a dietary supplement (and for distribution rather than for return to the supplier) to determine whether the received product meets specifications established in accordance with § 111.70(f), and as applicable, § 111.70(a); and
(e) Representative samples of each lot of packaged and labeled dietary supplements to determine whether the packaging and labeling of the finished packaged and labeled dietary supplements meet specifications established in accordance with § 111.70(g), and as applicable, § 111.70(a).

§ 111.83 What are the requirements for reserve samples?

(a) You must collect and hold reserve samples of each lot of packaged and labeled dietary supplements that you distribute.
(b) The reserve samples must:
   (1) Be held using the same container-closure system in which the packaged and labeled dietary supplement is distributed, or if distributing dietary supplements to be packaged and labeled, using a container-closure system that provides essentially the same characteristics to protect against contamination or deterioration as the one in which it is distributed for packaging and labeling elsewhere;
   (2) Be identified with the batch, lot, or control number;
   (3) Be retained for 1 year past the shelf life date (if shelf life dating is used), or for 2 years from the date of distribution of the last batch of

dietary supplements associated with the reserve sample, for use in appropriate investigations; and
(4) Consist of at least twice the quantity necessary for all tests or examinations to determine whether or not the dietary supplement meets product specifications.

§ 111.87 Who conducts a material review and makes a disposition decision?

Quality control personnel must conduct all required material reviews and make all required disposition decisions.

§ 111.90 What requirements apply to treatments, in-process adjustments, and reprocessing when there is a deviation or unanticipated occurrence or when a specification established in accordance with § 111.70 is not met?

(a) You must not reprocess a rejected dietary supplement or treat or provide an in-process adjustment to a component, packaging, or label to make it suitable for use in the manufacture of a dietary supplement unless:
   (1) Quality control personnel conduct a material review and make a disposition decision to approve the reprocessing, treatment, or in-process adjustment; and
   (2) The reprocessing, treatment, or in-process adjustment is permitted by § 111.77;
(b) You must not reprocess any dietary supplement or treat or provide an in-process adjustment to a component to make it suitable for use in the manufacture of a dietary supplement, unless:
   (1) Quality control personnel conduct a material review and make a disposition decision that is based on a scientifically valid reason and approves the reprocessing, treatment, or in-process adjustment; and
   (2) The reprocessing, treatment or in-process adjustment is permitted by § 111.77;
(c) Any batch of dietary supplement that is reprocessed, that contains components that you have treated, or to which you have made in-process adjustments to make them suitable for use in the manufacture of the dietary supplement must be approved by quality control personnel and comply with § 111.123(b) before releasing for distribution.

§ 111.95 Under this subpart E, what records must you make and keep?

(a) You must make and keep records required under this subpart E in accordance with subpart P of this part.
(b) Under this subpart E, you must make and keep the following records:
   (1) The specifications established;
   (2) Documentation of your qualification of a supplier for the purpose of relying on the supplier's certificate of analysis;
   (3) Documentation for why meeting in-process specifications, in combination with meeting component specifications, helps ensure that the

dietary supplement meets the specifications for identity, purity, strength, and composition; and for limits on those types of contamination that may adulterate or may lead to adulteration of the finished batch of the dietary supplement; and

(4) Documentation for why the results of appropriate tests or examinations for the product specifications selected under § 111.75(c)(1) ensure that the dietary supplement meets all product specifications;

(5) Documentation for why any component and in-process testing, examination, or monitoring, and any other information, will ensure that a product specification that is exempted under § 111.75(d) is met without verification through periodic testing of the finished batch, including documentation that the selected specifications tested or examined under § 111.75 (c)(1) are not able to verify that the production and process control system is producing a dietary supplement that meets the exempted product specification and there is no scientifically valid method for testing or examining such exempted product specification at the finished batch stage.

Subpart F—Production and Process Control System: Requirements for Quality Control

§ 111.103 What are the requirements under this subpart F for written procedures?

You must establish and follow written procedures for the responsibilities of the quality control operations, including written procedures for conducting a material review and making a disposition decision, and for approving or rejecting any reprocessing.

§ 111.105 What must quality control personnel do?

Quality control personnel must ensure that your manufacturing, packaging, labeling, and holding operations ensure the quality of the dietary supplement and that the dietary supplement is packaged and labeled as specified in the master manufacturing record. To do so, quality control personnel must perform operations that include:

(a) Approving or rejecting all processes, specifications, written procedures, controls, tests, and examinations, and deviations from or modifications to them, that may affect the identity, purity, strength, or composition of a dietary supplement;

(b) Reviewing and approving the documentation setting forth the basis for qualification of any supplier;

(c) Reviewing and approving the documentation setting forth the basis for why meeting in-process specifications, in combination with meeting component specifications, will help ensure that the identity, purity, strength, and composition of the dietary supplement are met;

(d) Reviewing and approving the documentation setting forth the basis for why the results of appropriate tests or examinations for each product specification selected under § 111.75(c)(1) will ensure that the finished batch of the dietary supplement meets product specifications;

APPENDIX FDA DIETARY SUPPLEMENT GMP REGULATIONS

(e) Reviewing and approving the basis and the documentation for why any product specification is exempted from the verification requirements in § 111.75(c)(1), and for why any component and in-process testing, examination, or monitoring, or other methods will ensure that such exempted product specification is met without verification through periodic testing of the finished batch;
(f) Ensuring that required representative samples are collected;
(g) Ensuring that required reserve samples are collected and held;
(h) Determining whether all specifications established under § 111.70(a) are met; and
(i) Performing other operations required under this subpart.

§ 111.110 What quality control operations are required for laboratory operations associated with the production and process control system?

Quality control operations for laboratory operations associated with the production and process control system must include:

(a) Reviewing and approving all laboratory control processes associated with the production and process control system;
(b) Ensuring that all tests and examinations required under § 111.75 are conducted; and
(c) Reviewing and approving the results of all tests and examinations required under § 111.75.

§ 111.113 What quality control operations are required for a material review and disposition decision?

(a) Quality control personnel must conduct a material review and make a disposition decision if:
  (1) A specification established in accordance with § 111.70 is not met;
  (2) A batch deviates from the master manufacturing record, including when any step established in the master manufacturing record is not completed and including any deviation from specifications;
  (3) There is any unanticipated occurrence during the manufacturing operations that adulterates or may lead to adulteration of the component, dietary supplement, or packaging, or could lead to the use of a label not specified in the master manufacturing record;
  (4) Calibration of an instrument or control suggests a problem that may have resulted in a failure to ensure the quality of a batch or batches of a dietary supplement; or
  (5) A dietary supplement is returned.
(b) (1) When there is a deviation or unanticipated occurrence during the production and in-process control system that results in or could lead to adulteration of a component, dietary supplement, or packaging, or could lead to the use of a label not specified in the master manufacturing record, quality control personnel must reject the component, dietary supplement, packaging, or label unless it approves a treatment,

an in-process adjustment, or reprocessing to correct the applicable deviation or occurrence.
(2) When a specification established in accordance with § 111.70 is not met, quality control personnel must reject the component, dietary supplement, package or label, unless quality control personnel approve a treatment, an in-process adjustment, or reprocessing, as permitted in § 111.77.
(c) The person who conducts a material review and makes the disposition decision must, at the time of performance, document that material review and disposition decision.

§ 111.117 What quality control operations are required for equipment, instruments, and controls?

Quality control operations for equipment, instruments, and controls must include:

(a) Reviewing and approving all processes for calibrating instruments and controls;
(b) Periodically reviewing all records for calibration of instruments and controls;
(c) Periodically reviewing all records for calibrations, inspections, and checks of automated, mechanical, or electronic equipment; and
(d) Reviewing and approving controls to ensure that automated, mechanical, or electronic equipment functions in accordance with its intended use.

§ 111.120 What quality control operations are required for components, packaging, and labels before use in the manufacture of a dietary supplement?

Quality control operations for components, packaging, and labels before use in the manufacture of a dietary supplement must include:

(a) Reviewing all receiving records for components, packaging, and labels;
(b) Determining whether all components, packaging, and labels conform to specifications established under § 111.70 (b) and (d);
(c) Conducting any required material review and making any required disposition decision;
(d) Approving or rejecting any treatment and in-process adjustments of components, packaging, or labels to make them suitable for use in the manufacture of a dietary supplement; and
(e) Approving, and releasing from quarantine, all components, packaging, and labels before they are used.

§ 111.123 What quality control operations are required for the master manufacturing record, the batch production record, and manufacturing operations?

(a) Quality control operations for the master manufacturing record, the batch production record, and manufacturing operations must include:
 (1) Reviewing and approving all master manufacturing records and all modifications to the master manufacturing records;

APPENDIX FDA DIETARY SUPPLEMENT GMP REGULATIONS 271

- (2) Reviewing and approving all batch production-related records;
- (3) Reviewing all monitoring required under subpart E;
- (4) Conducting any required material review and making any required disposition decision;
- (5) Approving or rejecting any reprocessing;
- (6) Determining whether all in-process specifications established in accordance with § 111.70(c) are met;
- (7) Determining whether each finished batch conforms to product specifications established in accordance with § 111.70(e); and
- (8) Approving and releasing, or rejecting, each finished batch for distribution, including any reprocessed finished batch.

(b) Quality control personnel must not approve and release for distribution:
- (1) Any batch of dietary supplement for which any component in the batch does not meet its identity specification;
- (2) Any batch of dietary supplement, including any reprocessed batch, that does not meet all product specifications established in accordance with § 111.70(e);
- (3) Any batch of dietary supplement, including any reprocessed batch, that has not been manufactured, packaged, labeled, and held under conditions to prevent adulteration under section 402(a)(1), (a)(2), (a)(3), and (a)(4) of the act; and
- (4) Any product received from a supplier for packaging or labeling as a dietary supplement (and for distribution rather than for return to the supplier) for which sufficient assurance is not provided to adequately identify the product and to determine that the product is consistent with your purchase order.

§ 111.127 What quality control operations are required for packaging and labeling operations?

Quality control operations for packaging and labeling operations must include:

- (a) Reviewing the results of any visual examination and documentation to ensure that specifications established under § 111.70(f) are met for all products that you receive for packaging and labeling as a dietary supplement (and for distribution rather than for return to the supplier);
- (b) Approving, and releasing from quarantine, all products that you receive for packaging or labeling as a dietary supplement (and for distribution rather than for return to the supplier) before they are used for packaging or labeling;
- (c) Reviewing and approving all records for packaging and label operations;
- (d) Determining whether the finished packaged and labeled dietary supplement conforms to specifications established in accordance with § 111.70(g);
- (e) Conducting any required material review and making any required disposition decision;
- (f) Approving or rejecting any repackaging of a packaged dietary supplement;
- (g) Approving or rejecting any relabeling of a packaged and labeled dietary supplement; and
- (h) Approving for release, or rejecting, any packaged and labeled dietary supplement (including a repackaged or relabeled dietary supplement) for distribution.

§ 111.130 What quality control operations are required for returned dietary supplements?

Quality control operations for returned dietary supplements must include:

(a) Conducting any required material review and making any required disposition decision; including:
   (1) Determining whether tests or examination are necessary to determine compliance with product specifications established in accordance with § 111.70(e); and
   (2) Reviewing the results of any tests or examinations that are conducted to determine compliance with product specifications established in accordance with § 111.70(e);
(b) Approving or rejecting any salvage and redistribution of any returned dietary supplement;
(c) Approving or rejecting any reprocessing of any returned dietary supplement; and
(d) Determining whether the reprocessed dietary supplement meets product specifications and either approving for release, or rejecting, any returned dietary supplement that is reprocessed.

§ 111.135 What quality control operations are required for product complaints?

Quality control operations for product complaints must include reviewing and approving decisions about whether to investigate a product complaint and reviewing and approving the findings and follow-up action of any investigation performed.

§ 111.140 Under this subpart F, what records must you make and keep?

(a) You must make and keep the records required under this subpart F in accordance with subpart P of this part.
(b) You must make and keep the following records:
   (1) Written procedures for the responsibilities of the quality control operations, including written procedures for conducting a material review and making a disposition decision and written procedures for approving or rejecting any reprocessing;
   (2) Written documentation, at the time of performance, that quality control personnel performed the review, approval, or rejection requirements by recording the following:
      (i) Date that the review, approval, or rejection was performed; and
      (ii) Signature of the person performing the review, approval, or rejection; and
   (3) Documentation of any material review and disposition decision and follow-up. Such documentation must be included in the appropriate batch production record and must include:
      (i) Identification of the specific deviation or the unanticipated occurrence;

(ii) Description of your investigation into the cause of the deviation from the specification or the unanticipated occurrence;
(iii) Evaluation of whether or not the deviation or unanticipated occurrence has resulted in or could lead to a failure to ensure the quality of the dietary supplement or a failure to package and label the dietary supplement as specified in the master manufacturing record;
(iv) Identification of the action(s) taken to correct, and prevent a recurrence of, the deviation or the unanticipated occurrence;
(v) Explanation of what you did with the component, dietary supplement, packaging, or label;
(vi) A scientifically valid reason for any reprocessing of a dietary supplement that is rejected or any treatment or in-process adjustment of a component that is rejected; and
(vii) The signature of the individual(s) designated to perform the quality control operation, who conducted the material review and made the disposition decision, and of each qualified individual who provides information relevant to that material review and disposition decision.

Subpart G—Production and Process Control System: Requirements for Components, Packaging, and Labels and for Product That You Receive for Packaging or Labeling as a Dietary Supplement

§ 111.153 What are the requirements under this subpart G for written procedures?

You must establish and follow written procedures for fulfilling the requirements of this subpart G.

§ 111.155 What requirements apply to components of dietary supplements?

(a) You must visually examine each immediate container or grouping of immediate containers in a shipment that you receive for appropriate content label, container damage, or broken seals to determine whether the container condition may have resulted in contamination or deterioration of the components;
(b) You must visually examine the supplier's invoice, guarantee, or certification in a shipment you receive to ensure the components are consistent with your purchase order;
(c) You must quarantine components before you use them in the manufacture of a dietary supplement until:
  (1) You collect representative samples of each unique lot of components (and, for components that you receive, of each unique shipment, and of each unique lot within each unique shipment);
  (2) Quality control personnel review and approve the results of any tests or examinations conducted on components; and
  (3) Quality control personnel approve the components for use in the manufacture of a dietary supplement, including approval of any

treatment (including in-process adjustments) of components to make them suitable for use in the manufacture of a dietary supplement, and releases them from quarantine.
(d) (1) You must identify each unique lot within each unique shipment of components that you receive and any lot of components that you produce in a manner that allows you to trace the lot to the supplier, the date received, the name of the component, the status of the component (e.g., quarantined, approved, or rejected); and to the dietary supplement that you manufactured and distributed.
 (2) You must use this unique identifier whenever you record the disposition of each unique lot within each unique shipment of components that you receive and any lot of components that you produce.
(e) You must hold components under conditions that will protect against contamination and deterioration, and avoid mixups.

§ 111.160 What requirements apply to packaging and labels received?

(a) You must visually examine each immediate container or grouping of immediate containers in a shipment for appropriate content label, container damage, or broken seals to determine whether the container condition may have resulted in contamination or deterioration of the packaging and labels.
(b) You must visually examine the supplier's invoice, guarantee, or certification in a shipment to ensure that the packaging or labels are consistent with your purchase order.
(c) You must quarantine packaging and labels before you use them in the manufacture of a dietary supplement until:
 (1) You collect representative samples of each unique shipment, and of each unique lot within each unique shipment, of packaging and labels and, at a minimum, conduct a visual identification of the immediate containers and closures;
 (2) **Quality** control personnel review and approve the results of any tests or examinations conducted on the packaging and labels; and
 (3) **Quality** control personnel approve the packaging and labels for use in the manufacture of a dietary supplement and release them from quarantine.
(d) (1) You must identify each unique lot within each unique shipment of packaging and labels in a manner that allows you to trace the lot to the supplier, the date received, the name of the packaging and label, the status of the packaging and label (e.g., quarantined, approved, or rejected); and to the dietary supplement that you distributed; and
 (2) You must use this unique identifier whenever you record the disposition of each unique lot within each unique shipment of packaging and labels.
(e) You must hold packaging and labels under conditions that will protect against contamination and deterioration, and avoid mixups.

APPENDIX FDA DIETARY SUPPLEMENT GMP REGULATIONS

§ 111.165 What requirements apply to a product received for packaging or labeling as a dietary supplement (and for distribution rather than for return to the supplier)?

(a) You must visually examine each immediate container or grouping of immediate containers in a shipment of product that you receive for packaging or labeling as a dietary supplement (and for distribution rather than for return to the supplier) for appropriate content label, container damage, or broken seals to determine whether the container condition may have resulted in contamination or deterioration of the received product.

(b) You must visually examine the supplier's invoice, guarantee, or certification in a shipment of the received product to ensure that the received product is consistent with your purchase order.

(c) You must quarantine the received product until:
  (1) You collect representative samples of each unique shipment, and of each unique lot within each unique shipment, of received product;
  (2) Quality control personnel review and approve the documentation to determine whether the received product meets the specifications that you established under § 111.70(f); and
  (3) Quality control personnel approve the received product for packaging or labeling as a dietary supplement and release the received product from quarantine.

(d) (1) You must identify each unique lot within each unique shipment of received product in a manner that allows you to trace the lot to the supplier, the date received, the name of the received product, the status of the received product (e.g., quarantined, approved, or rejected), and to the product that you packaged or labeled and distributed as a dietary supplement.
  (2) You must use this unique identifier whenever you record the disposition of each unique lot within each unique shipment of the received product.

(e) You must hold the received product under conditions that will protect against contamination and deterioration, and avoid mixups.

§ 111.170 What requirements apply to rejected components, packaging, and labels, and to rejected products that are received for packaging or labeling as a dietary supplement?

You must clearly identify, hold, and control under a quarantine system for appropriate disposition any component, packaging, and label, and any product that you receive for packaging or labeling as a dietary supplement (and for distribution rather than for return to the supplier), that is rejected and unsuitable for use in manufacturing, packaging, or labeling operations.

§ 111.180 Under this subpart G, what records must you make and keep?

(a) You must make and keep records required under this subpart G in accordance with subpart P of this part.
(b) You must make and keep the following records:
  (1) Written procedures for fulfilling the requirements of this subpart.

(2) Receiving records (including records such as certificates of analysis, suppliers' invoices, and suppliers' guarantees) for components, packaging, and labels and for products that you receive for packaging or labeling as a dietary supplement (and for distribution rather than for return to the supplier); and
(3) Documentation that the requirements of this subpart were met.
  (i) The person who performs the required operation must document, at the time of performance, that the required operation was performed.
  (ii) The documentation must include:
     (A) The date that the components, packaging, labels, or products that you receive for packaging or labeling as a dietary supplement were received;
     (B) The initials of the person performing the required operation;
     (C) The results of any tests or examinations conducted on components, packaging, or labels, and of any visual examination of product that you receive for packaging or labeling as a dietary supplement; and
     (D) Any material review and disposition decision conducted on components, packaging, labels, or products that you receive for packaging or labeling as a dietary supplement.

Subpart H—Production and Process Control System: Requirements for the Master Manufacturing Record

§ 111.205 What is the requirement to establish a master manufacturing record?

(a) You must prepare and follow a written master manufacturing record for each unique formulation of dietary supplement that you manufacture, and for each batch size, to ensure uniformity in the finished batch from batch to batch.
(b) The master manufacturing record must:
  (1) Identify specifications for the points, steps, or stages in the manufacturing process where control is necessary to ensure the quality of the dietary supplement and that the dietary supplement is packaged and labeled as specified in the master manufacturing record; and
  (2) Establish controls and procedures to ensure that each batch of dietary supplement that you manufacture meets the specifications identified in accordance with paragraph (b)(1) of this section.
(c) You must make and keep master manufacturing records in accordance with subpart P of this part.

§ 111.210 What must the master manufacturing record include?

The master manufacturing record must include:

(a) The name of the dietary supplement to be manufactured and the strength, concentration, weight, or measure of each dietary ingredient for each batch size;
(b) A complete list of components to be used;

APPENDIX FDA DIETARY SUPPLEMENT GMP REGULATIONS 277

(c) An accurate statement of the weight or measure of each component to be used;
(d) The identity and weight or measure of each dietary ingredient that will be declared on the Supplement Facts label and the identity of each ingredient that will be declared on the ingredients list of the dietary supplement;
(e) A statement of any intentional overage amount of a dietary ingredient;
(f) A statement of theoretical yield of a manufactured dietary supplement expected at each point, step, or stage of the manufacturing process where control is needed to ensure the quality of the dietary supplement, and the expected yield when you finish manufacturing the dietary supplement, including the maximum and minimum percentages of theoretical yield beyond which a deviation investigation of a batch is necessary and material review is conducted and disposition decision is made;
(g) A description of packaging and a representative label, or a cross-reference to the physical location of the actual or representative label;
(h) Written instructions, including the following:
  (1) Specifications for each point, step, or stage in the manufacturing process where control is necessary to ensure the quality of the dietary supplement and that the dietary supplement is packaged and labeled as specified in the master manufacturing record;
  (2) Procedures for sampling and a cross-reference to procedures for tests or examinations;
  (3) Specific actions necessary to perform and verify points, steps, or stages in the manufacturing process where control is necessary to ensure the quality of the dietary supplement and that the dietary supplement is packaged and labeled as specified in the master manufacturing record.
    (i) Such specific actions must include verifying the weight or measure of any component and verifying the addition of any component; and
    (ii) For manual operations, such specific actions must include:
      (A) One person weighing or measuring a component and another person verifying the weight or measure; and
      (B) One person adding the component and another person verifying the addition.
  (4) Special notations and precautions to be followed; and
  (5) Corrective action plans for use when a specification is not met.

Subpart I—Production and Process Control System: Requirements for the Batch Production Record
§ 111.255 What is the requirement to establish a batch production record?

(a) You must prepare a batch production record every time you manufacture a batch of a dietary supplement;
(b) Your batch production record must include complete information relating to the production and control of each batch;
(c) Your batch production record must accurately follow the appropriate master manufacturing record and you must perform each step in the production of the batch; and

(d) You must make and keep batch production records in accordance with subpart P of this part.

§ 111.260 What must the batch record include?
The batch production record must include the following:
(a) The batch, lot, or control number:
   (1) Of the finished batch of dietary supplement; and
   (2) That you assign in accordance with § 111.415(f) for the following:
      (i) Each lot of packaged and labeled dietary supplement from the finished batch of dietary supplement;
      (ii) Each lot of dietary supplement, from the finished batch of dietary supplement, that you distribute to another person for packaging or labeling;
(b) The identity of equipment and processing lines used in producing the batch;
(c) The date and time of the maintenance, cleaning, and sanitizing of the equipment and processing lines used in producing the batch, or a cross-reference to records, such as individual equipment logs, where this information is retained;
(d) The unique identifier that you assigned to each component (or, when applicable, to a product that you receive from a supplier for packaging or labeling as a dietary supplement), packaging, and label used;
(e) The identity and weight or measure of each component used;
(f) A statement of the actual yield and a statement of the percentage of theoretical yield at appropriate phases of processing;
(g) The actual results obtained during any monitoring operation;
(h) The results of any testing or examination performed during the batch production, or a cross-reference to such results;
(i) Documentation that the finished dietary supplement meets specifications established in accordance with § 111.70(e) and (g);
(j) Documentation, at the time of performance, of the manufacture of the batch, including:
   (1) The date on which each step of the master manufacturing record was performed; and
   (2) The initials of the persons performing each step, including:
      (i) The initials of the person responsible for weighing or measuring each component used in the batch;
      (ii) The initials of the person responsible for verifying the weight or measure of each component used in the batch;
      (iii) The initials of the person responsible for adding the component to the batch; and
      (iv) The initials of the person responsible for verifying the addition of components to the batch;
(k) Documentation, at the time of performance, of packaging and labeling operations, including:
   (1) The unique identifier that you assigned to packaging and labels used, the quantity of the packaging and labels used, and, when label reconciliation is required, reconciliation of any discrepancies between issuance and use of labels;

APPENDIX FDA DIETARY SUPPLEMENT GMP REGULATIONS

    (2) An actual or representative label, or a cross-reference to the physical location of the actual or representative label specified in the master manufacturing record; and
    (3) The results of any tests or examinations conducted on packaged and labeled dietary supplements (including repackaged or relabeled dietary supplements), or a crossreference to the physical location of such results;
(l) Documentation at the time of performance that quality control personnel:
    (1) Reviewed the batch production record, including:
        (i) Review of any monitoring operation required under subpart E of this part; and
        (ii) Review of the results of any tests and examinations, including tests and examinations conducted on components, in-process materials, finished batches of dietary supplements, and packaged and labeled dietary supplements;
    (2) Approved or rejected any reprocessing or repackaging; and
    (3) Approved and released, or rejected, the batch for distribution, including any reprocessed batch; and
    (4) Approved and released, or rejected, the packaged and labeled dietary supplement, including any repackaged or relabeled dietary supplement.
(m) Documentation at the time of performance of any required material review and disposition decision.
(n) Documentation at the time of performance of any reprocessing.

Subpart J—Production and Process Control System: Requirements for Laboratory Operations
§ 111.303 What are the requirements under this subpart J for written procedures?
You must establish and follow written procedures for laboratory operations, including written procedures for the tests and examinations that you conduct to determine whether specifications are met.

§ 111.310 What are the requirements for the laboratory facilities that you use?
You must use adequate laboratory facilities to perform whatever testing and examinations are necessary to determine whether:

(a) Components that you use meet specifications;
(b) In-process specifications are met as specified in the master manufacturing record; and
(c) Dietary supplements that you manufacture meet specifications.

§ 111.315 What are the requirements for laboratory control processes?
You must establish and follow laboratory control processes that are reviewed and approved by quality control personnel, including the following:

(a) Use of criteria for establishing appropriate specifications;
(b) Use of sampling plans for obtaining representative samples, in accordance with subpart E of this part, of:
   (1) Components, packaging, and labels;
   (2) In-process materials;
   (3) Finished batches of dietary supplements;
   (4) Product that you receive for packaging or labeling as a dietary supplement (and for distribution rather than for return to the supplier); and
   (5) Packaged and labeled dietary supplements.
(c) Use of criteria for selecting appropriate examination and testing methods;
(d) Use of criteria for selecting standard reference materials used in performing tests and examinations; and
(e) Use of test methods and examinations in accordance with established criteria.

§ 111.320 What requirements apply to laboratory methods for testing and examination?

(a) You must verify that the laboratory examination and testing methodologies are appropriate for their intended use.
(b) You must identify and use an appropriate scientifically valid method for each established specification for which testing or examination is required to determine whether the specification is met.

§ 111.325 Under this subpart J, what records must you make and keep?

(a) You must make and keep records required under this subpart J in accordance with subpart P of this part.
(b) You must make and keep the following records:
   (1) Written procedures for laboratory operations, including written procedures for the tests and examinations that you conduct to determine whether specifications are met;
   (2) Documentation that laboratory methodology established in accordance with this subpart J is followed.
      (i) The person who conducts the testing and examination must document, at the time of performance, that laboratory methodology established in accordance with this subpart J is followed.
      (ii) The documentation for laboratory tests and examinations must include the results of the testing and examination.

Subpart K—Production and Process Control System: Requirements for Manufacturing Operations

§ 111.353 What are the requirements under this subpart K for written procedures?

You must establish and follow written procedures for manufacturing operations.

APPENDIX FDA DIETARY SUPPLEMENT GMP REGULATIONS

§ 111.355 What are the design requirements for manufacturing operations?

You must design or select manufacturing processes to ensure that product specifications are consistently met.

§ 111.360 What are the requirements for sanitation?

You must conduct all manufacturing operations in accordance with adequate sanitation principles.

§ 111.365 What precautions must you take to prevent contamination?

You must take all the necessary precautions during the manufacture of a dietary supplement to prevent contamination of components or dietary supplements. These precautions include:

(a) Performing manufacturing operations under conditions and controls that protect against the potential for growth of microorganisms and the potential for contamination;
(b) Washing or cleaning components that contain soil or other contaminants;
(c) Using water that, at a minimum, complies with the applicable Federal, State, and local requirements and does not contaminate the dietary supplement when the water may become a component of the finished batch of dietary supplement;
(d) Performing chemical, microbiological, or other testing, as necessary to prevent the use of contaminated components;
(e) Sterilizing, pasteurizing, freezing, refrigerating, controlling hydrogen-ion concentration (pH), controlling humidity, controlling water activity (aw), or using any other effective means to remove, destroy, or prevent the growth of microorganisms and prevent decomposition;
(f) Holding components and dietary supplements that can support the rapid growth of microorganisms of public health significance in a manner that prevents the components and dietary supplements from becoming adulterated;
(g) Identifying and holding any components or dietary supplements, for which a material review and disposition decision is required, in a manner that protects components or dietary supplements that are not under a material review against contamination and mixups with those that are under a material review;
(h) Performing mechanical manufacturing steps (such as cutting, sorting, inspecting, shredding, drying, grinding, blending, and sifting) by any effective means to protect the dietary supplements against contamination, by, for example:
 (1) Cleaning and sanitizing contact surfaces;
 (2) Using temperature controls; and
 (3) Using time controls.
(i) Using effective measures to protect against the inclusion of metal or other foreign material in components or dietary supplements, by, for example:
 (1) Filters or strainers;
 (2) Traps;
 (3) Magnets, or
 (4) Electronic metal detectors.

(j) Segregating and identifying all containers for a specific batch of dietary supplements to identify their contents and, when necessary, the phase of manufacturing; and
(k) Identifying all processing lines and major equipment used during manufacturing to indicate their contents, including the name of the dietary supplement and the specific batch or lot number and, when necessary, the phase of manufacturing.

§ 111.370 What requirements apply to rejected dietary supplements?

You must clearly identify, hold, and control under a quarantine system for appropriate disposition any dietary supplement that is rejected and unsuitable for use in manufacturing, packaging, or labeling operations.

§ 111.375 Under this subpart K, what records must you make and keep?

(a) You must make and keep records required under this subpart K in accordance with subpart P of this part.
(b) You must make and keep records of the written procedures for manufacturing operations.

Subpart L—Production and Process Control System: Requirements for Packaging and Labeling Operations

§ 111.403 What are the requirements under this subpart L for written procedures?

You must establish and follow written procedures for packaging and labeling operations.

§ 111.410 What requirements apply to packaging and labels?

(a) You must take necessary actions to determine whether packaging for dietary supplements meets specifications so that the condition of the packaging will ensure the quality of your dietary supplements;
(b) You must control the issuance and use of packaging and labels and reconciliation of any issuance and use discrepancies. Label reconciliation is not required for cut or rolled labels if a 100-percent examination for correct labels is performed by appropriate electronic or electromechanical equipment during or after completion of finishing operations; and
(c) You must examine, before packaging and labeling operations, packaging and labels for each batch of dietary supplement to determine whether the packaging and labels conform to the master manufacturing record; and
(d) You must be able to determine the complete manufacturing history and control of the packaged and labeled dietary supplement through distribution.

§ 111.415 What requirements apply to filling, assembling, packaging, labeling, and related operations?

You must fill, assemble, package, label, and perform other related operations in a way that ensures the quality of the dietary supplement and that the dietary supplement is packaged and labeled as specified in the master manufacturing record. You must do this using any effective means, including the following:

(a) Cleaning and sanitizing all filling and packaging equipment, utensils, and dietary supplement packaging, as appropriate;
(b) Protecting manufactured dietary supplements from contamination, particularly airborne contamination;
(c) Using sanitary handling procedures;
(d) Establishing physical or spatial separation of packaging and label operations from operations on other components and dietary supplements to prevent mixups;
(e) Identifying, by any effective means, filled dietary supplement containers that are set aside and held in unlabeled condition for future label operations, to prevent mixups;
(f) Assigning a batch, lot, or control number to:
 (1) Each lot of packaged and labeled dietary supplement from a finished batch of dietary supplement; and
 (2) Each lot of dietary supplement, from a finished batch of dietary supplement, that you distribute to another person for packaging or labeling.
(g) Examining a representative sample of each batch of the packaged and labeled dietary supplement to determine whether the dietary supplement meets specifications established in accordance with § 111.70(g); and
(h) Suitably disposing of labels and packaging for dietary supplements that are obsolete or incorrect to ensure that they are not used in any future packaging and label operations.

§ 111.420 What requirements apply to repackaging and relabeling?

(a) You may repackage or relabel dietary supplements only after quality control personnel have approved such repackaging or relabeling.
(b) You must examine a representative sample of each batch of repackaged or relabeled dietary supplements to determine whether the repackaged or relabeled dietary supplements meet all specifications established in accordance with § 111.70(g).
(c) Quality control personnel must approve or reject each batch of repackaged or relabeled dietary supplement prior to its release for distribution.

§ 111.425 What requirements apply to a packaged and labeled dietary supplement that is rejected for distribution?

You must clearly identify, hold, and control under a quarantine system for appropriate disposition any packaged and labeled dietary supplement that is rejected for distribution.

§ 111.430 Under this subpart L, what records must you make and keep?

(a) You must make and keep records required under this subpart L in accordance with subpart P of this part.

(b) You must make and keep records of the written procedures for packaging and labeling operations.

Subpart M—Holding and Distributing
§ 111.453 What are the requirements under this subpart for M written procedures?
You must establish and follow written procedures for holding and distributing operations.

§ 111.455 What requirements apply to holding components, dietary supplements, packaging, and labels?

(a) You must hold components and dietary supplements under appropriate conditions of temperature, humidity, and light so that the identity, purity, strength, and composition of the components and dietary supplements are not affected.
(b) You must hold packaging and labels under appropriate conditions so that the packaging and labels are not adversely affected.
(c) You must hold components, dietary supplements, packaging, and labels under conditions that do not lead to the mixup, contamination, or deterioration of components, dietary supplements, packaging, and labels.

§ 111.460 What requirements apply to holding in-process material?
(a) You must identify and hold in-process material under conditions that protect against mixup, contamination, and deterioration.
(b) You must hold in-process material under appropriate conditions of temperature, humidity, and light.

§ 111.465 What requirements apply to holding reserve samples of dietary supplements?
(a) You must hold reserve samples of dietary supplements in a manner that protects against contamination and deterioration. This includes:
  (1) Holding the reserve samples under conditions consistent with product labels or, if no storage conditions are recommended on the label, under ordinary storage conditions; and
  (2) Using the same container-closure system in which the packaged and labeled dietary supplement is distributed, or if distributing dietary supplements to be packaged and labeled, using a container-closure system that provides essentially the same characteristics to protect against contamination or deterioration as the one in which you distribute the dietary supplement for packaging and labeling elsewhere.
(b) You must retain reserve samples for 1 year past the shelf life date (if shelf life dating is used), or for 2 years from the date of distribution of the last batch of dietary supplements associated with the reserve samples, for use in appropriate investigations.

APPENDIX   FDA DIETARY SUPPLEMENT GMP REGULATIONS

§ 111.470 What requirements apply to distributing dietary supplements?

You must distribute dietary supplements under conditions that will protect the dietary supplements against contamination and deterioration.

§ 111.475 Under this subpart M, what records must you make and keep?

(a) You must make and keep records required under this subpart M in accordance with subpart P of this part.
(b) You must make and keep the following records:
  (1) Written procedures for holding and distributing operations; and
  (2) Records of product distribution.

Subpart N—Returned Dietary Supplements

§ 111.503 What are the requirements under this subpart N for written procedures?

You must establish and follow written procedures to fulfill the requirements of this subpart.

§ 111.510 What requirements apply when a returned dietary supplement is received?

You must identify and quarantine returned dietary supplements until quality control personnel conduct a material review and make a disposition decision.

§ 111.515 When must a returned dietary supplement be destroyed, or otherwise suitably disposed of?

You must destroy, or otherwise suitably dispose of, any returned dietary supplement unless the outcome of a material review and disposition decision is that quality control personnel do the following:

(a) Approve the salvage of the returned dietary supplement for redistribution or
(b) Approve the returned dietary supplement for reprocessing.

§ 111.520 When may a returned dietary supplement be salvaged?

You may salvage a returned dietary supplement only if quality control personnel conduct a material review and make a disposition decision to allow the salvage.

§ 111.525 What requirements apply to a returned dietary supplement that quality control personnel approve for reprocessing?

(a) You must ensure that any returned dietary supplements that are reprocessed meet all product specifications established in accordance with § 111.70(e); and

(b) Quality control personnel must approve or reject the release for distribution of any returned dietary supplement that is reprocessed.

§ 111.530 When must an investigation be conducted of your manufacturing processes and other batches?

If the reason for a dietary supplement being returned implicates other batches, you must conduct an investigation of your manufacturing processes and each of those other batches to determine compliance with specifications.

§ 111.535 Under this subpart N, what records must you make and keep?

(a) You must make and keep records required under this subpart N in accordance with subpart P of this part.
(b) You must make and keep the following records:
  (1) Written procedures for fulfilling the requirements of this subpart N.
  (2) Any material review and disposition decision on a returned dietary supplement;
  (3) The results of any testing or examination conducted to determine compliance with product specifications established under § 111.70(e); and
  (4) Documentation of the reevaluation by quality control personnel of any dietary supplement that is reprocessed and the determination by quality control personnel of whether the reprocessed dietary supplement meets product specifications established in accordance with § 111.70(e).

Subpart O—Product Complaints

§ 111.553 What are the requirements under this subpart O for written procedures?

You must establish and follow written procedures to fulfill the requirements of this subpart O.

§ 111.560 What requirements apply to the review and investigation of a product complaint?

(a) A qualified person must:
  (1) Review all product complaints to determine whether the product complaint involves a possible failure of a dietary supplement to meet any of its specifications, or any other requirements of this part 111, including those specifications and other requirements that, if not met, may result in a risk of illness or injury; and
  (2) Investigate any product complaint that involves a possible failure of a dietary supplement to meet any of its specifications, or any other requirements of this part, including those specifications and other requirements that, if not met, may result in a risk of illness or injury.
(b) Quality control personnel must review and approve decisions about whether to investigate a product complaint and review and approve the findings and follow-up action of any investigation performed.

APPENDIX   FDA DIETARY SUPPLEMENT GMP REGULATIONS

(c) The review and investigation of the product complaint by a qualified person, and the review by quality control personnel about whether to investigate a product complaint, and the findings and follow-up action of any investigation performed, must extend to all relevant batches and records.

§ 111.570 Under this subpart O, what records must you make and keep?

(a) You must make and keep the records required under this subpart O in accordance with subpart P of this part.
(b) You must make and keep the following records:
: (1) Written procedures for fulfilling the requirements of this subpart,
: (2) A written record of every product complaint that is related to good manufacturing practice,
:: (i) The person who performs the requirements of this subpart must document, at the time of performance, that the requirement was performed.
:: (ii) The written record of the product complaint must include the following:
::: (A) The name and description of the dietary supplement;
::: (B) The batch, lot, or control number of the dietary supplement, if available;
::: (C) The date the complaint was received and the name, address, or telephone number of the complainant, if available;
::: (D) The nature of the complaint including, if known, how the product was used;
::: (E) The reply to the complainant, if any; and
::: (F) Findings of the investigation and follow-up action taken when an investigation is performed.

Subpart P—Records and Recordkeeping
§ 111.605 What requirements apply to the records that you make and keep?

(a) You must keep written records required by this part for 1 year past the shelf life date, if shelf life dating is used, or 2 years beyond the date of distribution of the last batch of dietary supplements associated with those records.
(b) Records must be kept as original records, as true copies (such as photocopies, microfilm, microfiche, or other accurate reproductions of the original records), or as electronic records.
(c) All electronic records must comply with part 11 of this chapter.

§ 111.610 What records must be made available to FDA?

(a) You must have all records required under this part, or copies of such records, readily available during the retention period for inspection and copying by FDA when requested.
(b) If you use reduction techniques, such as microfilming, you must make suitable reader and photocopying equipment readily available to FDA.

Index

Acceptance quality limit (AQL), 96–97, 100
Acceptance sampling, 95–96
Accuracy, defined, 68
Acid sanitizers, 60
ADA. *see* Americans with Disabilities Act (ADA)
Administrative Procedure Act (APA), 11, 12, 14
Adulteration, 200–210. *see also* Contamination/contaminants
   defined, 200
   of dietary supplements, 9
   intentional and unintentional, 200–201
Advance Notice of Proposed Rulemaking (ANPR), 10, 12
Adverse event reporting (AER), 197–198
   labeling requirements, 198
   record keeping, 198
   significance of, 198
Advisory committees, for FDA, 230–231
AER. *see* Adverse event reporting (AER)
Aflatoxin contamination, 207
AHPA. *see* American Herbal Products Association (AHPA)
AHU. *see* Air handling unit (AHU)
Airborne contamination, 208–209
Air handling unit (AHU), 52
AISI. *see* American Iron and Steel Institute (AISI)
Aisles, physical plant, 30
Aluminum, 138, 139
   as construction material for equipment/utensils, 47
AL-6XN (UNS N08367), 48
American Herbal Products Association (AHPA), 205, 220
American Iron and Steel Institute (AISI)
   numbering system, 48
American National Standards Institute (ANSI), 89–90
   standard 173, 221
American Society for Testing and Materials (ASTM), 43
   A380 "Standard Practice for Cleaning, Descaling, and Passivation of Stainless Steel Parts, Equipment, and Systems," 48
   A967, "Standard Specification for Chemical Passivation Treatments for Stainless Steel Parts," 48
Americans with Disabilities Act (ADA), 31
Analytical chemistry, laboratory operations, 154
Analytical chemists, 154
ANPR. *see* Advance Notice of Proposed Rulemaking (ANPR)
ANSI. *see* American National Standards Institute (ANSI)

Antiseptics, 59
AOAC. *see* Association of Official Analytical Chemists (AOAC)
AOAC International, 166
APA. *see* Administrative Procedure Act (APA)
AQL. *see* Acceptance quality limit (AQL)
Association of Official Analytical Chemists (AOAC), 88, 156
ASTM. *see* American Society for Testing and Materials (ASTM)
ASTM A380 "Standard Practice for Cleaning, Descaling, and Passivation of Stainless Steel Parts, Equipment, and Systems," 48
ASTM A967, "Standard Specification for Chemical Passivation Treatments for Stainless Steel Parts," 48
ATP bioluminescence, 58
Audits, 216–221
   ANSI standard 173, 221
   gap analysis and needs assessments, 219
   internal reports, FDA for, 218–219
   reasons for, 216
   second-party, 219–220
   self-inspection, 216–218
   third-party, 220–221
   types, 216

Batch number, 142
Batch production record (BPR), 113, 117–120
   information in, 118–119
   material review/disposition decisions, 119
   MMR and, 117
   QC and, 150
   reprocessing, 119
   review, 119–120
Bills of materials (BOM), 115
BIMs. *see* Botanical identification methods (BIMs)
Bioavailability, 128
Biofilms, 61–62
BioProcess Equipment Committee of the American Society of Mechanical Engineers, 42
Blend uniformity, 101
BOM. *see* Bills of materials (BOM)
Botanical identification methods (BIMs), 165
BPR. *see* Batch production record (BPR)
Brite stock, packaging/labeling, 141
Brushes, for cleaning/sanitation, 55
Buildings, maintenance of, 66
Bulk fermentation vessels, physical plant, 32

INDEX

Calcium hypochlorite, 60
Calibration, 68–72
  accuracy/precision, 68
  frequency, 69–70
  identification label, 69
  instruments, 71–72
  program, 69
  records/written procedures, 68–69
  requirement for, 68
  role of quality control personnel in, 70
  scales/balances, 71
  traceability, 70
CALs. *see* Contract analytical laboratories (CALs)
CAPA. *see* Corrective and preventive action (CAPA)
Cap liners, 83
Capsules, making of, 131–132
Causation, *vs.* correlation, 105
CCPs. *see* Critical control points (CCPs)
Ceilings, physical plant, 26–28
  cleaning, 53
CEN. *see* European Committee for Standardization (CEN)
Center for Food Safety and Applied Nutrition (CFSAN), 233
Certificate of Analysis for Dietary Supplement Components: A Voluntary Guideline, 213
Certificates of analysis (CofA), 80, 86–87, 161, 213
CFD technology. *see* Computational fluid dynamics (CFD) technology
CFR. *see* Code of Federal Regulations (CFR)
CFSAN. *see* Center for Food Safety and Applied Nutrition (CFSAN)
Change control, 194–196
  continual improvement, 195–196
Child-resistant packaging (CRP), 143
Chlorine-releasing sanitizers, 59–60
Chromatography, 158
CI. *see* Continual improvement (CI)
CIP. *see* Clean-in-place (CIP)
Citizen Petition, 14
Cleaning compounds, 56–57
  for physical plant, 33
  water and, 53
Cleaning/sanitation, 52–62
  biofilms, 61–62
  compounds (*see* Cleaning compounds)
  of equipment/utensils/machinery, 53–56
    brushes, 55
    CIP *vs.* COP, 54
    high-pressure water blasting, 56
    parts washers, 56
    soil and, 53
    steel wool, 56
    water and, 53–54
  frequency of, 54–55
  logs, 52
  outsourcing, 62
  in physical plant, 52–53
  procedures, 55
  verification of effectiveness, 57–58
Clean-in-place (CIP)
  vs. clean-out-of-place, 54

Clean-out-of-place (COP)
  clean-in-place *vs.*, 54
CM. *see* Corrective maintenance (CM)
CMMS. *see* Computerized maintenance management system (CMMS)
CMO. *see* Contract manufacturing operations (CMO)
Code of Federal Regulations (CFR), 13
CofA. *see* Certificates of analysis (CofA)
Column chromatography, 159
Complaints
  actions in handling, 171–172
  defined, 170–171
  importance of, 170
  information to management, 172
  record keeping, 172
  sources of, 171
Components
  defined, 79
  identity specification for, 79–80, 85–86
  incoming (*see* Incoming components)
  QC for, 151–152
Compressed air, equipment/utensils, 46
Compression coating, 131
Computational fluid dynamics (CFD) technology, 134
Computerized maintenance management system (CMMS), 66
Confidential informants, 247
Conflict resolution, by FDA, 235
Coning, 99
Consent decree, 238
Consultants, in outsourcing, 228
Consumer Complaint Reporting System, 171
Consumer Products Healthcare Association, 220
Consumer Products Safety Commission (CPSC), 83, 143
Consumer safety officers (CSO), 240
Contact surface, 20
Contamination/contaminants, 200–210.
  see also Adulteration
  airborne, 208–209
  clean hands in avoiding, 209
  defect action levels, 209
  defined, 201
  economically motivated, 202
  endotoxins in, 207
  equipment/utensils and, 42–45
  foreign material (*see* Foreign material contamination)
  heavy metal, 205
  limits on, setting, 201–202
  microbiological, 85, 206–207
  mycotoxins and aflatoxins in, 207
  natural toxins in, 205
  pesticide residue, 206
  residual solvent, 206
  testing for microorganisms as, 207–208
  water activity in, 208
Content uniformity, 101
Continual improvement (CI), 195–196
Contract acceptor, defined, 223

INDEX

Contract analytical laboratories (CALs), 153, 227–228
Contracting, 223
Contract manufacturing operations (CMO), 224
 offshore, 225
 quality agreements with, 226–227
 supply agreement with, 225–226
Contractors, defined, 223
Contract packagers/labelers
 specifications, 84–85
Contract research organizations (CRO), 227
Controlled room temperature (CRT), 174
Control number, 142
Coolants, and equipment/utensils, 46
COP. see Clean-out-of-place (COP)
Copper, as construction material for equipment/utensils, 47
Corporate social responsibility (CSR), 182
Corrective actions, 103–104
 vs. preventive action, 106
Corrective and preventive action (CAPA), 106
Corrective maintenance (CM), 65
Correlation vs. causation, 105
Council for Responsible Nutrition, 220
CPSC. see Consumer Products Safety Commission (CPSC)
Critical control points (CCPs), 2
CRO. see Contract research organizations (CRO)
CRP. see Child-resistant packaging (CRP)
CRT. see Controlled room temperature (CRT)
CSO. see Consumer safety officers (CSO)
CSR. see Corporate social responsibility (CSR)

DAL. see Defect action levels (DAL)
DDSP. see Division of Dietary Supplement Programs (DDSP)
Defect action levels (DAL), 209
DEG. see Diethylene glycol (DEG)
Department of Health and Human Services (DHHS), 231, 233
Deviations, 103–106
 and corrective actions, 103–104
 correlation vs. causation, 105
 human errors and, 105
 investigations, 104–105
 root cause analysis, 105
 and unanticipated occurrence, 103
DHHS. see Department of Health and Human Services (DHHS)
Dietary ingredients, sampling of, 98–100
Dietary Supplement Good Manufacturing Practice (DS GMP), 1
 subject to, 15
Dietary Supplement Health and Education Act (DSHEA), 9, 10–11, 137, 232
Diethylene glycol (DEG)
 in Elixir sulfanilamide, 231
 in glycerin, 202
Digitalis lanata, 80
Direct final rule, 14
Disgorgement of profits, defined, 238
Disinfectant, sanitizers, 59
Disintegration, of tablets/capsules, 127
Disputes, handling
 FDA for, 235
Distribution, of finished products, 177
Division of Dietary Supplement Programs (DDSP), 233
Documentation, 186–192
 inspections of, 243–244
 laboratory operation, 163–164
 management and control, 191
 of training, 18
Dosage form, 122–123
Dress, cleanliness of, 20
Dry-bulb thermometer, 174
DS GMP. see Dietary Supplement Good Manufacturing Practice (DS GMP)
DSHEA. see Dietary Supplement Health and Education Act (DSHEA)
Ducts, physical plant, 28–29

EBR. see Electronic Batch Record (EBR)
e-CFR. see Electronic Code of Federal Regulations (e-CFR)
Economically motivated adulteration (EMA)
 defined, 202
EHEDG. see European Hygiene Equipment Design Group (EHEDG)
EIRs. see Establishment inspection reports (EIRs)
Electronic Batch Record (EBR), 117
Electronic Code of Federal Regulations (e-CFR), 13
Electronic Freedom of Information Act, 236
Electronic reading rooms, of FDA, 236
Elixir sulfanilamide, 231
EMA. see Economically motivated adulteration (EMA)
Employees
 cleanliness of dress, 20
 personal hygiene, 19–20
 qualification requirements, 17
 training of, 18–19
Endotoxin contamination, 207
Enforcement powers, of FDA, 237–239
 consent decree, 238
 FD&C Act, 238–239
 seizures and injunctions and, 238
 WL in, 237–238
Environmental control equipment, physical plant, 29–30
Environmental Protection Agency (EPA), 61
EPA. see Environmental Protection Agency (EPA)
Equipment/utensils, 40–50
 automated/mechanical/electronic, 49–50
 cleaning of, 53–56, 58–59
 cold storage, 49
 compressed air, 46
 contamination and, 42–45
 coolants/heat transfer fluids, 46
 FDA-approved, 41
 GMP requirements for, 40–41
 identification of, 64
 installation, 47
 logs, 49, 63–64

[Equipment/utensils]
  lubricants and, 45
  maintenance, 47, 63
    program manual, 64
  materials of construction for, 47–49
  metal/glass fragments, 46
  overview, 40
  sanitary, standards for, 42
  suitability, 40
Establishment inspection reports (EIRs), 137, 234, 245–246
  classification, 246
  Turbo, 246
European Committee for Standardization (CEN), 42
European Hygiene Equipment Design Group (EHEDG), 42
Examinations Institute of the Division of Chemical Education of the American Chemical Society, 154
Excipients
  physical characteristics of, 81
  sampling, 98–100
  specifications for, 80–81
Expiration dating, 144

FACA. *see* Federal Advisory Committee Act (FACA)
Factory samples, for FDA inspections, 243
FACTS. *see* Field accomplishments and compliance tracking system (FACTS)
Fair Packaging and Labeling Act, 9
FDA. *see* Food and Drug Administration (FDA)
FD&C Act. *see* Federal Food, Drug, and Cosmetic (FD&C) Act
Federal Advisory Committee Act (FACA), 231
Federal Food, Drug, and Cosmetic (FD&C) Act, 8, 183, 231–232, 238–239
  foods for special dietary uses, 10
  prohibited acts by, 200
  Section 701(a) of, 11
  Section 201(ff) of, 122
  section 402(f) of, 200
  section 402(g) of, 200
  Section 402(g)(1) of, 9
  Section 331 of, 9
  Section 401 of, 10
  Section 403 of, 9
Federal Register, 12, 13
Federal Register Act of 1934, 13
Fiberglass reinforced polyester (FRP), 26
Field accomplishments and compliance tracking system (FACTS), 246
Film coating, 130
Final Rule (FR), 12
Finished product
  examining samples of, 142
  requirement to establish, 87–88
  sampling, 101–102
  specifications of, 84, 85
  testing of, 160–161
Firm(s)
  defined, 223
  reason to outsource, 223–224

Fixtures, physical plant, 28–29
Flexibility, 80
Floors, physical plant, 26–28
  cleaning, 53
Fluid-bed processing, 133
Fluidization, 133
FOIA. *see* Freedom of Information Act (FOIA)
Food Advisory Committee, 231
Food and Drug Administration (FDA), 1–2, 230–239
  advisory committees, 230–231
  Citizen Petition, 14
  conflict resolution by, 235
  DSHEA (*see* Dietary Supplement Health and Education Act (DSHEA))
  electronic reading rooms of, 236
  enforcement powers of, 237–239
  and equipment/utensils, 41
  establishment inspection reports, 137
  FOIA of (*see* Freedom of Information Act (FOIA))
  form 483, 234
  GMP and, 5
  guidance documents, 13–14, 234–235
  history, 231–233
  inspections, 240–247
    conduct of, 241
    confidential informants, 247
    of documentation, 243–244
    EIR and (*see* Establishment inspection reports (EIRs))
    factory samples for, 243
    form 482 (*see* Form 482)
    information, and FDA's entitlement, 242–243
    inspectional observations, 244–245
    opening stage, 240–241
    photography in, 243
    preparing to be inspected, 241–242
    WL by (*see* Warning letters (WL))
  for internal audit reports, 218–219
  investigators, qualification of, 240
  organizational structure, 233–234
  on outsourcing, 227
  overview, 230
  penalties imposed of, 9–10
  and recall process, 180–181
  Standards of Identity, 10
  transparency initative, 236–237
  United States Code and, 8–9
  and verification of cleaning effectiveness, 57–58
  web site, 235
Foreign material contamination, 202–205
  magnetic separators in, 202–203, 203f
  metal detectors in, 203–204, 204f
  shatter-resistant lamps in, 204, 205f
Form 482, FDA, 240, 244–245
  responding to, 245
FR. *see* Final Rule (FR)
Freedom of Information Act (FOIA), 236
Friabilator, 127
FRP. *see* Fiberglass reinforced polyester (FRP)
Fumigants usage, in physical plant, 34–35
Fungicides usage, in physical plant, 34–35

INDEX

Gap analysis, defined, 219
Gas chromatography (GC), 159
GC. *see* Gas chromatography (GC)
GDP. *see* Good documentation practices (GDP)
Gelatin, 132
Generally recognized as safe (GRAS), 81
Glass particle contamination, shatter-resistant lamps in, 204, 205f
Gloves, use as hygienic practice, 21
GMP. *see* Good Manufacturing Practice (GMP)
Good Distribution Practices, 81
Good documentation practices (GDP), 190
Good guidance practices (GGP), 13–14
  of FDA, 235
Good Manufacturing Practice (GMP)
  compliance, cost of, 7
  current, 4–5, 9
  goals of, 6
  as good business practice, 6
  history, overview of, 231–233
  hygienic practices, 20–22
  maintenance and, 63–67
  personnel issues (*see* Employees)
  principle of, 3–4
  regulations (*see* Regulations, GMP)
  requirements for equipment/utensils, 40–41
  top management and, 22–23
Granulation, 128–129
  dry method of, 128
  wet, 128
GRAS. *see* Generally recognized as safe (GRAS)
Grounds, of physical plant, 38–39
Guidance on Microbiology & Mycotoxins, 207

HACCP. *see* Hazard analysis and critical control points (HACCP)
Hands, clean
  in avoiding contamination, 209
Hand-washing
  facilities, in physical plant, 38
  hygienic practice, 21
HAZ. *see* Heat-affected zone (HAZ)
Hazard analysis and critical control points (HACCP), 2–3, 233
Heat-affected zone (HAZ), 48
Heating, ventilating, and airconditioning (HVAC), 29–30, 52, 66
Heat transfer fluids, and equipment/utensils, 46
Heavy metal contamination, 205
High-performance interior architectural wall coatings (HIPACs), 26
High-performance liquid chromatography (HPLC), 57, 159, 165
High-performance thin layer chromatography (HPTLC), 158
High-pressure water blasting, for cleaning/sanitation, 56
HIPACs. *see* High-performance interior architectural wall coatings (HIPACs)
Holding, 174–175

HPLC. *see* High-performance liquid chromatography (HPLC)
HPMC. *see* Hydroxypropyl methylcellulose (HPMC)
HPTLC. *see* High-performance thin layer chromatography (HPTLC)
Human errors
  and deviations, 105
  production/process controls, 76–77
Human relations management
  importance of, 17
Humidity, and holding of dietary supplements, 174
HVAC. *see* Heating, ventilating, and airconditioning (HVAC)
Hydroxypropyl methylcellulose (HPMC), 132

IBCs. *see* Intermediate bulk containers (IBCs)
ICH. *see* International Conference on Harmonization (ICH)
Identity testing, of dietary ingredients, 164–166
IEC. *see* International Electrotechnical Commission (IEC)
IFR. *see* Interim final rule (IFR)
Incoming components, 107–108
Injunctions, FDA's enforcement action and, 238
In-process sampling, 100–101
In-process specifications, 81–82
  requirement to establish, 87
In-process testing, 160
Insecticides, usage in physical plant, 34–35
Inspectional observations, by FDA, 244–245
Inspection walk through, defined, 241
Interim final rule (IFR), 14, 90
Intermediate bulk containers (IBCs), 101
International Conference on Harmonization (ICH), 75
International Electrotechnical Commission (IEC), 44
International Pharmaceutical Excipients Council (IPEC), 81
Interpretive rules, 11
Investigations, deviations, 104–105
Investigations Operations Manual (IOM), 5, 183
Investigators, FDA
  Notice of Inspection by, 240
  qualification of, 240
Iodophors, 60
IOM. *see* Investigations Operations Manual (IOM)
IPEC. *see* International Pharmaceutical Excipients Council (IPEC)
ISO 24153, 94
ISO-14159 Hygiene Requirements for the Design of Machinery, 42
ISO-21469 (Safety of Machinery—Lubricants and Incidental Product Contact), 45

Jewelry
  removal of unsecured, as personal hygiene practice, 21
Journal of the AOAC, 166
The Jungle, 231

INDEX

Kefauver–Harris Amendments of 1962, 9
Kelsey, Francis Oldham, 232
KM. *see* Knowledge management (KM)
Knowledge management (KM), 75
Labeling, 9, 137–144
  brite stock, 141
  defined, 110
  issuance/reconciliation of usage, 139–140
  labels *vs.*, 137
  line clearance procedures, 141
  manufacturing history, 140
  obsolete/incorrect, 142
  online controls, 143
  product received for, 110
  purpose of, 137
  QC for, 151
  requirements applicable to, 137–138
  risks of cross-contamination, 143–144
  SOPs for, 138
Labels
  defined, 110
  incoming, 108–110
  vs. labeling, 137
  sampling of, 100
  specification, 83–84
  visual examination prior to, 88
Laboratory Information Management Systems (LIMS), 164
Laboratory operations, 153–166
  analytical chemistry, 154
  certificates of analysis, 161
  certification programs for, 154
  equipment/instruments, care of, 164
  facilities for, 154–155
  finished product testing, 160–161
  identity testing, 164–166
  in-process testing, 160
  management of, 153–154
  out-of-specification, 161–162
  quality control of, 155
  quality control personnel and, 148–149, 153
  sampling plans, 156
  SOPs, 163–164
  specifications, 155–156
  SRMs, 161
  stability testing, 162–163
  system suitability testing, 160
  trends, 162
LAL. *see* Limulus amebocyte lysate (LAL) technique
Liability cases, defective products, 182
Lighting, physical plant, 30–31
LIMS. *see* Laboratory Information Management Systems (LIMS)
Limulus amebocyte lysate (LAL) technique, 207
Line clearance procedures, packaging/labeling, 141
List of CFR Sections Affected (LSA), 13
Logs
  cleaning/sanitation, 52
  equipment/utensils, 49, 63–64
Lot number, 142

Lot records, 117
LSA. *See* List of CFR Sections Affected (LSA)
Lubricants, and equipment/utensils, 45

Magnetic separators, 202–203, 203*f*
Maintenance, and GMP, 63–67
  buildings, 66
  cleanliness in, 64
  computerized maintenance management system, 66
  equipment, program manual for, 64
  importance of, 63
  organization, 64
  outsourcing/out-tasking of, 66–67
  predictive, 65
  preventive, 65
  regulatory aspects of, 63
  role of top management, 63
Management. *see* Top management
Management of Change (MOC) program, 194
Manufacturing operations, 121–135
  capsules (*see* Capsules)
  fluid-bed processing, 133
  for liquid products, 133–134
  mixing equipment, 125–126
  particulate technology, 123–125
  SOPs for, 121–122
  tablets (*see* Tablets)
  weighing/dispensing of ingredients and, 134–135
Market withdrawal, 179
Master batching, 125
Master manufacturing record (MMR), 74, 78, 81, 112–115
  bills of materials, 115
  BPR and, 117
  information included in, 113–115
  language used in, 113
  purpose of, 112
  QC and, 151
  SOP for preparing, 112–113
  theoretical yield statement, 114
  universality, 115
MedWatch Form 3500A, 197–198
MedWatch program, 171
Metal detectors, 203–204, 204*f*
Metallic particulate contamination
  magnetic separators for, 202–203, 203*f*
Metrology, 68
Microbiological contamination, 85, 206–207
Misbranding, of dietary supplements, 9–10
MMR. *see* Master manufacturing record (MMR)
MOC program. *see* Management of Change (MOC) program
Mycotoxin contamination, 207

National Electrical Manufactures Association (NEMA), 44
National Institute of Standards and Technology (NIST), 70, 161
Natural Products Association, 220
Natural toxins, in contamination, 205
Needs assessments, defined, 219

INDEX

NEMA. *see* National Electrical Manufactures Association (NEMA)
NIST. *see* National Institute of Standards and Technology (NIST)
Notice-and-comment procedure, 12
NSF International, 42, 45
Nutrition Labeling and Education Act, 10

OC. *see* Office of the Commissioner (OC), of FDA
Occupational Safety and Health Administration (OSHA), 31
ODS. *see* Office of Dietary Supplements (ODS)
Office of Dietary Supplements (ODS), 161
Office of Management and Budget (OMB), 235
Office of Regulatory Affairs (ORA), of FDA, 233–234
Office of the Commissioner (OC), of FDA, 233
Offshore outsourcing, 225
OMB. *see* Office of Management and Budget (OMB)
Online controls, packaging/labeling, 143
OOS. *see* Out of specification (OOS)
OOS laboratory test. *see* Out-of-specification (OOS) laboratory test
ORA. *see* Office of Regulatory Affairs (ORA), of FDA
Organic volatile impurities (OVI), 206
OSHA. *see* Occupational Safety and Health Administration (OSHA)
OTC drugs. *see* Overthe-counter (OTC) drugs
Out of specification (OOS), 103
Out-of-specification (OOS) laboratory test, 161–162
Outsourcing/outsource, 223–228
  CAL in (*see* Contract analytical laboratories (CALs))
  cleaning/sanitation, 62
  CMO in (*see* Contract manufacturing operations (CMO))
  consultants in, 228
  CRO in (*see* Contract research organizations (CRO))
  defined, 223
  FDA on, 227
  maintenance, 66–67
  offshore, 225
  quality agreements in, 226–227
  reason to, 223–224
  supply agreements in, 225–226
  virtual companies in, 224–225
Overthe-counter (OTC) drugs, 10
OVI. *see* Organic volatile impurities (OVI)
Ozone, sanitizer, 61

Packaging, 137–144
  brite stock, 141
  child-resistant, 143
  issuance/reconciliation of usage, 139–140
  line clearance procedures, 141
  manufacturing history, 140
  obsolete/incorrect, 142
  online controls, 143
  purpose of, 137
  QC for, 151
  requirements applicable to, 137–138
  risks of cross-contamination, 143–144
  SOPs for, 138

[Packaging]
  tamper-evident, 142–143
  types of, 138–139
Packaging material
  incoming, 108–110
  sampling of, 100
  specifications of, 82–83
  visual identification of incoming, 88–89
Pallets, in warehouse, 176–177
Pantone Matching System (PMS) Color, 100
Paper-based records, 191
Paper chromatography, 158
Park Doctrine, 239
Parts washers, for cleaning/sanitation, 56
Passivation, 47
PdM. *see* Predictive maintenance (PdM)
Penalties
  imposed of FDA, 9–10
Personal hygiene, employees, 19–20
Pest control, in physical plant, 34
Pesticide residue contamination, 206
Photography, in FDA inspections, 243
Physical plant, 25–39
  cleaning compounds/sanitizing agents, 33
  cleaning in, 52–53
  design/construction, 26–39
    aisles/working spaces, 30
    bulk fermentation vessels, 32
    clean and sanitary condition, 32–33
    fixtures/ducts/pipes, 28–29
    hand-washing facilities, 38
    lighting, 30–31
    protection against pests, 32
    ventilation, 29–30
    walls/floors/ceilings, 26–28
  grounds of, 38–39
  insecticides/fumigants/fungicides/rodenticides, usage of, 34–35
  maintenance of, 33
  pest control in, 34
  plumbing in, 36–37
  restrooms in, 37
  space requirements, 25–26
  storage of toxic materials in, 33
  trash disposal in, 38
  water supply in, 35–36
Pipes
  deadlegs in, 44
  physical plant, 28–29
Plantain, 80
Plumbing, in physical plant, 36–37
PM. *see* Preventive maintenance (PM)
PMS Color. *see* Pantone Matching System (PMS) Color
Poison Prevention Packaging Act, 143
Polymorphism, 123
Powdered solids
  characteristics of, 124
  and manufacturing operations, 123–125
PR. *see* Proposed Rule (PR)
Precision, defined, 68
Predictive maintenance (PdM), 65

Preventive action, *vs.* corrective action, 106
Preventive maintenance (PM), 65
Process, defined, 73
Production/process controls, 73–77
　design requirements for, 73–75
　human errors, 76–77
　improvement, 75
　knowledge management, 75
　variation in, 75–76
Prohibited acts, by FD&C Act, 200
Proposed Rule (PR), 12
Pure Food and Drug Act (1906), 231
Purified Water USP, 35

QA. *see* Quality assurance (QA)
QACs. *see* Quaternary ammonium compounds (QACs)
QC. *see* Quality control (QC)
QMS. *see* Quality management system (QMS)
Qualified person, 171–172
Quality
　cost of, 7
　defined, 1, 200
　manuals, 3
Quality agreements
　in outsourcing, 226–227
　in supply chain, 214–215
Quality assurance (QA)
　vs. quality control, 146
Quality control personnel, 23, 82, 145, 147–148
　and complaints handling, 171–172
　and laboratory operations, 148–149, 153
　and material review/disposition decisions, 149
　role of
　　in calibration, 70
　　in receiving, 111
Quality control (QC), 145–152
　and BPR, 150
　and MMR, 151
　operations, 148
　　equipment/instruments/controls, 149
　product complaint, 150
　vs. quality assurance, 146
　and reprocessing, 150
　returned goods and, 150
　statistical quality control, 146–147
Quality management system (QMS), 1–2, 194
　HACCP, 2–3
　quality manual and, 3
Quartering, 99
Quaternary ammonium compounds (QACs), 60

Radio-frequency identification (RFID) tracking, 177
Random sampling, 94
RCA. *see* Root cause analysis (RCA)
RCM. *see* Reliability-centered maintenance (RCM)
Recalls
　classification, 180
　defined, 179
　market withdrawal, 179
　reasons for, 179–180
　strategy, 181

Receiving report (RR), 107–108
　warehouse, 175–176
Record keeping, 186–192
　adverse event reporting, 198
　complaints, 172
　receiving items and, 111, 169
Records and information management (RIM), 191
Records, calibration, 68–69
Regulations, GMP, 8–15
　availability of, 14
　Citizen Petition, 14
　direct final rule, 14
　DSHEA, 10–11
　foods for special dietary uses, 10
　interim final rule, 14
　and maintenance, 63
　origin of, 11–12
　prohibited acts/penalties, 9–10
　Standards of Identity, 10
Regulatory Flexibility Act, 12
Regulatory science, defined, 230
Rejected items, handling, 110–111
Relative humidity (RH), 174
Reliability-centered maintenance (RCM), 65
Repackaging/relabeling, QC for, 151
Representative sample, 92
Reprocessing, in quality control, 150
Request for proposal (RFP), 224
Reserve samples, 93
　of dietary supplements, 175
Residual solvent contamination, 206
Responsible Corporate Officer (RCO) Doctrine, 239
Restrooms, in physical plant, 37
Returned dietary supplement
　impact on other batches, 169
　receiving/handling, 168–169
　records of, 169
RFID tracking. *see* Radio-frequency identification (RFID) tracking
RFP. *see* Request for proposal (RFP)
RH. *see* Relative humidity (RH)
RIM. *see* Records and information management (RIM)
Rodenticides, use in physical plant, 34–35
Root cause analysis (RCA), 105
RR. *see* Receiving report (RR)
Rulemaking
　categories of, 11–12
　defined, 11
Rx-360, 220

SAEs. *see* Serious adverse events (SAEs)
Sampling, 92–102
　acceptance, 95–96
　acceptance quality limit, 96–97
　dietary ingredients/excipients, 98–100
　goals of, 93
　in-process, 100–101
　laboratory operations, 156
　personnel aspects of, 98
　plans, 93–94
　random, 94

INDEX

[Sampling]
  representative sample, 92
  reserve sample, 93
  size of, 94–95
  square root of (N) plus one, 97–98
  systematic, 102
Sanitary equipment, standards for, 42
Sanitation
  physical plant, supervision of, 34
  warehouse, 176
Sanitation supervisors, 23
Sanitizers, 59–61
  acid, 60
  chlorine-releasing, 59–60
  iodophors, 60
  ozone, 61
  for physical plant, 33
  quaternary ammonium compounds, 60
  usage direcetions, 61
  water and, 53
Scales, calibration, 71
SCAR. *see* Supplier corrective action request (SCAR)
Science Board, of FDA, 231
SDOs. *see* Standards-setting organizations (SDOs)
Second-party audits, 219–220
Section 701(a) of FD&C Act, 11
Section 201(ff) of FD&C Act, 122
Section 402(g)(1) of FD&C Act, 9
Section 331 of FD&C Act, 9
Section 401 of FD&C Act, 10
Section 403 of FD&C Act, 9
Seizures, FDA's enforcement action and, 238
Self-inspection audits, 216–218
Serious adverse events (SAEs), 197
Shatter-resistant lamps, 204, 205f
Shelf life, 144
Show and shadow facility, audited plant and, 215
SIDI. *see* Standardized information on dietary ingredients (SIDI) method
Sinclair, Upton, 231
Sling psychrometer, 174
Soils, cleaning of, 53
SOP. *see* Standard Operating Procedure (SOP)
Specifications, 78–90
  archives, 89
  components, 80
    identification of, 79–80
  contract packagers/labelers, 84–85
  definition of, 78–79
  determination of, 85–89
    Certificate of Analysis, 86–87
    component identity, 85–86
    finished product, 87–88
    in-process specifications, 87
  for excipients, 80–81
  finished product, 84
  in-process, 81–82
  label, 83–84
  laboratory operations, 155–156
  packaging material, 82–83
  required, 79
  standards *vs.*, 89–90

SQC. *see* Statistical quality control (SQC)
Square root of (N) plus one method, and sampling, 97–98
SRMs. *see* Standard reference materials (SRMs)
SRS. *see* Supplier rating systems (SRS)
Standardized information on dietary ingredients (SIDI) method, 212, 220
Standard Operating Procedure (SOP), 73, 186–192, 194
  on complaint handling, 170
  deviations from, 104–105
  laboratory operations, 163–164
  for manufacturing operations, 121–122
  for packaging/labeling, 138
  for preparing MMR, 112–113
Standard reference materials (SRMs), 156, 161
Standards of Identity, FDA, 10
Standards-setting organizations (SDOs), 89
Standards *vs.* specifications, 89–90
Statistical quality control (SQC), 146–147
Steel, as construction material for equipment/utensils, 47
Steel wool, for cleaning/sanitation, 56
Stock recoveries, 179
Stock, rotation of, 111
Substantive regulations, 11–12
Sugar coating, 129–130
Suitability, equipment/utensils, 40
Sulfanilamide, 8
Supervisors
  and personal hygiene of employees, 19–20
  qualification requirements, 17–18
  sanitation, 23
Supplier corrective action request (SCAR), 214
Supplier rating systems (SRS), 214
Supplier(s), 211
  controls, 214
  qualification of, 213–214
  selection of, 212
    risk management in, 212–213
Supply agreements, in outsourcing, 225–226
Supply chain, defined, 211
Supply chain integrity, 211–215
  defined, 211
  quality agreements in, 214–215
  show and shadow facility and, 215
  suppliers in (*see* Supplier(s))
Supply chain management, defined, 211
Systematic sampling, 102
System suitability testing, 160

Tablets
  coating of, 129–131
  making of, 127–129
Tamper-evident packaging, 142–143
TBA. *see* Tribromoanisole (TBA)
TBP. *see* Tribromophenol (TBP)
Technical agreements, in outsourcing. *see* Quality agreements, in outsourcing
Temperature, and holding of dietary supplements, 174
Thalidomide, 8
  disaster, 231–232

Theoretical yield, MMR, 114
Thin layer chromatography (TLC), 158, 159
Third-party audits, 220–221
3-A Sanitary Standards, 42
   20-20, 49
   20-23, 49
Title 21 of the Code of Federal Regulations, 13
TLC. *see* Thin layer chromatography (TLC)
TOC technique. *see* Total organic carbon (TOC) technique
Top management
   additional duties, 184
   availability of resources, 184
   complaints information to, 172
   defined, 183
   legal responsibility, 183
   and maintenance, 63
   role in GMP, 22–23, 182–184
Total organic carbon (TOC) technique, 57
Total productive maintenance (TPM), 65
Toxic materials, storage in physical plant, 33
TPM. *see* Total productive maintenance (TPM)
Traceability, 178
   calibration, 70
Training
   documentation of, 18
   of employees, 18–19
Transparency initative, FDA, 236–237
Trash, disposal of
   in physical plant, 38
Tribromoanisole (TBA), 176
Tribromophenol (TBP), 176
Tumble blenders, 125
Turbo EIR, 246

Unanticipated occurrence, deviations and, 103
Unique identity code, 107
United States Code (U.S.C.), 8
United States Pharmacopoeia–National Formulary, 81
U.S.C. *see* United States Code (U.S.C.)
USDA. *see* U.S. Department of Agriculture (USDA)
U.S. Department of Agriculture (USDA), 42

USP. *see* U.S. Pharmacopoeia (USP)
U.S. Pharmacopoeia (USP)
   for Purified Water, 35
U.S. vs Dotterweich, 183
U.S. vs Park, 183
Utensils. *see* Equipment/utensils

Variation, in production/process controls, 75–76
Vendor, 211. *see also* Supplier(s)
   audits, 219–220
Ventilation, physical plant, 29–30
Virtual companies, in outsourcing, 224–225
Visible-residue limit (VRL) method, 58
VRL method. *see* Visible-residue limit (VRL) method

Walls, physical plant, 26–28
   cleaning, 52
Warehouse
   distribution of finished products, 177
   humidity, 175
   palletizing, 176–177
   receiving records, 175–176
   sanitation, 176
   temprature, 175
   traceability, 178
Warning letters (WL)
   FDA's enforcement action and, 237–238, 246–247
Water
   activity, in contamination, 208
   and cleaning process, 53–54
   supply, in physical plant, 35–36
Weighing/dispensing, of ingredients
   and manufacturing operations, 134–135
Wet-bulb thermometer, 174
WL. *see* Warning letters (WL)
Working spaces, physical plant, 30
World Health Organization
   and HACCP, 3
Written procedures, calibration, 68–69

Z1.4, 95
Z1.9, 95